PLC

AUTOMATION WITH PROGRAMMABLE LOGIC CONTROLLERS

PETER ROHNER

A TEXTBOOK FOR ENGINEERS AND TECHNICIANS

WARNING!

The author and publishers of this book accept no responsibility for injury to persons or damage to machinery or property or any arising liability when design suggestions and concepts made in the text of this book are followed.

MACMILLAN PRESS LTD
Houndmills, Basingstoke, Hampshire RG21 6XS
and London
Companies and representatives
throughout the world

ISBN 0–333–67485–5 paperback

A catalogue record for this book is available
from the British Library.

CONTENTS

FOREWORD

Knowledge means power is the latest catch-cry of the 1990s. While one may baulk at such a sweeping generalisation, there is little doubt that increasing competition has led many industries to embrace the fervent pursuit of knowledge. Hence the rapid expension of training facilities and course offerings.

There are few places today where this push for knowledge is so prevalent as in the automation/manufacturing industry, which faces competition from a vast number of national and international sources; the recent lifting of international trading restrictions will only serve to increase this level of competition.

'Mechatronics', the integration of mechanical and electronic systems, is yet another recent buzz word. Machine designers now rightfully recognise the strong association between these facets, which for too long was neglected by those in the control/automation industry.

PLC: Automation with programmable logic controllers wisely treats the mechanical systems (hydraulic and pneumatic) as being one entity with the electronic programmable controllers. This text, with its many illustrations and practical examples, provides a thorough grounding in elementary circuit concepts, design and programming, as well as sequential control systems.

It is my pleasure to recommend this course book, which is simple enough for the novice to understand, yet sufficiently detailed to be an invaluable tool for both education and industry personnel.

Harold Mulder
Product manager for FA (Flexible Automation) Products
OMRON Electronics Australia

ACKNOWLEDGMENTS

I am very grateful to my wife, Heidi, for patiently typing the entire manuscript and putting up with me, working often long hours on the text and numerous illustrations and circuits for this book.

I am also thanking Graham Williams (fellow lecturer, RMIT University, Australia), for giving constructive comments on many of the circuits included in this text.

Omron (Australia) deserves a special thank you for providing PLC equipment and making numerous constructive suggestions as to how certain complex functions on their PLCs work and must be understood. Omron (Australia) has also given me permission to use textual material from their PLC manuals.

The Royal Melbourne Institute of Technology (RMIT University) has created over many years an environment that lends itself to research and development of control algorithms. I am fortunate to be part of the lecturing staff of this famous university and I am grateful to have been given the opportunity to use campus equipment for developing and testing new automation control concepts and control circuits.

Peter Rohner

PREFACE

Automation with programmable logic controllers has opened up a fascinating, completely new and virtually limitless control world. Therefore, mastering the programming and designing of PLC control software is of paramount importance to everyone involved with automated and PLC-controlled machinery. This need not be difficult, if one learns to identify a seemingly complex control problem as a circuit solution consisting of numerous basic logic functions. Such functions are fundamental to all binary digital controls and embrace, by concept, the logic "OR" function, the logic "AND" function, the logic "NOT" function, the logic "YES" function and the logic "INHIBITION" function. With some more complex switching elements, such as flip-flops, timers, counters, shift registers, edge flags, arithmetic functions and internal output relays, one can assemble a control system to solve both simple and enormously complex control tasks.

Attempting this task, this book therefore amply explains and illustrates all of these logic functions and switching elements with a simple yet profound and methodical manner, and shows the reader how to assemble these functions into various and typical industrial control circuits. The book also presents deep insights into first principles of sequential and combinational machine control design, and applies such principles by presenting numerous solved case problems as they may arise in today's industrial control systems and machinery.

Flexible automation with PLCs receives special attention through the presentation of a design algorithm for sequentially operating machines, using the well known step-counter modular design concept. Machine routines, such as automatic or manual cycling, emergency stop (cycle interrupt), stepping or running selection and program selector blocks are exhaustively explained, illustrated and applied.

The book frequently makes comparison to existing pneumatic or hydraulic control circuits, as refurbishing or conversion of old circuits to new and often modified PLC control is common practice in industry these days. Pneumatic control has firmly been entrenched for at least 40 years and is still famous for its unique ability to control even complex sequentially and combinationally operating machinery. But with the upsurge of smart, inexpensive and fast responding PLC controllers, pure pneumatic control has lost much ground and will eventually have to make room altogether for entirely PLC-controlled, electropneumatic and electrohydraulic control systems. This is tough luck, but a fact and stark reality in this fast changing and electronically controlled world.

As revealed through the pages of this book, my concern is mainly to teach micro- rather than macroautomation. This implies that PLC control applications for small to medium-size control systems are highlighted in this book. Most PLCs do control machinery with some form of pneumatic or hydraulic actuators as motion and force generators, and thus use fluid-power solenoid valves to serve as an electrical (PLC) to mechanical interface. Control problems applying such fluid power actuators and solenoid valves are therefore frequently used throughout the pages of this text.

PLCs of various brands and makes obviously and inherently have their programming peculiarities. For these reasons I refrain from specific PLC programming language, and present all control problems and explanations with the universally popular ladder circuit presentation method and its resulting mnemonic programming statements. Programming a PLC with a ladder or mnemonic statement list is commonplace and should therefore be mastered by personnel wishing to program PLCs or to diagnose faults in PLC-controlled machines.

My teaching and consultancy experience have compelled me to include important concepts and methods of approach to fault-finding and diagnosing PLC-controlled machine systems. In several places the text therefore shows how to debug a system, and how a thorough mechatronic training and cybernetic knowledge helps with system diagnosis.

As a lecturer of 21 years at the Royal Melbourne Institute of Technology (now RMIT University) in control engineering, and a consultant to industry, I am uniquely positioned to pass on years of experience in PLC programming and interfacing the PLC with fluid power control circuits. For these reasons, the book is didactically unique and an elementary resource work, as well as an essential teaching and research text for students enrolled in industrial control courses and subjects dealing with PLC programming. The text teaches a strong methodical approach to circuit design and PLC programming, and as such it may therefore also be regarded as fundamental and essential reading for electrical, electronic, mechanical and production engineers, and also for electrical and fluid power technicians retraining for and upgrading their qualifications in PLC control design and PLC programming.

Peter Rohner (Dip.Tech. Teaching), Senior Lecturer, Electrical and Control, Royal Melbourne Institute of Technology (RMIT University).

Briar Hill Victoria 3088, 2 August 1995

1 PLC PROGRAMMING AND HARDWARE FUNDAMENTALS

Over recent years electronic programmable logic controllers (PLCs) have sprung up like mushrooms, and over three hundred different types and brands are now available. Electronic programmable logic controllers have undoubtedly become an indispensable feature of the ever advancing industrial automation.

These controllers evolved as industry sought more economical ways of automating production lines, particularly those involved in the manufacturing of equipment, consumer goods and heavy industry products. Thus the electronic programmable controller has replaced relay-based, hard-wired electrical systems, and more recently it has also made significant inroads into the traditional domain of pneumatic logic circuitry. This encroachment unfortunately has brought about some problems:

- Programming difficulties caused through non-standardised and varying programming techniques.
- Programming difficulties caused through the use of varied logic element names and element description for hardware and software components within the controller.
- Demarcation problems in factories between electrical and metal worker unions when maintenance on machines with fluid power control and PLC control is required.
- Extreme shortage of skilled personnel who can program and install such electronic PLC controllers and interface them with fluid power circuitry.

PLCs operate by monitoring input signals from such sources as push-button switches, proximity sensors, heat sensors, liquid level sensors, limit switches, and pressure and flow rate sensors. When binary logic changes are detected from these input signals, the PLC reacts through a user-programmed, internal, highly sophisticated, logic switching network and produces appropriate output signals. These output signals may then be used to operate external loads and switching functions of the attached fluid power control system (figure 1-01).

PLCs eliminate much of the wiring and rewiring that was necessary with conventional relay-based systems. Instead, the programmed "logic" network replaces the previously "hard-wired" network. This logic network may be altered as required by simply deleting, inserting or changing certain sections in the PLC's program. Thus, the automated processes of a production line or complex manufacturing machine can be controlled and modified at will for highly economical adaptability to a rapidly changing manufacturing environment.

Some people use the abbreviation PC for programmable logic controller, but PC normally stands for personal computer. To avoid confusion, therefore, throughout this book PLC is used for programmable logic controller.

WHAT IS AN ELECTRONIC PROGRAMMABLE LOGIC CONTROLLER?

Programmable logic controllers are purpose-built computers. A typical PLC has four separate yet interlinked components. These are:

- an input/output section, which connects the PLC to the outside world (the machine with its sensors, solenoid valves and switches, lamps, heaters and electric motors).
- a central processing unit (CPU), which is microprocessor-based. This may be an octal or hexadecimal microprocessor.
- a programming device, which may be a hand-held programming console, a special PLC desk-type programmer, similar to a lap-top computer, or a desk-top

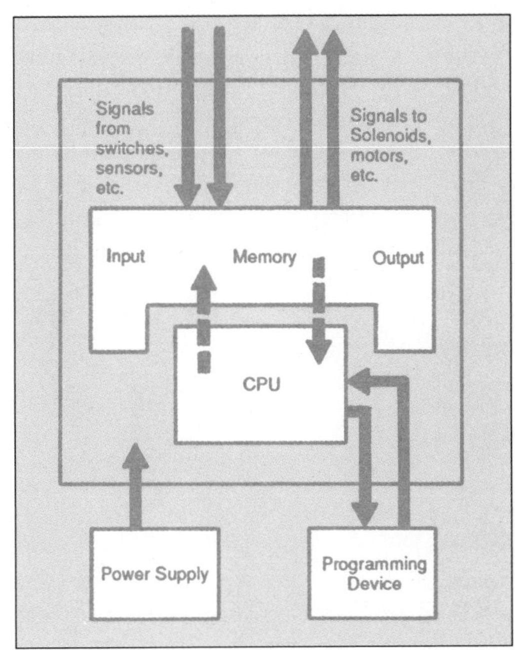

Figure 1-01. A PLC's internal configuration and interface with the machine and peripheral equipment.

computer with monitor (figures 1-04, 1-05 and 1-06).

- a power supply to power input sensors and output signals leading to lamps, motors, heaters and solenoids on the fluid power valves (usually 24V DC).

The CPU (central processing unit) is microprocessor based and may be regarded as the brain of the controller. It scans and reads all the on/off conditions of all input terminals and stores them in its input image memory before executing the program. The CPU then processes that information according to the control plan programmed into the user memory (UM). Such an internal control plan may include numerous memory functions, logic "AND", "OR" and "INHIBITION" functions, Arithmetic computation instructions, Timers and Counter functions. The CPU also continuously scans (monitors) the status of all output signals (bits) and thus constantly updates the contents of the input image memory according to changes made to the output image memory (because outputs may also serve as inputs). The CPU also organises its internal operation (watchdog timer, initialising program etc.) (figure 1-02). Larger PLCs also employ additional microprocessors to execute complex, time-consuming functions such as mathematical data processing and PID ( ) control.

The program is entered with ladder diagram or graphic logic symbols through a computer and monitor. It may also be entered by statement list or mnemonics via a hand-held or on-board programming console. It then remains in the RAM (random access memory) of the CPU. In practically all cases RAM is used for the initial program configuration (UM). RAM permits changes to be easily made during the initial programming stages. Current trends in PLC design are the use of CMOS (complementary metal-oxide silicon) RAM chips because of their extremely low power consumption, and providing battery backup to the chip to maintain logic status to flip-flops during interruptions to PLC power supply. RAM by nature is a volatile memory, which loses stored data during power failure! Execution results based upon the logic combinations rendering internal or external output signals are then written internally (electronically) into the element image memory. The element image memory then drives the output relays of the PLC.

THE PLC's INTERNAL ORGANISATION PROGRAM

When a programmable logic controller (PLC) operates, that is, when it executes its program to control an external system with, for example, fluid power valves and their actuators, a series of operations are automatically performed within the CPU. These automatic operations are regarded as the PLC's internal organisation program. These internal operations can broadly be grouped into four categories:

- common internal processes such as resetting the watchdog scan-cycle timer and checking the user program memory
- data input/output refreshing
- instruction execution
- peripheral device command servicing.

Of these four internal operations, the only one easily visible is the third, instruction execution (driving valves and their actuators and starting or stopping motors, turning lamps on or off and actuating heaters and fans).

Immediately after power application, the first three operations shown in figure 1-02 are performed once only. All other operations shown in figure 1-02 are performed in cyclic order, with each cycle forming one scan. The scan time is the time required for the CPU to complete one of these cycles. This scan cycle includes the four types of operation listed above.

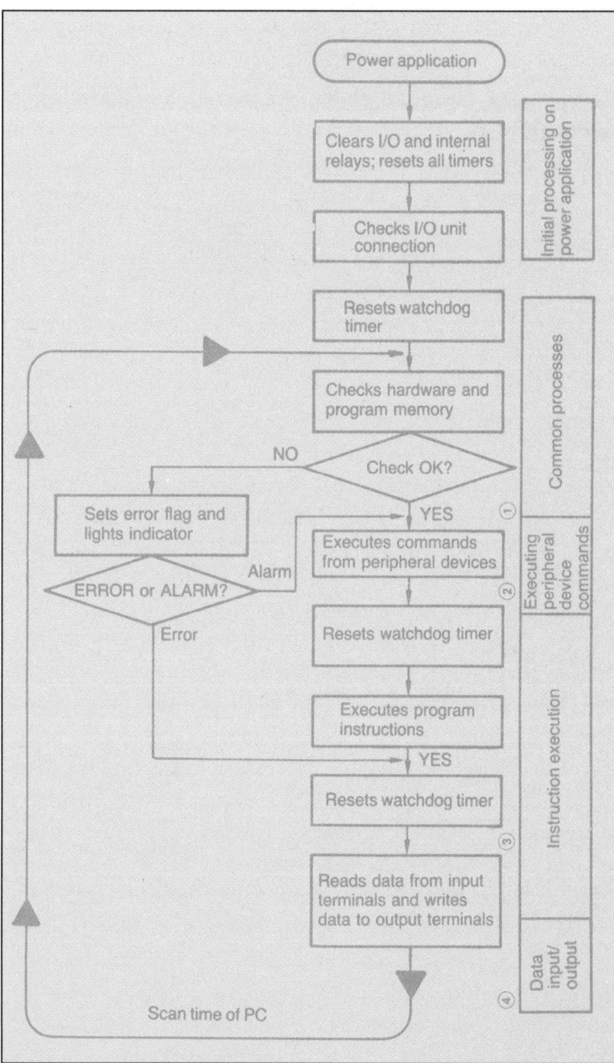

Figure 1-02. The internal organisation program of a PLC and its scan cycle.

The PLC's internal organisation program cannot be influenced by the person programming the PLC. This organisation program is exclusively designed by the PLC manufacturer to automatically execute these four operations, shown in the vertical boxes numbered 1, 2, 3 and 4 in figure 1-02.

Summary

A programmable controller is an aggregate control mechanism made up of multiple electronic relays, timers and counters and numerous special functions, used to execute the internal logical wiring by the programming panel. A conventional, hard-wired relay panel differs from a PLC in the sequence execution method. Its sequences are executed in parallel but the PLC executes its sequences in the order of the program and cyclically as the scanning reveals any input changes.

PLC INPUT/OUTPUT TERMINOLOGY

Although listed in the glossary at the back of this book, the following terms are essential to understanding and mastering the programming of PLCs.

PLCs are designed to receive signals on their inputs (input terminals), and send signals via their outputs (output terminals). Hence, a device connected to the PLC that sends a signal is called an "input device" (switch or sensor). The signal this device sends is called an "input signal". An input signal enters the PLC through terminals or connector pins on a multi-point detachable connector rail. The precise point where such an input signal enters the PLC is usually called an "input point". This input point is allocated a defined location in the PLC's memory, which reflects the incoming signal's logic status, being either "ON" (true) or "OFF" (not true). This memory location where this incoming signal is recorded is called an "input bit". The CPU (central processing unit) in its normal processing cycle monitors the logic status of all input points and turns "ON" or "OFF" corresponding input bits accordingly (figure 1-03).

Within the PLC's CPU, logic decisions are being made, based upon the program written into its user memory (UM). For such actions to be carried out, they need to be translated into signals. To accomplish and initiate such action, the PLC manipulates "output bits" by turning these output bits either "ON" or "OFF". These output bits in the PLC memory are allocated and connected to "output points" on a multi-point connector rail or a multiple-point terminal (figures 1-03 and 2-07). Signals emanating from such output points are called "output signals". The signal, when conducted through electrical wiring, ultimately reaches an "output device" (solenoid, lamp, motor starter, relay, buzzer, heater etc.), to turn such a device either on or off.

A signal can logically be "ON" or "OFF", according to its binary characteristics. When there is no signal leaving the PLC's output point, a lamp connected to that output point, for example, would be "OFF", but if the same signal were connected to a solenoid valve that is spring-returned, the absence of the PLC's output signal could mean action on the pneumatic or hydraulic actuator. Figure 2-08 shows that the absence of the output signal causes actuator piston motion. The PLC programmer therefore needs to know the peculiarities of components and devices controlled by the PLC.

Various terms regarding PLC inputs and outputs are used when describing different aspects of PLC operation, programming and installation. For example, when a program is being designed, written and discussed, one is primarily concerned with whatever information is held in the PLC's memory, and hence reference is made to **input and output bits** and their logic connections within the PLC.

For discussions and explanation of input and output devices being connected to the PLC, reference is frequently made to the PLC's **input and output points**. These are the very points where such devices are electrically wired in and out of the PLC.

For discussions pertaining to fault-finding and total electrical system diagnostics, one refers to the electrical **input and output signals** leading to and going from the PLC. Such entering and exiting electrical signals (sometimes simply referred to as "inputs" and "outputs") can be tested for their presence or absence with multimeters or similar electrical testing apparatus, or one may simply check for the "ON" light of the input or output signal LED (light-emitting diode) on the PLC's diagnostics panel. Most PLCs provide such input and output LEDs to facilitate fault-finding (figure 1-03).

The **total electrical control system** includes the PLC with all its connected input and output devices and the necessary power supplies. The input and output devices are connected to the machine, whereas the PLC is usually located in an electrical control cabinet nearby, and protected from dust, vibration, moisture and electrical interference.

PROGRAMMING DEVICES FOR PLC

PLCs may be programmed with a variety of devices. The three most common devices are:

- the hand-held programmer
- the desktop PLC programming unit
- the computer with keyboard, monitor and appropriate software.

The hand-held programmer is predominantly used for small to medium-sized PLCs, or for minor program alterations during machine and PLC commissioning.

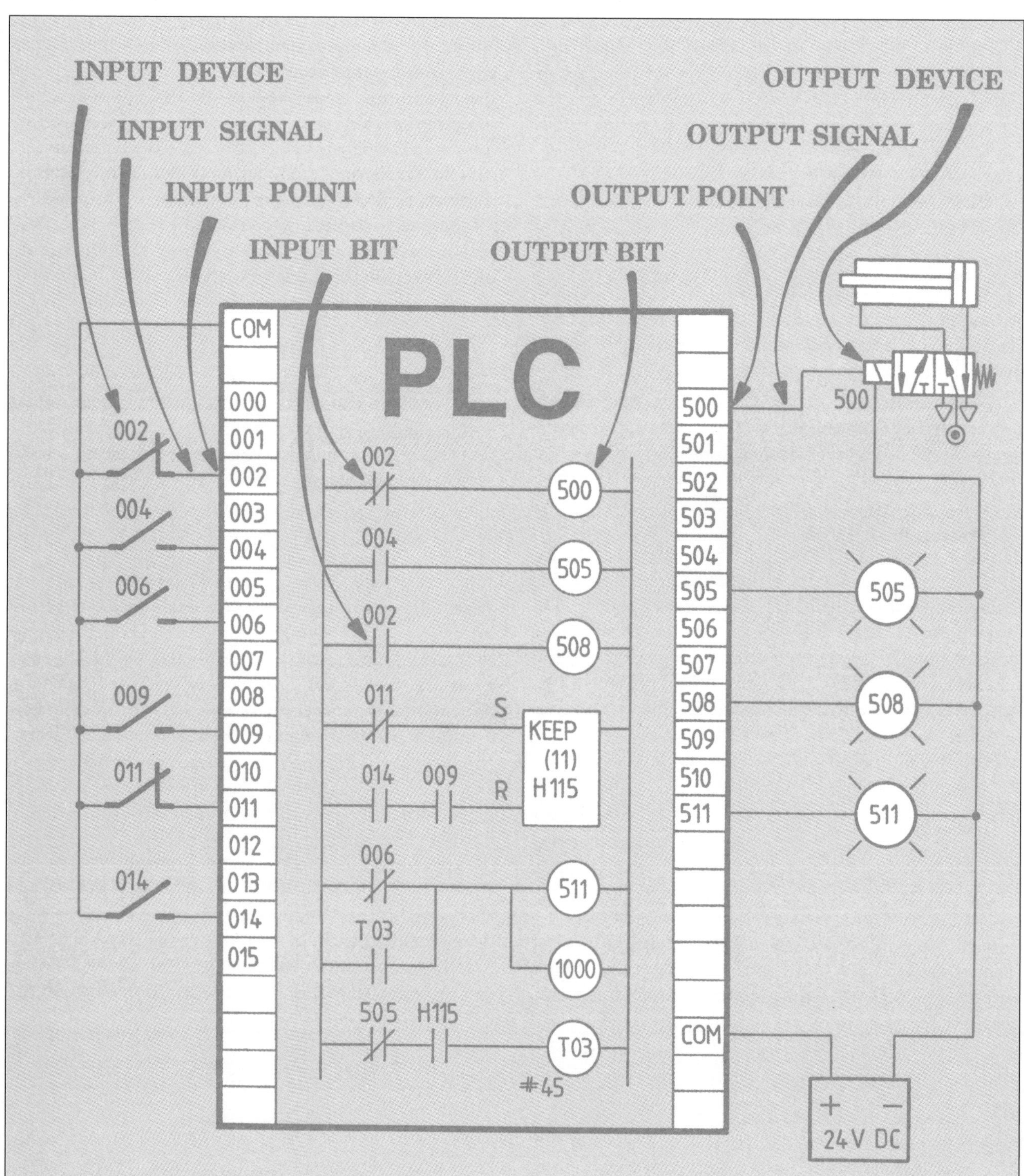

Figure 1-03. Input/output terminology for a PLC's integration with the machine.

The hand-held programmer is very popular and inexpensive (figure 1-04). Entering a program into a PLC with a hand-held programmer and mnemonic codes is usually much faster than with any other programming method. The hand-held programmer has a number of keys that allow the person programming the PLC to enter a program in mnemonic code form, similar to a Boolean equation expression, or sometimes also in ladder form. Hand-held programmers have a limited built-in diagnostics and debugging facility. Some hand-held programmers provide a small monitor window, which displays ladder rungs to enter the program in graphic form, similar to the computer monitor.

This book concentrates on the use of hand-held programmers.

Hand-held programmers for some PLC brands may also be fastened to the PLC with knurled screws for ease of removal. The hand-held Omron programming console is detachable and uniform for all Omron PLCs. Omron calls this hand-held programmer a programming console (figure 1-05). The programming console also functions as an interface to transfer programs between a standard cassette tape recorder and the PLC (figure 1-07).

Desktop PLC programmers and computer-based programmers provide usually extensive debugging and diagnostics features. Desktop PLC programmers often cost as much as a laptop computer, and come with in-built monitor (figure 1-06).

Unfortunately, programming software differs significantly from manufacturer to manufacturer of PLCs, and may come in menu or window selection and instruction form. Owing to the vast differences in software, this book does not show any graphic approach. For more information on PLC programming with PCs and programming software, contact the suppliers of the PLC brand in use. Omron now offers a Windows-based program called Syswin, which should make computer-based programming much simpler.

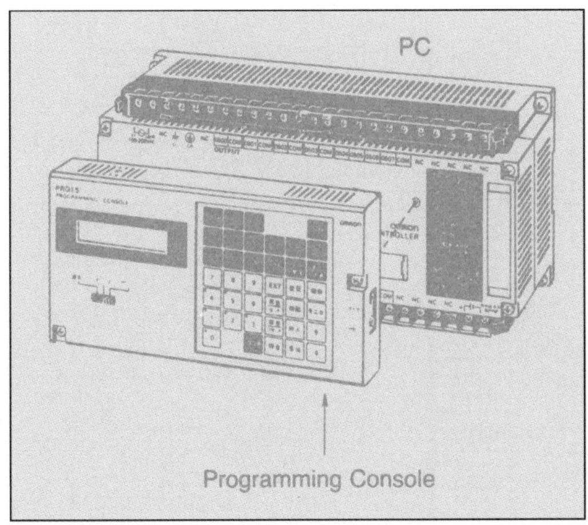

Figure 1-04. Programming console mounted to PLC (this can be an attached programming console or cable-linked, hand-held programming console).

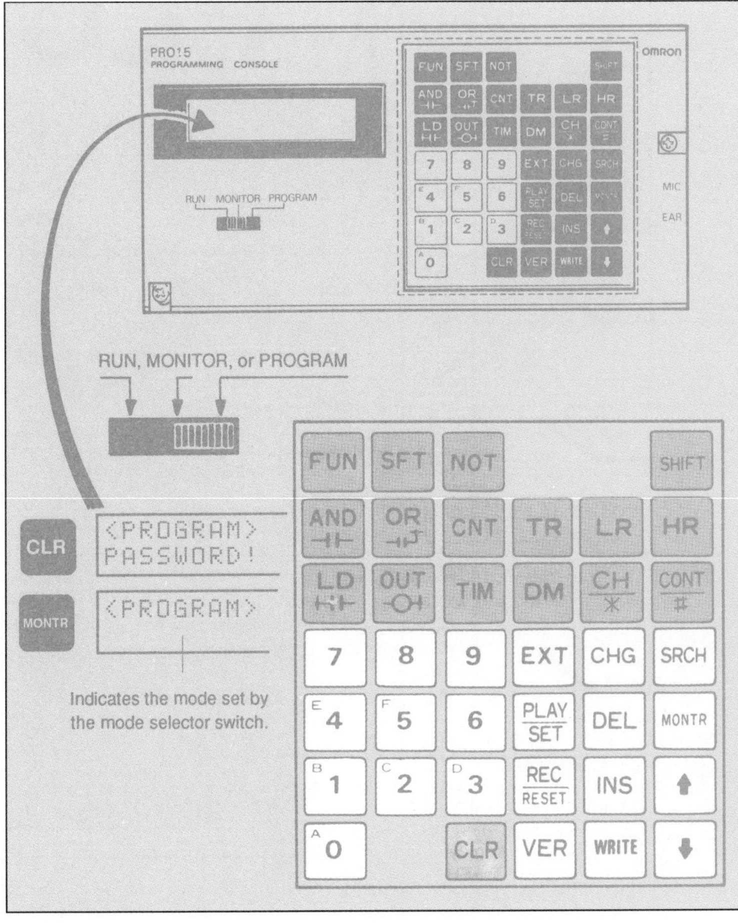

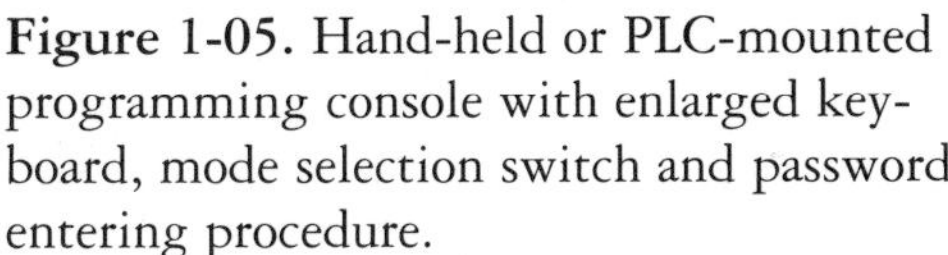

Figure 1-05. Hand-held or PLC-mounted programming console with enlarged keyboard, mode selection switch and password entering procedure.

BIT, BYTE, DIGIT, WORD

Program information or program instructions are stored within the PLC's data memory, also called user memory. The CPU makes logic conclusions, comparisons, data manipulations and arithmetic computations based solely on the program loaded and stored in the user memory. The processed results are then used to direct and control the machine to which the PLC is attached. Data in the user memory are organised and deposited in "words". A word has 16 bits or two bytes or four digits (figure 1-08). The bit is the smallest item in the data memory. A bit may assume the logic 1 or "ON" state or the logic 0 or "OFF" state. Since the PLC primarily operates as a digital binary controller, these control bits are binary—they are either logic 1 or logic 0. Some PLCs operate with an "octal" microprocessor, in which a word consists of eight bits. Omron and most other leading brand PLCs operate with a "hexadecimal" microprocessor, which has 16 bits to a word. Following international standards, the rightmost bit of the word is bit 00 and the leftmost bit is bit 15 (figure 1-08).

Program instructions are entered into the PLC's user memory in alphanumerical form or machine language. This form of entering programming instructions is also called

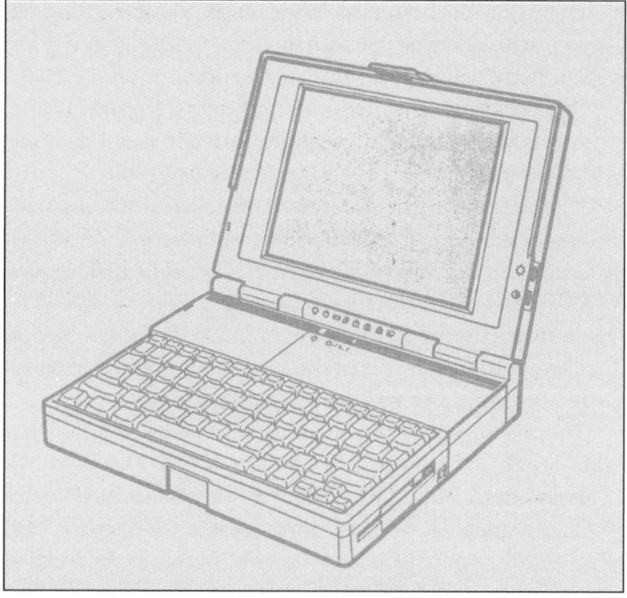

Figure 1-06. Desktop PLC programmer.

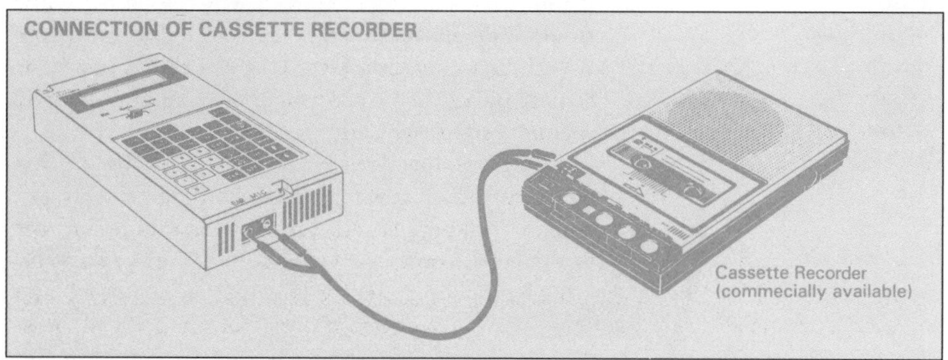

Figure 1-07. Saving the program onto cassette tape with a standard cassette recorder attached to the programming console.

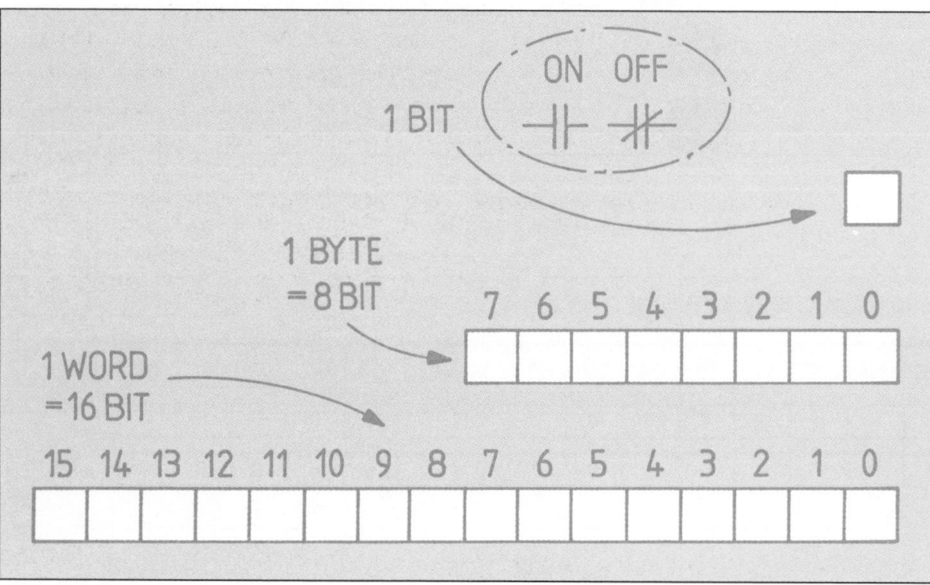

Figure 1-08. Bit, byte and word structure, used by PLCs to store programming instructions. The bit is the smallest unit in that structure and may take on either the logic 1 or logic 0 state. Four bits form a digit, eight bits form a byte and sixteen bits form a word.

statement list or mnemonic instruction programming. A single instruction line consists usually of three items:

- the instruction's location in the memory of the PLC. This location is called the "address" (figure 1-09). This is where the PLC with its in-built search facility can locate a stored programming instruction.
- the relationship of the instruction to other instructions, which is called the "mnemonic". Such a mnemonic could be LD, OR, AND, LD NOT, AND NOT, OR NOT, OUT etc. (figure 1-11).
- the instruction's word definer and bit definer, which is called the "operand". An operand therefore consists of the word number and the bit number within that word (figure 1-10) or other operand definers as shown later in other chapters and in figure 1-11. The operand always designates where the bit is found or filed within the data area (user memory; figures 1-12 and 1-14).

ADDRESS	MNEMONIC	OPERAND
0000	LD	0007
0001	AND	0009
0002	OUT	0503
¦	¦	¦

Figure 1-09. Programming instructions commonly used for mnemonic or statement list programming. A single instruction is grouped into the address section, mnemonic section and operand section.

DATA AREA STRUCTURE

A PLC has several data areas (see figure 1-12). When designating one of these, the acronym for the area is always required for all but the IR and SR areas. Although the acronyms for the internal and special relay areas (IR and SR areas) are often given for clarity, they are not required and not entered when programming. Any data area designation without an acronym is assumed to be in either the IR or the SR area. Because IR and SR areas run consecutively, the word or bit definers are sufficient to differentiate these two areas (see figure 1-12; IR operand 1807 [the last operand in IR] is followed by SR operand 1808 [the first operand in SR]).

An actual data location within any data area except the timer and counter area (TC area) is designated by its operand. The operand designates the bit or word within the area where the desired data are located. The TC area, as an exception, consists of TC numbers. Each of these TC numbers is used for a specific timer or counter in the user program (see Chapters 5 and 6 for timers and counters and figures 1-11 and 1-12). Hence timers are not organised in words and one therefore often refers to the timer as a "flag". An Omron C28K PLC has 48 such timers (see also figure 1-14).

Data areas except for timers and counters consist of words, each of which again consists of 16 bits numbered 00 to 15 from right to left (figures 1-08, 1-10 and 7-06). IR words 00 and 01 are shown with their bit numbers in figure 1-10. Here, the content of each word is shown as all zeros. Bit 00 is called the rightmost bit; bit 15 is the leftmost bit within that word.

Note: The term "least significant bit" is often used for the rightmost bit and the term "most significant bit" is used for the leftmost bit. These terms so far have not been used in this chapter because a single word is often split into two or more sections in which each section is used for different parameters or operands, sometimes even with bits in an other word. When this is done, the rightmost bit in such a compounded word may actually be the most significant bit.

To designate one of these data areas by word, all that is necessary is the acronym (if required) and the one- or two-digit word definers. To designate an area by bit, the word definer is combined with the bit number to a single three- or four-digit operand (see figures 1-11 and 1-12). The two rightmost digits of an operand must therefore indicate a bit between 00 and 15. The examples given in figure 1-11 should make this sufficiently clear.

The Data Memory area (DM area, figure 1-12) as a section within the PLC's many memory areas is accessible by word only; one cannot designate an individual bit within a DM word. Data in the IR, SR and HR areas, however, are accessible by either bit or word, depending on the instruction in which the data are being used.

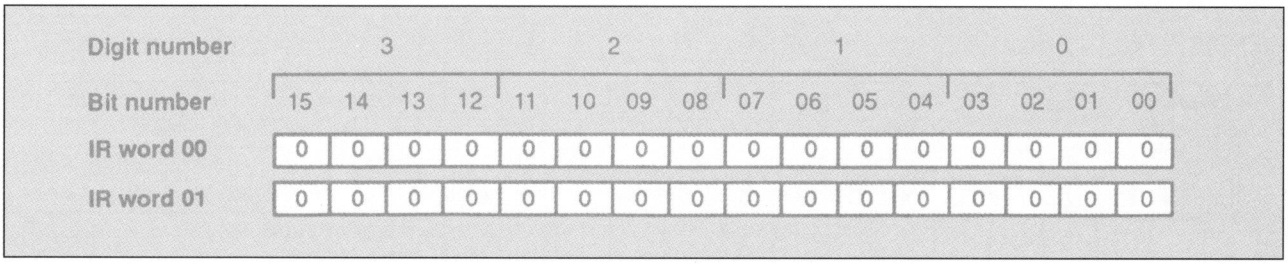

Digit number	3				2				1				0			
Bit number	15	14	13	12	11	10	09	08	07	06	05	04	03	02	01	00
IR word 00	0	0	0	0	0	0	0	0	0	0	0	0	0	0	0	0
IR word 01	0	0	0	0	0	0	0	0	0	0	0	0	0	0	0	0

Figure 1-10. Word, byte, digit and bit structure for data location and organisation.

Address	Mnemonic	Operand
0217	LD	0006
0218	AND	1000
0219	AND	1902
0220	LD	HR 0007
0221	OR	0505
0222	OUT	1003
0223	OUT	0511
0224	LD NOT	0015
0225	OR	TIM 47
0226	LD	0504
0227	KEEP (11)	HR 801
0228	END (01)	

Figure 1-11. Samples of operand configurations, consisting of acronym, word and bit label. Not all data specified in the operand column require an acronym (see also figures 7-18, 9-12 and 10-34).

Example

For an Omron C20 or Omron C28K PLC, the instruction operand labelled 1006 may be broken down as follows:

- The two leftmost digits in this operand label indicate the word, which in this case is word 10.
- The two rightmost digits indicate the bit location within that word, which in this case is bit 06.
- Together, the word indication (10) and the bit location (06) formulate the operand label, which now is 1006.

On an Omron C28K, if mixed expansion input/output units are added, even numbered words, such as 00, 02, 04, 06, 08, are used as input words. Odd numbered words, such as 01, 03, 05, 07, 09, are used as output words. Smaller to medium-size PLCs have their input and output words usually preallocated, and larger PLCs with rack-mounted input and output cards leave the allocation up to the user.

On an Omron C20 PLC, if expansion input/output units are added, the words 00, 01, 02, 03, 04 are used as input words, whereas words 05, 06, 07, 08, 09 are used as output words. Each input word uses all 16 bits. This amounts to a total of 80 input bits (5 words × 16 bits = 80 input points), which means that 80 switches or sensors may be attached to an Omron C20 PLC when fully expanded. Each output word on this controller uses only 12 bits (bit 00 to bit 11). This amounts to a total of 60 output bits, which means that 60 output devices (solenoids, lamps, motors, heaters, alarms, etc.) may be attached (figure 1-14).

Words HR 00 to HR 09 (see figure 1-14), when combined with FUN 11, serve as a KEEP function (the battery connection instruction is FUN 11). They then become KEEP relays or so-called memory-retentive relays (see Chapter 4).

PLC words may be used as complete units of 16 bits as a block, where all 16 bits are occupied (e.g. in shift registers, Chapter 7), or they may be used as fragmented words, by using only some of the bits within that word.

Figure 1-12. Various PLC data areas with their respective Acronym, Word and Bit range and control function.

Area	Acronym	Range	Function
Internal Relay area	IR	Words: 00 to 18 (bits 00 to 07) Bits: 0000 to 1807	Used to manage I/O points, control other bits, timers, and counters, to temporarily store data.
Special Relay area	SR	Words: 18 (bits 08 to 15) and 19 (bits 00 to 07) Bits: 1808 to 1907	Contains system clocks, flags, control bits, and status information.
Data Memory area	DM	DM 00 to DM 63 (words only)	Used for internal data storage and manipulation.
Holding Relay area	HR	Words: HR 0 to HR 9 Bits: HR 000 to HR 915	Used to store data and to retain the data values when the power to the PC is turned off.
Timer/Counter area	TC	TC 00 to TC 47 (TC numbers are used to access other information)	Used to define timers and counters and to access completion flags, PV, and SV for them.
Temporary Relay area	TR	TR 00 to TR 07 (bits only)	Used to temporarily store execution conditions.
Program Memory	UM	UM: 1,194 words.	Contains the program executed by the CPU.

WORK BITS AND WORDS

When some bits and words in certain data areas are not used for their intended purpose, they can be used in programming as required to control other bits. Words and bits available for use in this fashion are called work bits and work words. Most unused bits can be used as work bits. For further information regarding work bits and work words, check with the manual for the PLC in use.

USING THE PROGRAMMING CONSOLE FOR MNEMONIC PROGRAMMING

The keyboard of the programming console for Omron PLCs (figure 1-05) is functionally divided by colour into white numeric keys, the red CLR key, the yellow operation keys and the grey instruction and data area keys.

White numeric keys

The ten white keys numbered 0 to 9 are used to enter numeric program data such as program addresses, data area definers and operand values. These numeric keys are also used in combination with the function key (FUN) to enter instructions with function codes. For example, the KEEP function as used for the programming of holding relays with memory-retentive characteristics during power failure uses function 11 (FUN 11; figures 4-03 and 4-04). These function keys are also used when the end instruction (FUN 01) is programmed in mnemonic code (see figure 4-03).

Red CLR key

The red CLR (clear) key clears the display and cancels current programming console operations. It is also used for entering the password at the beginning of programming operations. Any programming console operation can be cancelled by pressing the CLR key, although it may have to be pressed two or three times to cancel the operation and clear the display (with monitored and displayed data, for example).

Yellow operation keys

The twelve yellow operation keys located in the lower right-hand corner are used for writing and correcting programs. Detailed explanations of their functions are given in the User Manual for Omron PLCs.

Grey instruction and data area keys

Except for the SHIFT key on the upper right of the keyboard, the sixteen grey keys are used to enter instructions and designate data area prefixes when entering or changing a program (figure 1-13). The SHIFT key is similar to the SHIFT key of a typewriter or the second function key on calculators, and is used to alter the function of the next key pressed. Note: It is not necessary to hold the SHIFT key down, just to press it once and then press the key to be used with it.

The grey keys other than the SHIFT key have either the mnemonic name of the instruction or the abbreviation (acronym) of the data area written on them. The functions of these keys are described in figure 1-13.

Clarification

Internal auxiliary relays, where mentioned in this book as such, designate internal outputs not falling into the HR, TIM, CNT, DM or special relay area. These internal auxiliary relays are listed in figure 1-14, bits 1000 to 1807. For an Omron C20 PLC, these relays constitute all relays from word 10 to word 18, in which only bits 00 to 07 of word 18 are accessible.

This chapter is not intended to promote any particular brand or type of programmable controller and, therefore, great care has been taken to present the necessary information in as unbiased a way as possible. Nevertheless, to make the applications meaningful, some practical programming procedures are given, and have been based on the Omron C20, Omron C28K and Omron C200H programmable controllers. These first two PLC types are essentially identical to the Festo FPC 201 and FPC 202, and are closely related to the smaller types of Mitsubishi, Izumi, Texas Instruments, Siemens and Allen Bradley programmable controllers. These are all small controllers, ideally suited for small to medium-size, sequential and combinational, fluid power controls. The Omron C200H PLC is an extremely powerful, medium-range PLC with sophisticated, state-of-the-art functions.

Key	Function
FUN	Used to select and enter instructions with function codes. To enter an instruction with function code, press the FUN key and then the appropriate numerical value. Instructions and their function codes are listed in Appendix C.
SFT	Enters a shift register instruction.
NOT	Inverts the instruction before it. Often used to form a normally closed input or output. It is also used to change instructions from differentiated to non-differentiated and vice versa.
AND	Enters a logical AND instruction.
OR	Enters a logical OR instruction.
CNT	Enters counter instructions. After **CNT,** enter the counter number and data.
LD	Enters load instructions.
OUT	Enters output instructions.
TIM	Enters timer instructions. After **TIM,** enter the timer number and data.
TR	Used to specify a TR bit.
LR	Used to specify the LR area.
SHIFT HR	Used to specify the AR area. (both keys need to be pressed)
HR	Used to specify the HR area.
DM	Used to specify the DM area.
CH / *	Used to specify a channel.
CONT / #	Used to search for a bit.

Figure 1-13. The sixteen grey programming console keys and their function for the programming and diagnosing of PLC programs (see also figure 1-05).

Figure 1-14. Assignment of input–output channels (16 Bit Words) and relay numbers for an "OMRON C20" programmable controller.

Name	No. of points	Relay number									
Input relay	80	0000 to 0415									
		00CH		01CH		02CH		03CH		04CH	
		00	08	00	08	00	08	00	08	00	08
		01	09	01	09	01	09	01	09	01	09
		02	10	02	10	02	10	02	10	02	10
		03	11	03	11	03	11	03	11	03	11
		04	12	04	12	04	12	04	12	04	12
		05	13	05	13	05	13	05	13	05	13
		06	14	06	14	06	14	06	14	06	14
		07	15	07	15	07	15	07	15	07	15
Output relay	60	0500 to 0915									
		05CH		06CH		07CH		08CH		09CH	
		00	08	00	08	00	08	00	08	00	08
		01	09	01	09	01	09	01	09	01	09
		02	10	02	10	02	10	02	10	02	10
		03	11	03	11	03	11	03	11	03	11
		04	12	04	12	04	12	04	12	04	12
		05	13	05	13	05	13	05	13	05	13
		06	14	06	14	06	14	06	14	06	14
		07	15	07	15	07	15	07	15	07	15
Internal auxiliary relay	136	1000 to 1807									
		10CH		11CH		12CH		13CH		14CH	
		00	08	00	08	00	08	00	08	00	08
		01	09	01	09	01	09	01	09	01	09
		02	10	02	10	02	10	02	10	02	10
		03	11	03	11	03	11	03	11	03	11
		04	12	04	12	04	12	04	12	04	12
		05	13	05	13	05	13	05	13	05	13
		06	14	06	14	06	14	06	14	06	14
		07	15	07	15	07	15	07	15	07	15
		15CH		16CH		17CH		18CH			
		00	08	00	08	00	08	00			
		01	09	01	09	01	09	01			
		02	10	02	10	02	10	02			
		03	11	03	11	03	11	03			
		04	12	04	12	04	12	04			
		05	13	05	13	05	13	05			
		06	14	06	14	06	14	06			
		07	15	07	15	07	15	07			
Holding relay (retentive relay)	160	HR000 to 915									
		00CH		01CH		02CH		03CH		04CH	
		00	08	00	08	00	08	00	08	00	08
		01	09	01	09	01	09	01	09	01	09
		02	10	02	10	02	10	02	10	02	10
		03	11	03	11	03	11	03	11	03	11
		04	12	04	12	04	12	04	12	04	12
		05	13	05	13	05	13	05	13	05	13
		06	14	06	14	06	14	06	14	06	14
		07	15	07	15	07	15	07	15	07	15
		05CH		06CH		07CH		08CH		09CH	
		00	08	00	08	00	08	00	08	00	08
		01	09	01	09	01	09	01	09	01	09
		02	10	02	10	02	10	02	10	02	10
		03	11	03	11	03	11	03	11	03	11
		04	12	04	12	04	12	04	12	04	12
		05	13	05	13	05	13	05	13	05	13
		06	14	06	14	06	14	06	14	06	14
		07	15	07	15	07	15	07	15	07	15

Name	No. of points	Timer/counter number					
Timer/counter	48	TIM/CNT00 to 47					
		00	08	16	24	32	40
		01	09	17	25	33	41
		02	10	18	26	34	42
		03	11	19	27	35	43
		04	12	20	28	36	44
		05	13	21	29	37	45
		06	14	22	30	38	46
		07	15	23	31	39	47

2 PLC PROGRAMMING CONCEPTS

A most misleading impression for many control engineers and electricians when confronted with a programmable logic controller is that PLCs can easily be programmed with the "ladder logic" method. Ladder logic *is not a method of circuit design*; it is a method of circuit presentation, like an electrical, pneumatic or hydraulic circuit. For this reason numerous "amateur" programmers have insurmountable difficulties and frustration while programming their little electronic "wizard"; some to such an extent that they give up before they ever see their controller work! In the same way as one has to learn to design a properly functioning hydraulic or pneumatic control circuit one needs also to know how to design a PLC ladder logic circuit. It is therefore important and mandatory that the designer of PLC circuits possess fundamental knowledge and understanding of general switching logic, combinational and sequential circuit design concepts, and the function of the five basic logic concepts "NOT", "YES", "AND", "OR" and "INHIBITION". Other equally essential switching concepts include timer function, counter function, shift register function, auxiliary relay function, memory function as applied to flip-flops, BCD number systems and data conversion.

This book, therefore, presents a method for programming combinational and sequential type industrial controls. Combinational controls are presented and illustrated in Chapters 7 and 10. Sequential controls, using the well known step-counter circuit design concept, are explained and illustrated in Chapters 8 and 9.

PLC PROGRAMMING LANGUAGES

A PLC may be programmed in three language forms or ways. That is what most PLC "experts" say, but this statement can be misleading. PLCs understand and can be programmed with only one programming language. This language is sometimes called machine language or a statement list, but more correctly, it is called Boolean instruction language or mnemonic code language. This mnemonic code language consists of mnemonic coded input instructions embracing the four standardised mnemonic words:

"LOAD" (yes), "AND", "OR", "NOT"

and mnemonic coded output instructions. These may be instructions for an output relay, a flip-flop, a timer or a counter, or an arithmetic or special instruction (figure 2-06). Some of these outputs require an acronym to complement the operand, such as TIM, CNT, FUN 11 and HR. Mnemonic instructions are entered into the addresses of the PLC's user memory and each mnemonic instruction is usually followed by an operand label, which contains the word and the bit within that word (figures 2-02 and 1-10).

To design a PLC control circuit directly in Boolean or mnemonic format is often difficult, if not impossible. For this reason, most PLC program designers create their program first on paper sheets in ladder diagram form (see figures 2-01 and 7-13), or system flow chart form (see figures 2-03 and 9-11). Since electricians and electronic engineers usually program PLCs, and electricians are used to ladder diagram circuit presentation, it comes as no surprise that ladder programming is so popular.

As PLCs cannot be directly programmed from ladder or system flow chart circuits, the ladder diagram and the system flow chart need to be translated and fragmented into mnemonic code instructions followed by the necessary operand for each instruction (see figures 2-02 and 7-13). These instructions are then entered into the PLC via the programming console (hand-held programmer). Alternatively, one may create the ladder diagram or system flow chart directly on the screen of a computer or desktop PLC programmer, and then transfer it to the PLC with the aid of a host link and an appropriate PLC module.

Figure 2-01 shows the ladder diagram for a simple "AND" logic function, consisting of input 007 and 009, driving output relay 503 when the logic function is true. The order of listing the two inputs is unimportant.

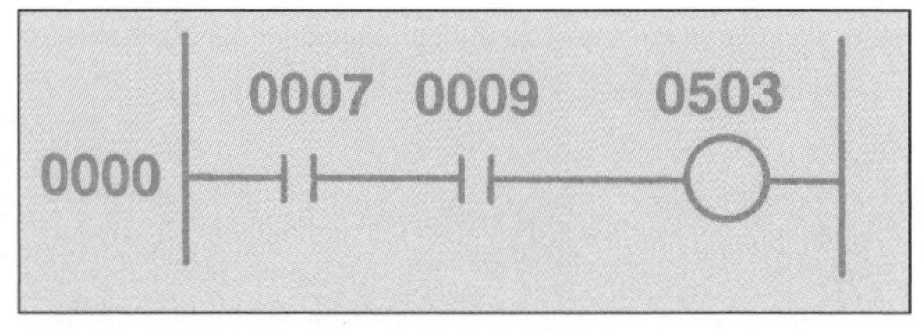

Figure 2-01. Ladder image for a logic line.

Hence, 009 could be on the left and 007 on the right of the "AND" function on the rung. Note: PLCs do not series-connect, they connect with "AND" gates.

ADDRESS	MNEMONIC	OPERAND
0000	LD	0007
0001	AND	0009
0002	OUT	0503

Figure 2-02. Address–mnemonic–operand Boolean-type PLC programming image (mnemonic list) for the ladder diagram shown in figure 2-01.

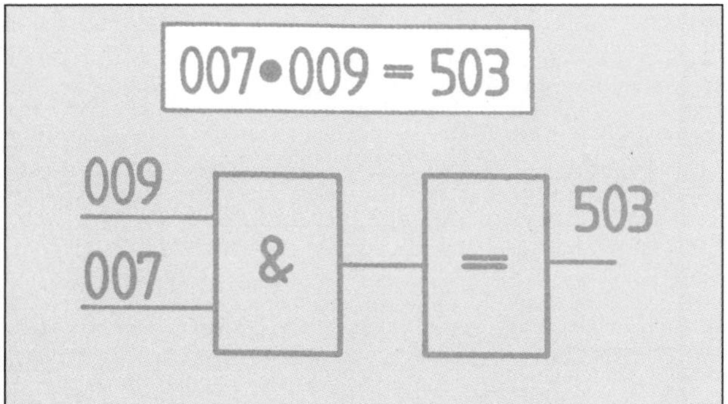

Figure 2-03. System flow chart presentation for logic line, shown in figures 2-01 and 2-02.

Figure 2-02 shows the mnemonic list and order of entering the "AND" function shown in figure 2-01 into the PLC with a hand-held programming console. Each instruction is given an address in the user memory of the PLC, and each mnemonic (instruction) is followed by an operand. To write the address line just programmed into the user memory, and advance from one address to the next, one simply needs to press the "WRITE" key on the programmer after having entered the operand (see also the section "Logic Line (Network)" in this chapter).

Figure 2-03 shows the system flow chart circuit presentation for the "AND" function depicted in figures 2-01 and 2-02. The symbols used here reflect the international logic symbols shown in figure 3-15. Some European PLC manufacturers prefer and recommend system flow chart programming over ladder programming.

BOOLEAN SWITCHING EQUATIONS FOR PLC PROGRAMMING

Logic switching functions can be expressed in equation form. This applies to all sorts of control, mechanical, pneumatic, hydraulic, electrical or electronic. The fundamental principles of logic switching expression use the notations of Boolean algebra, which was devised and published by George Boole, professor of mathematics at Queens College, County Cork, Ireland, in 1854. His laws and theorems are still valid today and are the foundation of logic switching, used in many areas, including telephones and computers.

Such switching equations consist of signal labels, which for PLCs are usually numbers and are called operands. They also include "OR" function signs, "AND" function signs and brackets (parentheses) where necessary, and an equals sign before the output designation. For example:

$$(0003 \bullet 0501) + (1006 \bullet 0008) + 0006 = 0503$$

These signal labels (called operands) are made up of the bit number, which is the two rightmost digits, and the word number, which is the two leftmost digits. A dash above the signal label means the inverse of the signal. The inverse of the signal means the signal is "NOT ON" or "NOT PRESENT". The logic Boolean expression for connecting inputs by the "AND" function is denoted by a dot sign (•). The logic Boolean expression for connecting inputs by the "OR" function is denoted by a plus sign (+). These two unique signs may also be used to logically connect switching expressions enclosed by brackets. For example:

$$(0006 \bullet \overline{0003}) + (1000 \bullet 502)$$

for "AND" connecting within brackets and connecting the brackets with logic "OR"; and:

$$(0011 + 1003) \bullet (\overline{507} + \mathrm{T}\ 03)$$

for "OR" connecting within brackets and connecting the brackets with logic "AND".

Switching equations may be used directly to program a PLC controller by translating these equations into mnemonic codes (figures 2-15 and 2-16). Alternatively, one may start with a ladder diagram and then enter instructions into the PLC by translating the ladder diagram graphics into mnemonic codes. Experienced programmers, however, program the PLC directly from the ladder, by making the translation to mnemonic coding a mental process.

Whatever programming language the program designer is using, it must be made clear that the program

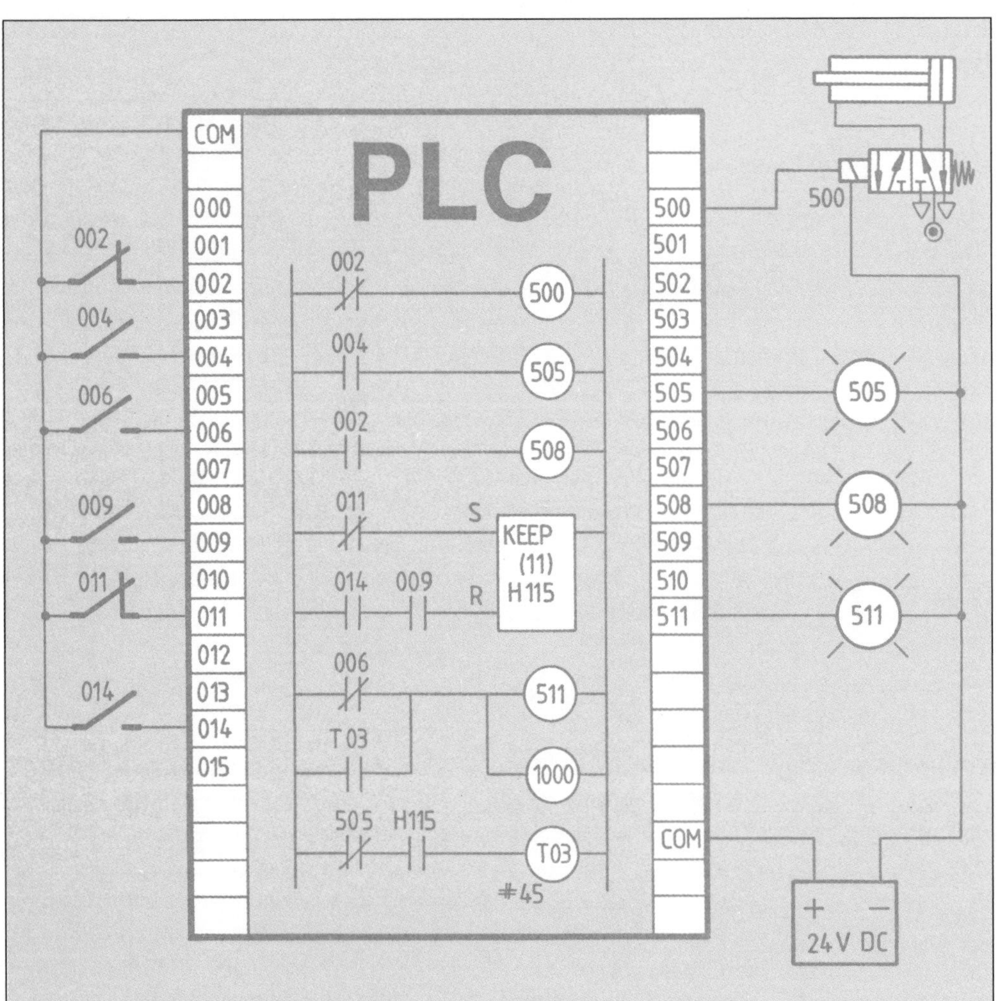

Figure 2-04. The ladder diagram circuit reflects the structure of the program written into the user memory of the PLC.

previously designed is written into the user memory of the PLC and is therefore located and operating inside the PLC (figure 2-04). The sensors and switches are hard-wired into the PLC and do not influence the circuit written into the user memory of the PLC, other than what is shown in figures 2-07 and 2-08.

LADDER DIAGRAM CIRCUIT PRESENTATION

A ladder diagram consists of two vertical lines called bus bars or power rails, and horizontal lines, called rungs or logic lines (figures 2-04 and 2-06). The rungs contain one or several normally open contacts (⊣ ⊢), and normally closed contacts (⊣/⊢), and at their right hand end, immediately before the right hand bus bar, an instruction that could be one of the commonly used PLC instructions shown in figure 2-05. For further instructions, see Chapters 4 to 10 and the Appendix.

The horizontal rungs are sometimes also called branching lines. These branching lines may be a singular rung (figure 2-06 F), or they may be used in "OR" connection as parallel rungs (figure 2-06 A), or they may form interim blocks of "OR" circuits (figure 2-06 C). Where necessary, such "OR" blocks may be "AND" connected into complex subcircuits as is shown in figures 2-16 and 2-19. As shown in figure 2-06 E, the rungs can branch apart, and they can join back together. Furthermore, such rungs may branch out into several output instructions, as is shown in figures 2-06 D and 2-22.

When logically "true", these subcircuits enable the output instruction on the right hand side and determine when and how the output instruction is executed. The complete circuit arrangement ending in an output instruction is called a logic line or network (see logic lines in figure 2-06).

LOGIC LINE (NETWORK)

Ladder diagrams consist of one or a number of individual logic lines (sometimes also called networks or function blocks). Each logic line starts from the left hand bus bar of the ladder diagram and ultimately terminates on the right hand bus bar with an output instruction, as shown in figures 2-04, 2-05 and 2-06. It is not absolutely essential to draw the right hand bus bar in a ladder diagram; it is, however, customary to do so.

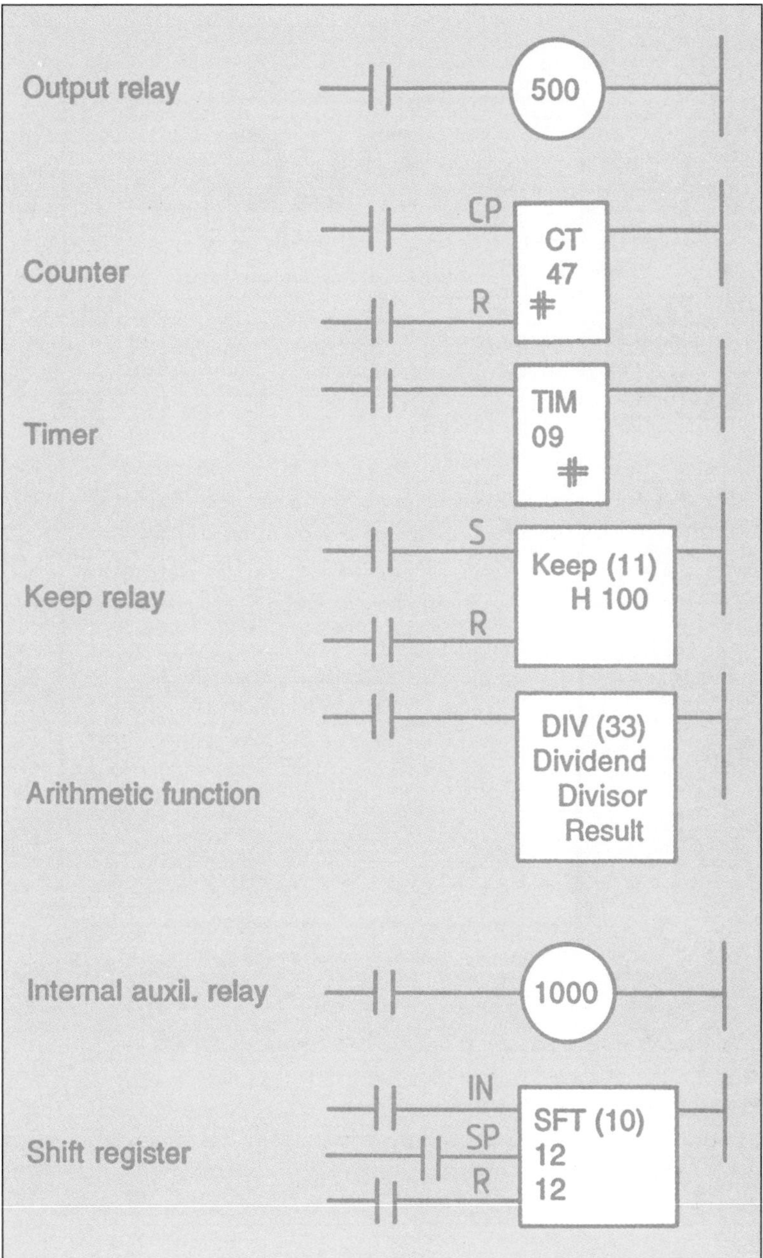

Figure 2-05. Various PLC output instructions (internal as well as external outputs).

Contacts placed in the rungs of these logic lines are either normally closed or normally open (sometimes called normally closed or normally open conditions). They may logically be arranged with other contacts, in logic "AND" connections, in logic "OR" connections, or when attached immediately adjacent to the left hand bus bar as a "load" connection (figure 2-06). The number above each condition (contact) indicates the operand given to this condition (bit). The logic status of these conditions (bits), either 0 or 1, determines the execution of the output instruction at the right hand end of the logic line (see figure 2-06).

Programming examples

When a logic line starts with, say, input 0015 as a normally open contact, the programming instruction in the mnemonic list will be LD 0015. If the logic line starts with a normally closed contact, say 0007, the instruction in the mnemonic list would then be LD NOT 0007. The load instruction ("LD") is also used to start a new block within a logic line (figure 2-06, address 33). Figure 2-06 shows six typical logic lines as they may frequently be found in ladder diagrams for PLC.

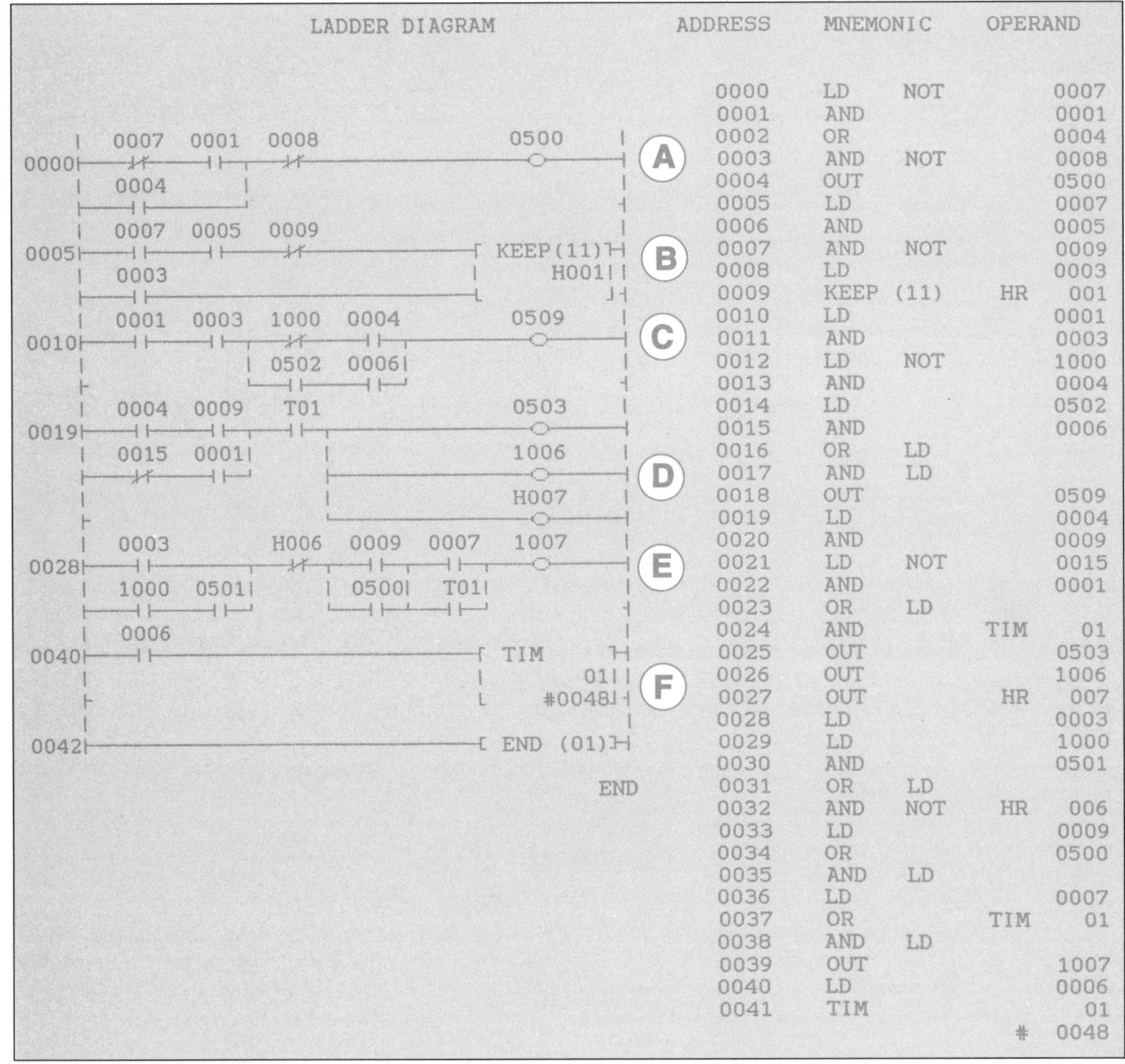

ADDRESS	MNEMONIC		OPERAND	
0000	LD	NOT		0007
0001	AND			0001
0002	OR			0004
0003	AND	NOT		0008
0004	OUT			0500
0005	LD			0007
0006	AND			0005
0007	AND	NOT		0009
0008	LD			0003
0009	KEEP	(11)	HR	001
0010	LD			0001
0011	AND			0003
0012	LD	NOT		1000
0013	AND			0004
0014	LD			0502
0015	AND			0006
0016	OR	LD		
0017	AND	LD		
0018	OUT			0509
0019	LD			0004
0020	AND			0009
0021	LD	NOT		0015
0022	AND			0001
0023	OR	LD		
0024	AND		TIM	01
0025	OUT			0503
0026	OUT			1006
0027	OUT		HR	007
0028	LD			0003
0029	LD			1000
0030	AND			0501
0031	OR	LD		
0032	AND	NOT	HR	006
0033	LD			0009
0034	OR			0500
0035	AND	LD		
0036	LD			0007
0037	OR		TIM	01
0038	AND	LD		
0039	OUT			1007
0040	LD			0006
0041	TIM			01
			#	0048

Figure 2-06. Six typical logic lines, as used for PLC ladder circuit programming.

- Logic line Ⓐ in figure 2-06 drives output relay 500 if its logic conditions are true:

$$[(\overline{0007} \bullet 0001) + (0004)] \bullet \overline{0008} = 500$$

- Logic line Ⓑ is a keep relay (memory-retentive relay):

$$0007 \bullet 0005 \bullet \overline{0009} = \text{SET HR } 001$$
$$0003 = \text{RESET HR } 001$$

- Logic line Ⓒ drives output relay 509 if its logic is true (consisting of a complex "OR" and an "AND" function, including an inhibition function):

$$(0001 \bullet 0003) \bullet [(\overline{1000} \bullet 0004) + (502 \bullet 0006)] = 509$$

- Logic line Ⓓ illustrates how a complex logic including a timer contact drives several output relays.
- Logic line Ⓔ illustrates how a complex logic, consisting of an "OR" block (B1) inhibited by the presence of HR006 and then "AND" connected to two more "OR" blocks (Block 2 and Block 3), ultimately drives auxiliary relay 1007, if its complex logic is true. It must be noted that inhibiting contact HR006 could also be placed in front of block 1 or behind block 3, or between blocks 2 and 3, giving the same logic effect.
- Logic line Ⓕ shows a timer relay (T01) with its driving contact 0006. The timer starts to time-out when contact 0006 is logic 1 ("ON"). (See also figure 5-02).

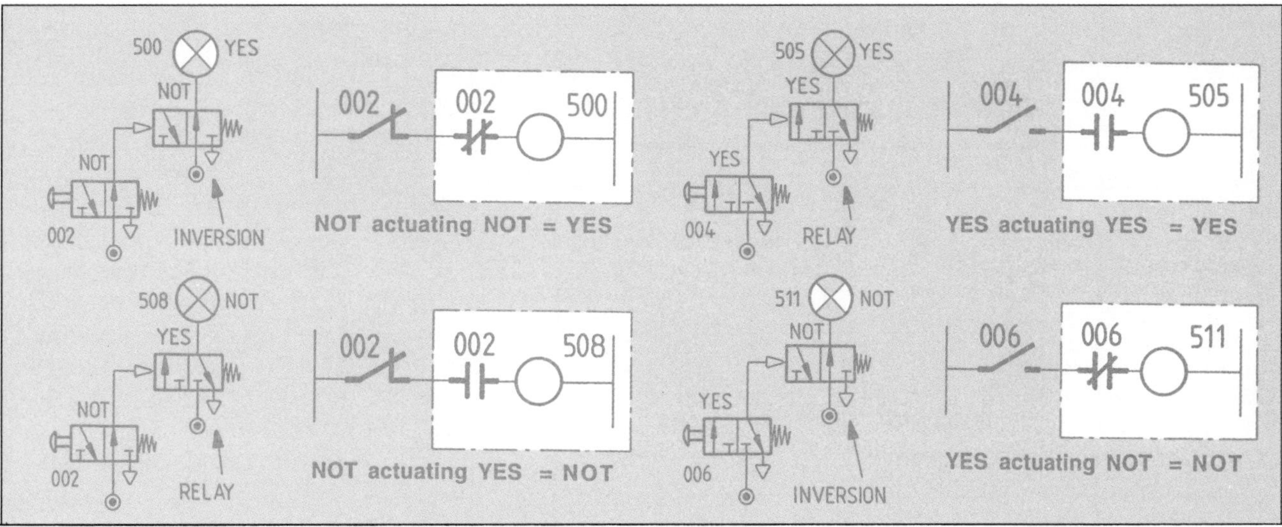

Figure 2-07. Switch-to-PLC bit contact relationship, to obtain internal or external output signals or to invert input signals for internal processing.

BINARY CONTACT LOGIC

Each condition (contact) in a ladder diagram may be either logically "ON" or "OFF", depending on the logic status of the operand bit assigned to the contact. A normally open condition is "ON" if its operand bit is "ON", and "OFF" if its operand bit is "OFF". The operand bit is "ON" if its internal relay or external switch or sensor on the machine driving it is "ON". Conversely, a normally closed condition is "ON" if its operand bit is "OFF". The operand bit is "OFF" if its internal relay or external switch or sensor on the machine is "OFF". To qualify, "OFF" in a logic sense means not actuated. Generally speaking, one uses a normally open condition (contact), when something must happen, when the PLC bit is "ON". Conversely, one uses a normally closed condition (contact), when something has to happen, when the PLC bit is "OFF" (figure 2-04).

Note: For the signalling of emergency action, one usually uses normally closed switches on the machine, so that, if a wire between switch and PLC breaks, the emergency action is started. This is illustrated in figure 2-07 with "NOT" actuating "NOT". (See also next section, "Input Device to Input Bit Relationship".) Such action may be the start of an emergency machine sequence, or the switching "OFF" of only some machine functions (partial disablement) or, where essential, the disconnection of power to all machine actuators. For such requirements, check with your local or national machine safety regulations.

INPUT DEVICE TO INPUT BIT RELATIONSHIP

The boxed circuit sections in figure 2-07 reflect the ladder program written into the user memory of the PLC (the central processing unit). The switches on the left hand side of the boxes are located outside the PLC, and are usually mounted on the machine controlled by the PLC. The illustration, therefore, explains the logic relationship and logic output result obtained from each of the four electrical connections.

- Picture 1 demonstrates how a normally closed switch, driving a normally closed bit contact within the PLC, causes the output to go "ON" when the N/C switch is actuated, or the wire between switch and PLC breaks.
- Picture 2 demonstrates how a normally closed switch, driving a normally open bit contact within the PLC, causes the output to go "OFF" when the N/C switch is actuated, or the wire between switch and PLC breaks.

Picture 1, NOT actuating NOT = YES
Picture 2, NOT actuating YES = NOT
Picture 3, YES actuating YES = YES
Picture 4, YES actuating NOT = NOT

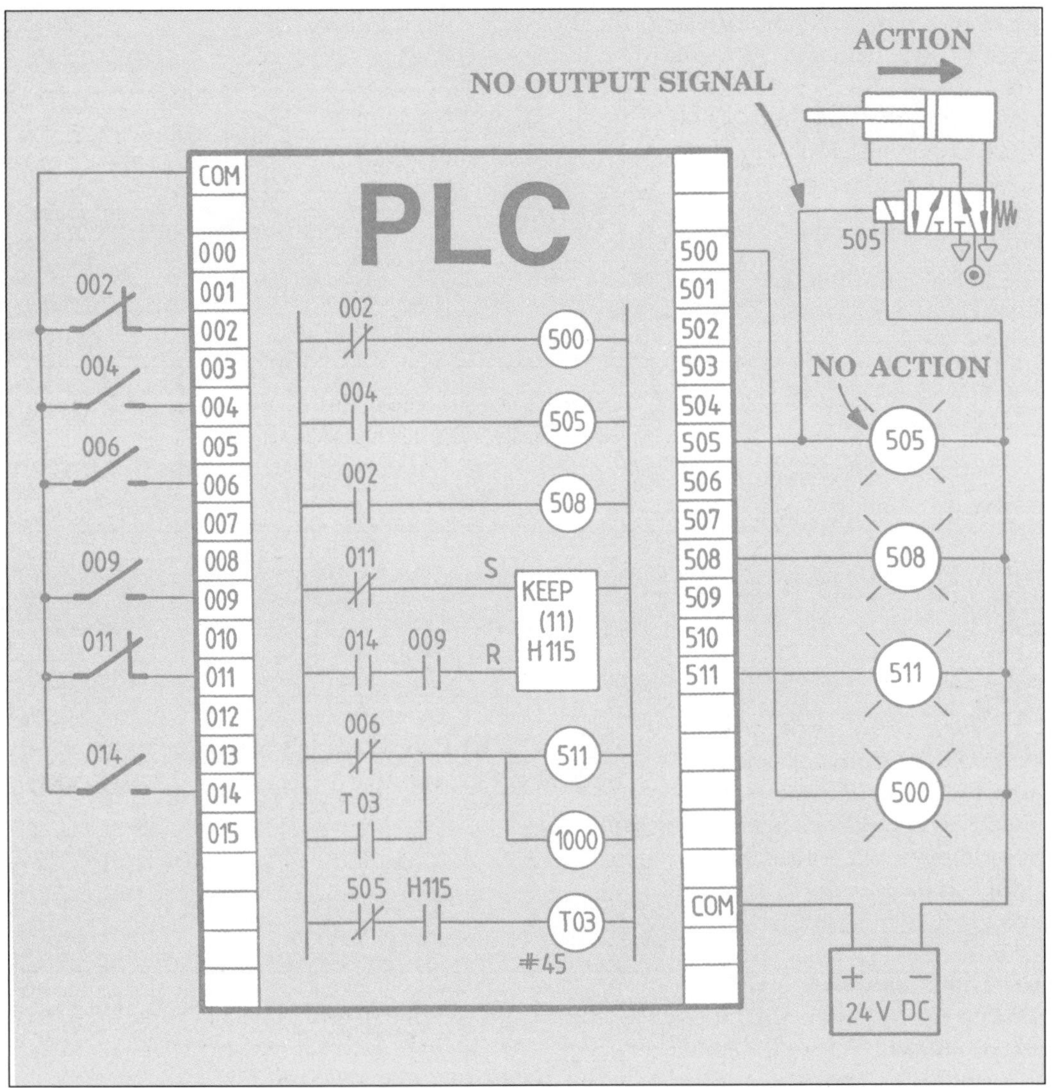

Figure 2-08. PLC with normally open and normally closed input devices, internal software circuit (ladder diagram) and output devices, explaining the relationship of input devices actuating and inverting bits, and finally actuating outputs inside and outside the PLC (see also figures 2-04 and 2-05).

- Picture 3 demonstrates how a normally open switch, driving a normally open bit contact within the PLC, causes the output to go "ON" when the N/O switch is actuated.
- Picture 4 demonstrates how a normally open switch, driving a normally closed bit contact within the PLC, causes the output to go "OFF" when the N/O switch is actuated.

OUTPUT DEVICE CONSIDERATIONS

Depending entirely on the output device being electrically connected to the PLC output point, no output signal being present on that point could mean action or no action on the machine under control (figure 2-07). The absence of output signal 505 on the PLC schematic shown in figure 2-07 causes the lamp connected to that output point to be "OFF" (not illuminated), whereas the absence of the same 505 signal also connected to the solenoid valve prompts the spring to reset the solenoid valve to its "normal" position, thus causing the linear pneumatic actuator to retract (the normal position for this valve is the spring-selected position). It is for this reason that machine designers need to be careful when choosing pneumatic and hydraulic valves, and consideration must always be given to the consequences of a potential power loss to the PLC, or severing of its output and input signal wires! The author strongly recommends the exclusive use of double solenoid valves, which maintain their presently selected valve spool position regardless of PLC power loss or damage to PLC input and output signal wires.

OPERAND BITS

Operand bits, given to conditions and instructions (inputs and outputs) in the ladder diagram, can be any bit in the internal relay area (IR), special relay area (SR), holding relay area (HR), auxiliary relay area (AR), timer/counter relay area (TC), or link relay area (LR). This means that an output relay operand bit may also serve in a logic line as an input bit for an N/C or N/O contact (see figure 4-05, operand bit 506, and figure 7-13, operand bits 505 and 506 in addresses 32 to 42). The same is true for internal outputs such as flip-flops (HR), which are always used as input bits driving either other internal outputs or external outputs (see figure 4-06).

PROGRAMMING WITH MNEMONIC CODES (BOOLEAN PROGRAMMING)

If the logic line shown in figure 2-01 had to be entered (programmed) into the PLC with a hand-held or PLC-fastened programming console, then the following programming keys had to be pressed:

LD 7 WR AND 9 WR OUT 5 0 3 WR

The zeros in front of the operands 0007, 0009 and 0503 need not be entered when programming, as the PLC puts them in automatically.

> **Note:** WR means the WRITE key, shown in figure 1-05. From now on, this key is abbreviated as WR. By pressing the WR key, one confirms the completion of an address instruction. The address instruction consists of a mnemonic code and its operand. By pressing the WR key, one also advances the address counter to the next address (see figure 1-11).

The program is entered into addresses in the program memory. Program memory addresses start at 0000 and run until the program memory is full. For an Omron C28K PLC this memory capacity is 1194 addresses. During programming, addresses are automatically displayed and do not have to be entered. When converting a ladder diagram to mnemonic code, and entering this into the user memory of the PLC, it is best to start at program memory address 0000, unless there is a specific reason for starting elsewhere.

For the programming of the logic line with an "OR" function in figure 2-09, press the following thirteen keys:

LD 1 2 WR OR 4 WR OUT 1 0 0 0 WR

For the programming of the logic line in figure 2-10, which includes an inhibition function (NOT 003 AND 005) and an "OR" function, press the following sixteen keys on the programming console:

LD NOT 3 WR AND 5 WR
OR 1 4 WR OUT 5 0 6 WR

For the programming of further logic lines and circuits, see mnemonic lists in figures 2-06, 4-03, 4-05, 4-07, 4-10, 5-04, 5-08, 5-12, 6-08, 6-12, 7-06, 7-13 and 10-34.

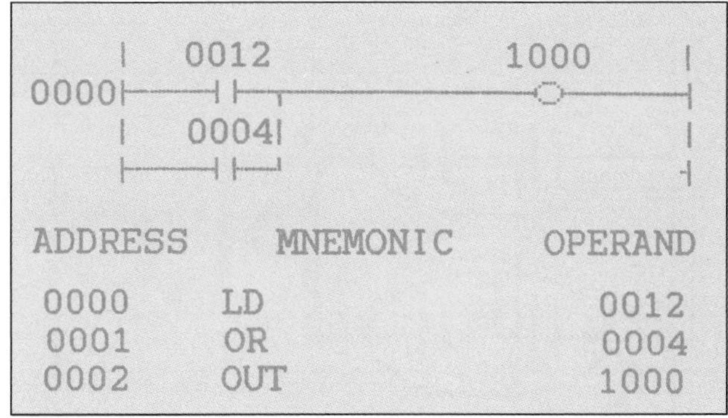

Figure 2-09. Mnemonic instructions plus ladder diagram for an "OR" logic line with output.

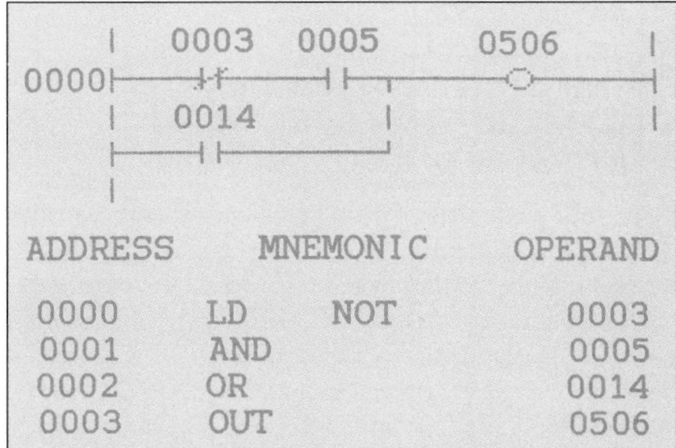

Figure 2-10. Mnemonic instructions plus ladder diagram for an "INHIBITION" rung, as well as an "OR" rung, terminating in output 506.

PROGRAMMING BRACKET-ENCLOSED BOOLEAN EQUATIONS

Most PLC switching functions may be expressed in Boolean equation format. Such Boolean equations invariably contain bit conditions (contacts) that are "AND" connected, "OR" connected, or often both (figures 2-09 to 2-12 and 2-15).

Single logic blocks, such as shown in figures 2-01 and 2-09, are very simple to program, since they would need no logic sum brackets or logic product brackets if expressed as a Boolean equation. Logic Boolean equations that are of mixed nature with logic "OR" blocks being "AND" connected, or logic "AND" blocks being "OR" connected, however, need brackets around these blocks and, therefore, need special programming procedures (figures 2-12 and 2-15).

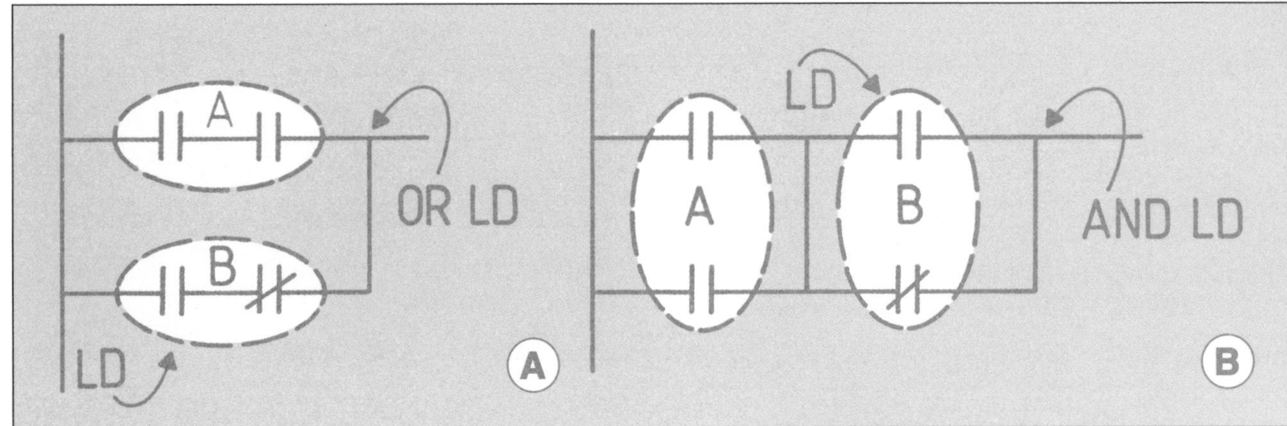

Figure 2-11. "AND" blocks joined in "OR" connection (vertical joining) and "OR" blocks joined with "AND" connection (horizontal joining).

CONNECTING "AND" BLOCKS WITH LOGIC "OR" USING "OR LD" (VERTICAL BLOCK CONNECTION)

If an "OR" connection has more than one contact in its rungs below the top rung, then the programming of such rungs must be started with a load ("LD") mnemonic instruction, instead of the usual "OR" instruction (see figure 2-11 and compare it with figure 2-10). If a normally closed contact starts such a rung, then the "LD NOT" instruction is used (figure 2-18, address 0006). Such rungs are terminated with the additional "OR LD" instruction, as is shown in figures 2-11 A, 2-13 and 2-18. "OR LD" is a mnemonic instruction that is not followed by an operand. However, one needs to press the WR key to finish the instruction and advance the address counter. The "OR LD" instruction and the "LD" instruction assigned to the first contact in that rung become a substitute for setting logic product brackets (some PLC brands require the setting of opening and closing logic product brackets). The "OR LD" load instruction is a mnemonic code that needs no operand. It is, however, followed by pressing the WRITE key to advance the address counter to the next address (figures 2-13 and 2-18).

Why the need for this "OR LD" instruction? To answer that question, it is best to investigate the programming of this function by expressing its logic combination with its Boolean equation, which reads:

$$003 + (004 \bullet \overline{007}) = 501$$

If no logic product brackets were set for this equation, the PLC, by executing the program cyclically from top to bottom during the scan, could turn output 501 "ON", when input 003 or input 004 are "ON", and if input 007 is not "ON". This is best shown with logic system flow chart symbols in figure 2-12 A and the mnemonic code programming for this equation in figure 2-13.

Figure 2-12. Vertical joining of "AND" blocks with wrong and correct circuit presentation using system flow chart symbols to illustrate the wrong and correct connection of this logic line. The Boolean equation shows the correct circuit expression.

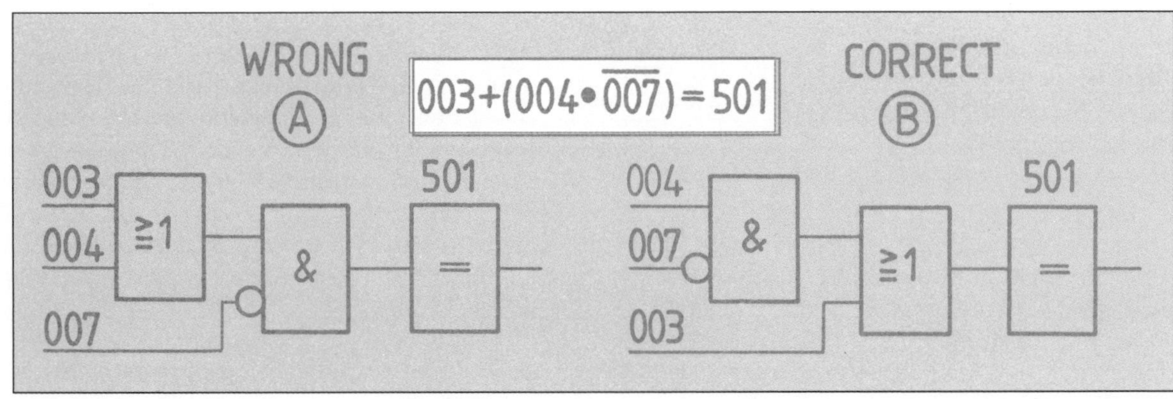

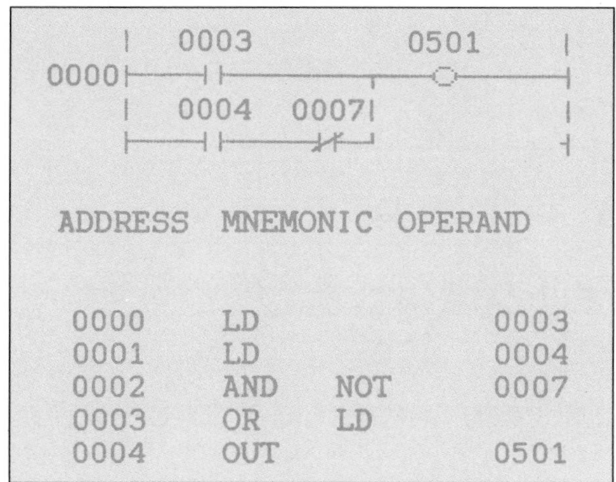

ADDRESS	MNEMONIC		OPERAND
0000	LD		0003
0001	LD		0004
0002	AND	NOT	0007
0003	OR	LD	
0004	OUT		0501

Figure 2-13. Ladder diagram presentation and mnemonic PLC programming instructions for the circuit and equation shown in figure 2-12.

By setting logic product brackets, the program emphasises the closer logic coherence of bits 004 and 007, as is shown in the equations given in figure 2-12. The PLC now turns output 501 "ON" if either 003 or the logic product in the brackets is present. This is shown in figures 2-12 B and 2-13.

If the "OR" connection in this illustration were rearranged so that contacts 004 and 007 were located in the top rung and contact 003 was located in the bottom rung, the mnemonic code "OR LD" would then no longer be necessary, and the bottom rung contact 003 would be entered into the PLC as OR 003, instead of LD 003 (compare figure 2-13 with 2-14).

Note: The top rung of such an "OR" connection may contain one or several contacts without the need to use "OR LD" (see figure 7-18, logic line for internal auxiliary relay 1001, address 0049).

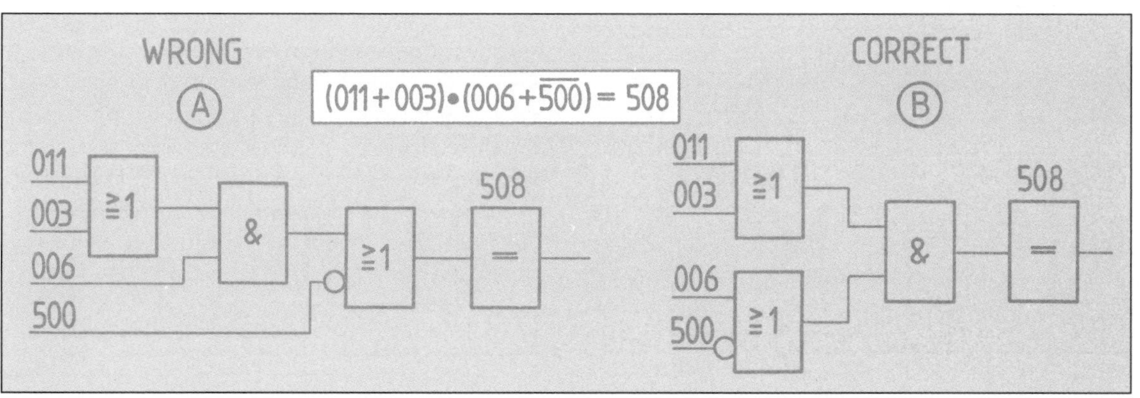

Figure 2-15. Boolean equation with wrong and correct system-flow-chart circuit presentations for the ladder diagram circuit shown in figure 2-17.

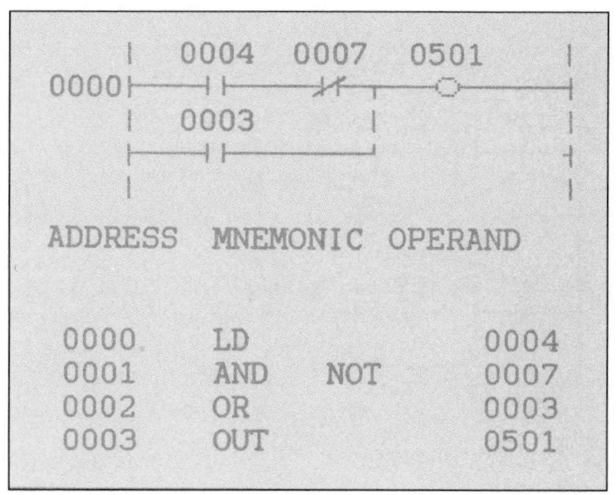

ADDRESS	MNEMONIC		OPERAND
0000	LD		0004
0001	AND	NOT	0007
0002	OR		0003
0003	OUT		0501

Figure 2-14. Rearranging the ladder circuit for the "OR" blocks in figure 2-13 simplifies programming instructions, and the mnemonic code "OR LD" is no longer required.

CONNECTING "OR" BLOCKS WITH LOGIC "AND" USING "AND LD" (HORIZONTAL BLOCK CONNECTION)

Where two "OR" blocks, such as shown in figures 2-11 B and 2-06 E, need to be "AND" connected (horizontal block connection), such connections require an "AND LD" mnemonic instruction at the end of programming the second "OR" block. This "AND LD" mnemonic code is a substitute for the setting of logic sum brackets (figures 2-15 B and 2-16). Being a mnemonic code only, the "AND LD" instruction needs no operand, but it is always followed by pressing the WRITE key to advance the address counter. The first contact for that block, in figure 2-16, is entered as "LD"

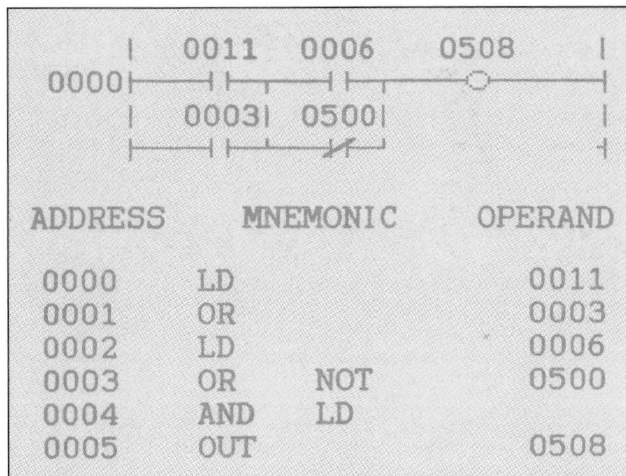

ADDRESS	MNEMONIC		OPERAND
0000	LD		0011
0001	OR		0003
0002	LD		0006
0003	OR	NOT	0500
0004	AND	LD	
0005	OUT		0508

Figure 2-16. "AND" connected "OR" blocks, using the "AND LD" mnemonic instruction to join the "OR" blocks horizontally.

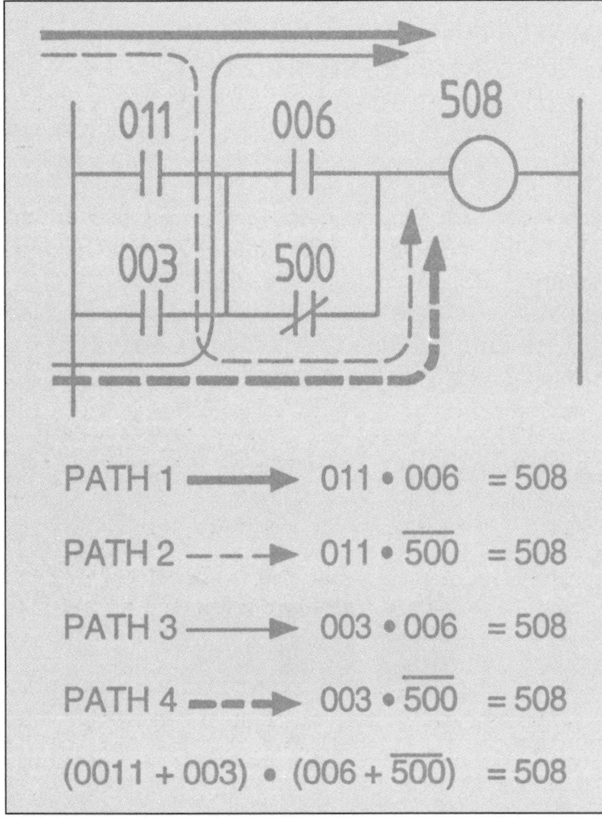

Figure 2-17. Four individual logic paths may render output 508.

(LOAD), followed by the contact's operand, or with "LD NOT", if that contact is normally closed (figures 2-06 C and 2-06 E). Several examples for the programming of "OR" blocks and connecting blocks with "OR LD" or "AND LD" are given in figures 2-06, 2-16, 2-18, 2-19, 2-22, 2-23 and 9-12.

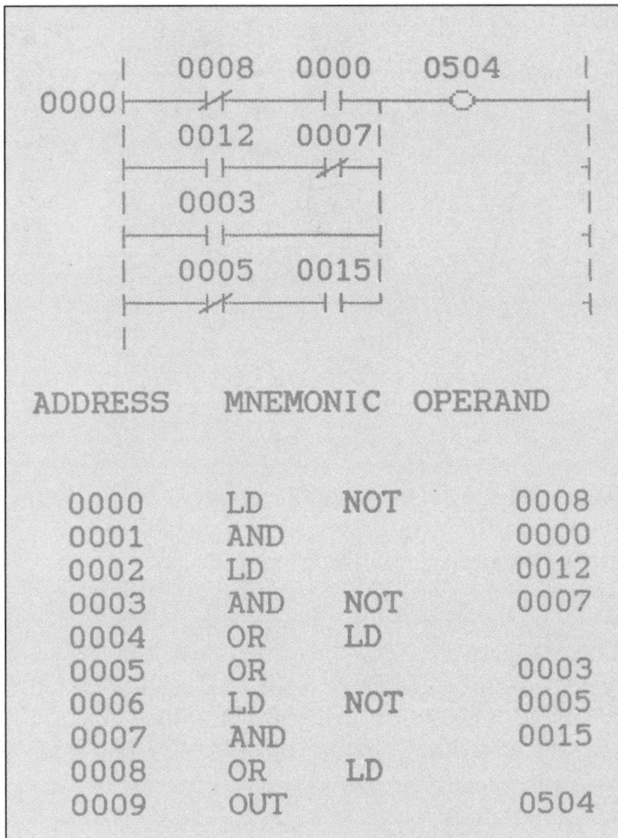

ADDRESS	MNEMONIC		OPERAND
0000	LD	NOT	0008
0001	AND		0000
0002	LD		0012
0003	AND	NOT	0007
0004	OR	LD	
0005	OR		0003
0006	LD	NOT	0005
0007	AND		0015
0008	OR	LD	
0009	OUT		0504

Figure 2-18. "OR" connected "AND" blocks, using the "OR LD" mnemonic instruction to join the "AND" blocks vertically.

Why the need for this "AND LD" instruction? To answer that question, it is again best to demonstrate its necessity by using a Boolean equation and its equivalent system flow chart symbols, as shown in illustration 2-15.

By setting logic sum brackets to emphasise the coherence of 011 and 003 and of 006 and $\overline{500}$, the PLC is instructed to turn output 508 "ON" if the "OR" function and the "INHIBITION" function are true. (For logic functions, see Chapter 3.)

The previous few paragraphs have extensively explained how "OR" blocks can be "AND" connected, and how "AND" blocks can be "OR" connected. Some PLCs use bracket instructions for the programming of complex Boolean logic equations to ensure that the program is scanned and executed in the correct logic order (figures 2-12 and 2-15). It is often regarded as more user-friendly if a PLC operation and management program avoids bracketing instructions and offers the programmer the block connection instructions instead ("OR LD" and "AND LD"), as illustrated and explained in this chapter.

Figure 2-19. Complex logic line, rendering output 506. Eight different logic paths may, if true, activate output 506.

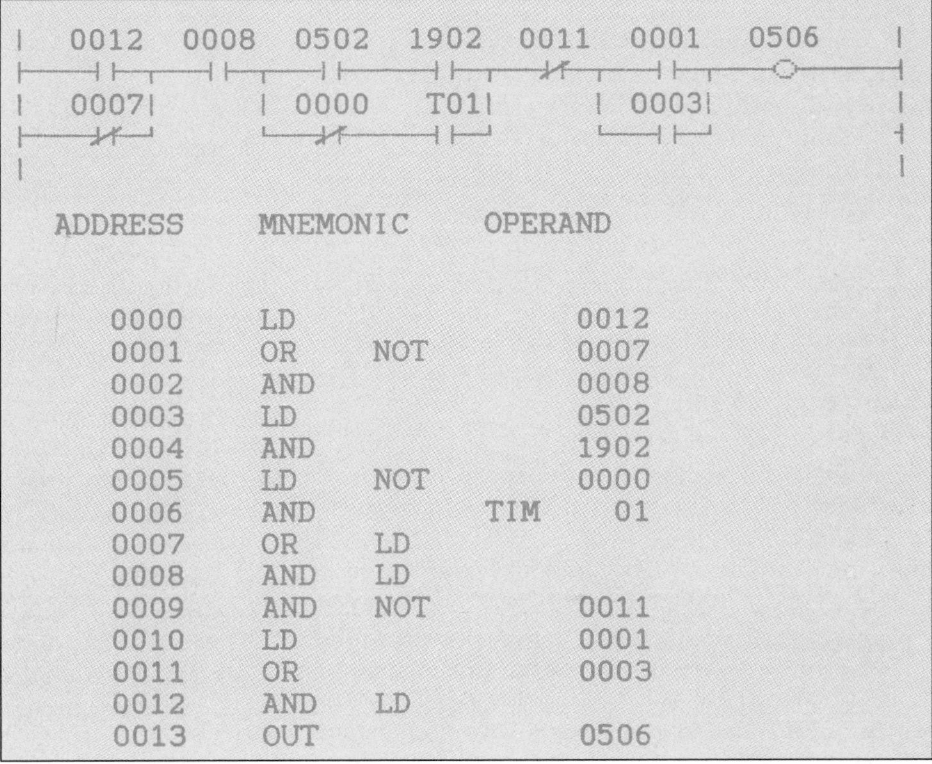

ADDRESS	MNEMONIC		OPERAND
0000	LD		0012
0001	OR	NOT	0007
0002	AND		0008
0003	LD		0502
0004	AND		1902
0005	LD	NOT	0000
0006	AND		TIM 01
0007	OR	LD	
0008	AND	LD	
0009	AND	NOT	0011
0010	LD		0001
0011	OR		0003
0012	AND	LD	
0013	OUT		0506

LOGIC PATH POSSIBILITIES

Ladder diagrams provide a unique opportunity to demonstrate logic path possibilities for two "OR" brackets being "AND" connected (horizontally joined), as is shown in figures 2-15 and 2-17. The programming with mnemonic codes for this circuit is shown in figure 2-16.

The same holds true when "AND" blocks are "OR" connected, as is shown in figure 2-18, where output 504 may be turned "ON" via four different "OR" rungs. Figure 2-19 shows a logic line that renders output 506 if one of its eight possible logic paths becomes true. These eight paths are shown in the solution of control problem 1 at the end of this chapter, in which the Boolean switching equation is also given for each logic path.

END INSTRUCTION (FUN 01)

The last instruction in any program must be the "END" instruction. When the CPU scans the program, it executes all instructions, up to the first END instruction, before returning to the beginning of the program and beginning scanning and execution again. Although an END instruction can be placed at any point in a program, which is sometimes done to subdivide a program for the purpose of debugging, no instruction past the first END instruction will be executed until the END instruction is removed. If there is no END instruction anywhere in the program, the program will not be executed at all. The PLC usually sounds a warning and gives an error message if the END instruction is missing. The END instruction requires no operand, and no conditions (contacts) can be placed on the same instruction line with it (figure 2-22).

PLC PROGRAMMING AND MACHINE INTEGRATION CONCEPTS

There are several basic steps involved in designing and writing a satisfactory and economically working PLC program. It is therefore recommended that, before programming, the program writer should study:

- the concepts of Boolean logic functions with their ladder diagram interpretation (see figures 3-01 to 3-15)
- the concept of sequential machine control notation by means of step-action diagrams (step-motion diagrams; see figures 8-03 and 8-28)
- the concept of step-counter sequencing as a means of effective, economical and flexible programming of asynchronously operating sequential machine controls (see figures 8-06, 8-07, 8-14 and 8-15)
- the concepts of pneumatic and hydraulic control design, for which most PLCs are used in industrial automation applications (see figures 8-12 and 8-27).

All these concepts are explained in detail in Chapters 18 and 19 of the book *Industrial Hydraulic Control* by Peter Rohner and in Chapters 7, 8 and 10 in the book *Pneumatic Control for Industrial Automation* by

Peter Rohner and Gordon Smith. Designing a PLC circuit, programming the PLC and connecting it to a machine should ideally be attempted with the following sixteen steps:

1. If the circuit to be controlled by the PLC is of predominantly sequential nature (and most industrial circuits are), then draw a step-action diagram for all actuators involved (hydraulic and pneumatic linear actuators, hydraulic and pneumatic motors, electric motors, pneumatic vacuum generators etc.). For typical step-action diagrams, including a timer, see figures 8-14 and 9-16.

2. If the circuit is of predominantly a combinational nature, such as a traffic control system, a combination lock circuit or a sorting operation control, based on a multitude of input parameter combinations, then draw a truth table and, where necessary, a Karnaugh-Veitch map for contact minimisation, and establish Boolean equations for the designing of the PLC circuit (see the book *Pneumatic Control for Industrial Automation* by Peter Rohner and Gordon Smith, Chapter 10).

3. Draw the electrical, electrohydraulic or electropneumatic circuit with all its directional control valves, linear actuators, motors etc. (figures 8-14 and 8-27). Assign PLC output point numbers to all the hydraulic, pneumatic and electric elements that require a PLC signal for their operation (valve solenoids, prime movers, buzzers, control panel indication lamps, heating elements etc., as shown in figure 9-16). However, if the circuit to be designed is of an almost pure combinational nature (few circuits are), then skip steps 1, 4, 6, 7 and 8.

4. Now transfer the assigned output point numbers to the sequence step time-lines of the step-action diagram developed in step 1 (figures 8-27 and 8-28).

5. Identify all input devices that may influence the operation of the PLC control circuit and assign to them PLC input point numbers. Such input devices may include limit switches, push button switches, heat sensors, pressure switches, flow rate sensors and proximity sensors. (See input/output assignment list for control problem 1 in Chapter 8.)

6. Transfer all these input numbers that directly affect the sequencing of the machine into the step-action diagram established in step 1 (figure 8-28). Some input signals such as program selection, machine interrupt and automatic or manual cycling do not directly affect the normal sequencing. These signals are, however, required in step 8.

7. Design and draw a ladder diagram (or system flow chart, if this is preferred; see figure 9-11), firstly for the sequential step-counter circuit only, and following on to that also for all output commands, leading to the fluid power and electrical devices on the machine (see figure 8-29). Procedures for sequential circuit design, using the step-counter circuit design method, are given in Chapters 8 and 9.

8. Design and draw a ladder diagram for all fringe condition modules, if such modules are required (machine interrupt module, cycle selection module, stepping module, program selector module etc.). Place the ladder diagram developed in this step in front of the sequential circuit diagram of step 7. This makes the circuit orderly and groups it into segments for ease of trouble-shooting (figures 9-12 and 9-14). Procedures for the design of fringe condition modules are given in Chapter 9.

9. Design and draw a ladder diagram for all other machine control functions that are not typically of sequential or modular nature. Such machine control functions are usually of pure or semi-combinational nature and may require the knowledge of Boolean algebra, truth tables and Karnaugh-Veitch maps, and they often require extensive circuit design experience (see figures 7-13 and 10-34). This is particularly true if special interrupt sequences, sequence jumps, arithmetic functions and alternative sequence programs are required to control the automated machine (see Chapters 8 to 10 in the book *Pneumatic Control for Industrial Automation* by Peter Rohner and Gordon Smith). Procedures for the design of such circuits are given in Chapters 7 and 10 in this book.

10. Enter the program established in steps 6 to 9 into the CPU. When using a programming console or a hand-held programmer, this will involve converting the ladder diagram to mnemonic PLC instruction form. Alternatively, if a PC with the appropriate software is used, the ladder diagram is established via the keyboard on the monitor screen, and then transferred via the host link into the PLC.

11. Now check the entered program visually by pressing the CLR button and then the DOWN button on the programming console for each step. This permits the program to be stepped through each address to verify that the entered program instructions agree with the ladder diagram designed initially on paper or on

the monitor. Such a test is called a "visual test". It is the first of four tests the program designer has to execute for a successful and trouble-free PLC programming and commissioning procedure. This visual test is best performed in the program mode so that if any programming errors are detected, they can immediately be rectified.

12. Check the program with the PLC's syntax check facility for syntax errors. This is the second of the four tests and is called the "syntax test". If the PLC gives an error message, then correct the errors. To check for syntax errors on an Omron C20, one may press the keys CLR, CLR, FUN, MONTR in any of the three selectable modes (RUN, MONITOR or PROGRAM). For further debugging checks and error messages, consult the operation manual of the PLC in use.

13. Execute the program now with the so-called "dry-run test". This is the third of the four tests. Execute the program with simulation input switches to check for any basic programming and execution errors and correct these immediately. These errors do not fall into the syntax error category. Such basic programming errors might creep in relatively easily. For example, an actuator is meant to extend and reach a particular limit switch, but instead of extending, the actuator retracts and hence does not reach the limit switch for which the sequential logic is programmed to look. The step-counter therefore stops at this point until the programming error is corrected. This dry-run test is described in detail in Chapter 8 under the heading "Testing the PLC Program with the 'Dry-Run Test'".

$$012 \cdot 008 \cdot 502 \cdot 1902 \cdot \overline{011} \cdot 001 = 506$$

$$012 \cdot 008 \cdot 502 \cdot 1902 \cdot \overline{011} \cdot 003 = 506$$

$$012 \cdot 008 \cdot \overline{000} \cdot \mathrm{TIM01} \cdot \overline{011} \cdot 001 = 506$$

$$012 \cdot 008 \cdot \overline{000} \cdot \mathrm{TIM01} \cdot \overline{011} \cdot 003 = 506$$

$$\overline{007} \cdot 008 \cdot 502 \cdot 1902 \cdot \overline{011} \cdot 001 = 506$$

$$\overline{007} \cdot 008 \cdot 502 \cdot 1902 \cdot \overline{011} \cdot 003 = 506$$

$$\overline{007} \cdot 008 \cdot \overline{000} \cdot \mathrm{TIM01} \cdot \overline{011} \cdot 001 = 506$$

$$\overline{007} \cdot 008 \cdot \overline{000} \cdot \mathrm{TIM01} \cdot \overline{011} \cdot 003 = 506$$

Figure 2-20. Eight Boolean switching equations for the eight possible logic switching paths for the ladder diagram shown in figure 2-19.

14. Once the PLC is programmed, the syntax is checked and execution errors are eliminated, one may electrically connect the PLC to the machine and run the PLC in conjunction with the fluid-power-controlled machine. This is "the hot-run test". Any time-dependent functions can now be fine-tuned, if required (see "Changing Timer or Counter Set Values (SV Change)" in Chapter 11 and figure 11-10).

15. Once the machine runs satisfactorily, it is wise to make a final backup copy of the program. This copy contains all logic, timing and counting data for future reference. The program may now be stored on disk, magnetic tape, paper or EEPROM for safe-keeping (see "Saving the Program to Cassette Tape" and "Loading and Verifying the Program from Cassette Tape to PLC" in Chapter 11).

16. Finally, one needs to document all input and output numbers in an input/output assignment list. It is also advisable to document the keep relays, which need to be "set" for program start-up (see control problem 2 in Chapter 9).

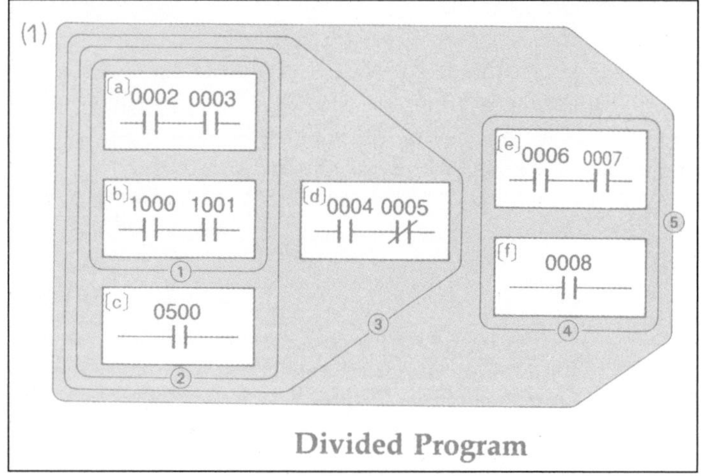

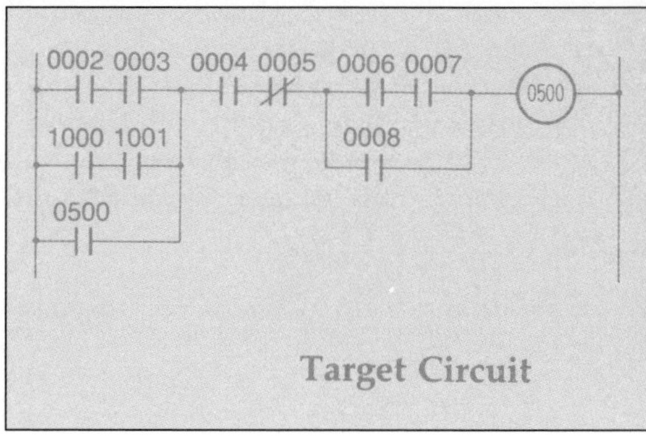

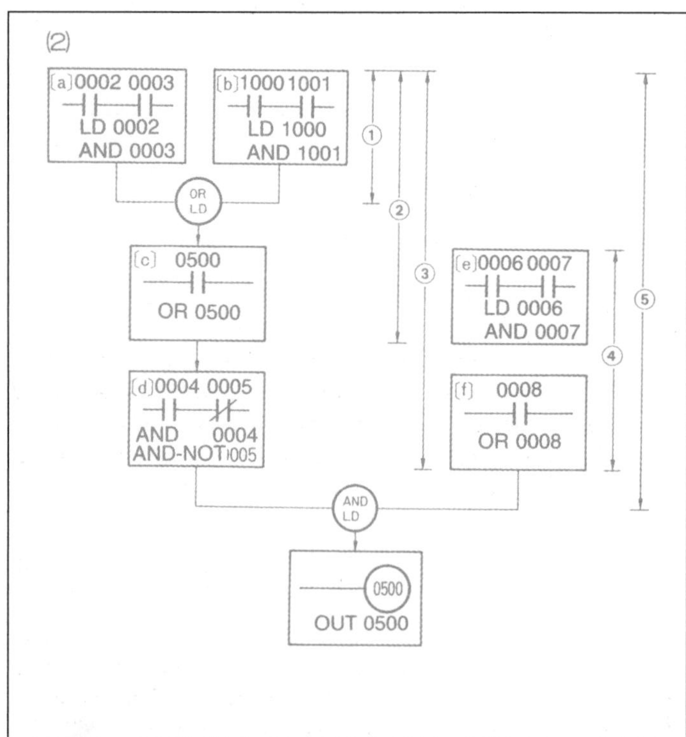

Address	Instruction	Data
0200	LD	0002
0201	AND	0003
0202	LD	1000
0203	AND	1001
0204	OR-LD	-
0205	OR	0500
0206	AND	0004
0207	AND-NOT	0005
0208	LD	0006
0209	AND	0007
0210	OR	0008
0211	AND-LD	-
0212	OUT	0500

Figure 2-21. Development stages to reach small circuit blocks and final mnemonic list for the target circuit given.

PLC PROGRAMMING CONCEPTS — CONTROL PROBLEM 1

For the control circuit given in figure 2-19, eight different logic flow paths may render output signal 506. With a pen, mark these eight paths on the circuit given, and show the Boolean switching equation for each logic path (solution in figure 2-20).

PLC PROGRAMMING CONCEPTS — CONTROL PROBLEM 2

Programming can be very simple or complicated, depending on the procedure used. Generally, a seemingly complicated circuit can be divided into several simple program blocks.

Divide the target circuit given in figure 2-21 into six small program blocks, labelled a, b, c, d, e and f. Now, program each block from top to bottom and left to right. Always combine blocks vertically and from left to right. Therefore, the order in which the six

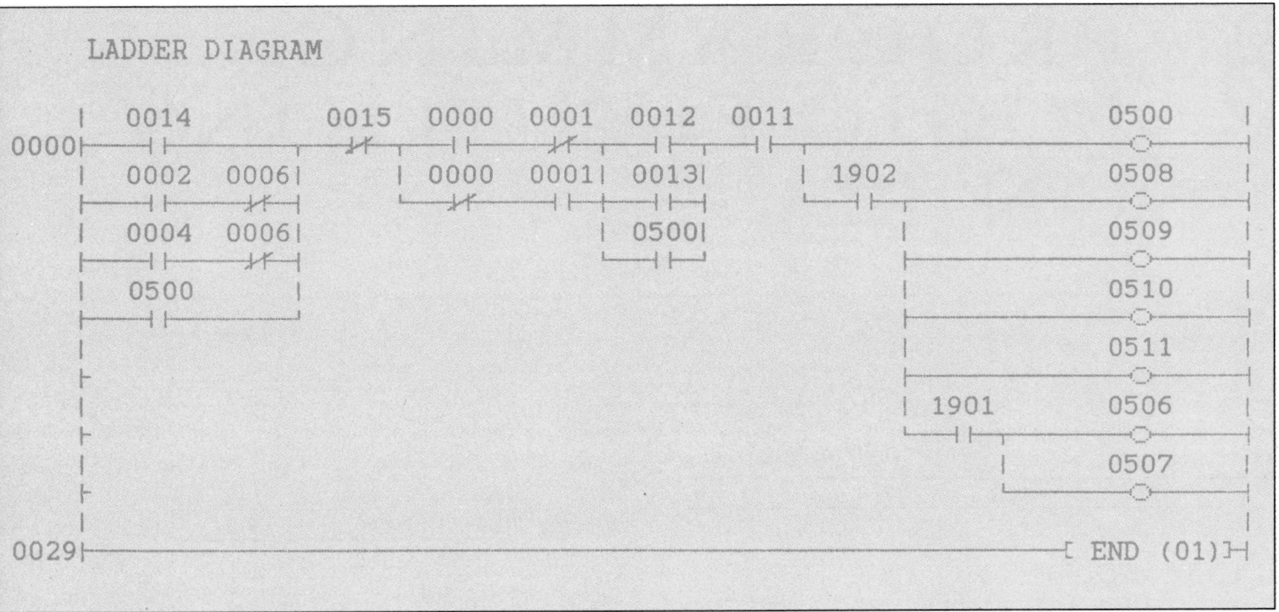

Figure 2-22. Control problem 3 ladder diagram to be translated into a mnemonic list.

blocks are programmed and combined is from 1 to 5. Picture 1 shows the subdivided circuit, and picture 2 shows how the subdivided circuit blocks interact with each other. On the right hand side of picture 2 is the final mnemonic list for the target circuit.

PLC PROGRAMMING CONCEPTS — CONTROL PROBLEM 3

Translate the ladder diagram given in figure 2-22 into mnemonic codes followed by their respective operand. Such a coding list is called a "mnemonic list" or statement list. Enter the program into an Omron C20 or C28K PLC, or a PLC of your choice.

For an Omron C28K all output operands starting with word 05 must be changed to word 01 (for an Omron C28K PLC, the output word is word 01, whereas for an Omron C20 PLC, the output word is 05). Hence, output 0500 becomes 0100, and output 0508 becomes 0108. All inputs using word 00 remain the same, and so do the clock pulse bits 1902 and 1901.

ADDRESS	MNEMONIC		OPERAND
0000	LD		0014
0001	LD		0002
0002	AND	NOT	0006
0003	OR	LD	
0004	LD		0004
0005	AND	NOT	0006
0006	OR	LD	
0007	OR		0500
0008	AND	NOT	0015
0009	LD		0000
0010	AND	NOT	0001
0011	LD	NOT	0000
0012	AND		0001
0013	OR	LD	
0014	AND	LD	
0015	LD		0012
0016	OR		0013
0017	OR		0500
0018	AND	LD	
0019	AND		0011
0020	OUT		0500
0021	AND		1902
0022	OUT		0508
0023	OUT		0509
0024	OUT		0510
0025	OUT		0511
0026	AND		1901
0027	OUT		0506
0028	OUT		0507
0029	END	(01)	

Figure 2-23. Mnemonic list translation for ladder diagram given in control problem 3 (see figure 2-22).

3 PROGRAMMING LOGIC FUNCTIONS ON PLC

To be able to program a PLC successfully, one needs intricate knowledge of the basic logic concepts required for all kinds of control programming. These logic switching concepts embrace the following five basic logic functions:

- "YES" (identity)
- "NOT" (negation)
- "AND" (conjunction)
- "OR" (disjunction)
- "INHIBITION" (impedance)

and the following five compound logic functions:

- "EXCLUSIVE OR" (antivalence)
- "NAND" (exclusion)
- "NOR" (rejection)
- "INHIBITED AND" (nonjunction)
- "INHIBITED OR" (nisjunction)

The following paragraphs extensively explain and illustrate these five basic and five compound logic functions and the "Law of Duality", which is also well known as "De Morgan's theorem".

BASIC LOGIC FUNCTIONS FOR PLC

The five most important logic functions are shown in figures 3-01 to 3-05. From these five basic logic functions one can construct further logic compound functions and even a latching type RS flip-flop. Figure 3-15 summarises all of these functions in a comparison table. Column 1 shows the name of the logic function. Column 2 shows the truth table. Column 3 shows the internationally standardised logic symbol, which is also used for flow-chart PLC programming. Column 4 shows the equivalent fluid power symbol. Column 5 shows the equivalent ladder diagram symbol circuit as used in PLC ladder diagram programming. Column 6 shows the Boolean switching equation for that logic function. Since various PLC brands use differing input and output signal numbering concepts, names for input and output signals have been generalised in figure 3-15. The letter "A" stands for signal input 1, the letter "B" for signal input 2, and the letter "S" for the output signal of the logic function.

Chapter 2 already briefly explained that a bar above the signal label means the inverse of the signal. The inverse of a signal means the signal is "NOT ON", or the signal is "NOT PRESENT". The logic Boolean expression for connecting inputs by the "AND" function (logic product) is denoted by a dot sign (•). The logic Boolean expression for connecting inputs by "OR" function (logic sum) is denoted by a plus sign (+). These two unique signs may also be used to logically connect switching expressions enclosed by brackets, as is demonstrated in Chapter 2 in figures 2-12 and 2-15.

DESCRIPTION OF LOGIC PLC FUNCTIONS

"YES" function (Identity)

With the "YES" function, shown in figure 3-01, the central processing unit (CPU) of the PLC interrogates a designated signal input point for the presence of its signal. For example, if input 0006 is "ON", then output 504 is turned "ON".

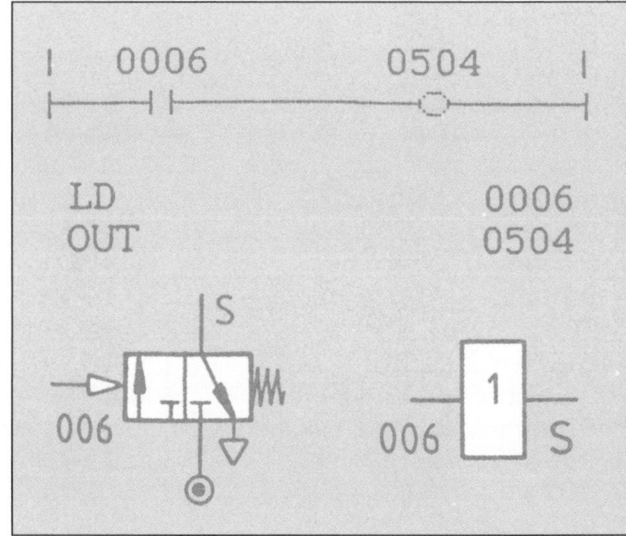

Figure 3-01. Ladder diagram and mnemonic list for the "YES" function. Output 504 serves the purpose of verifying the logic function.

Here, for the purpose of illustration only, output relay 504 is used to show the presence of input 0006. The "YES" function itself often requires no immediate output, as it may also be embedded with an "OR" logic connection or an "AND" logic connection, or may contribute to an "INHIBITION" function (figures 3-03 to 3-05). Most PLCs have LED input and output monitoring lights, which may be used for system diagnostics as well as output monitoring of the logic function.

Conclusion

Output relay 504 is "ON" when input point 0006 receives a signal from its connected sensor or switch. Alternatively, one may say that output 504 is "ON" if the "YES" contact of bit 0006 is "ON". That is the case when input switch 0006 is actuated and is sending an input signal to input point 0006 (see also figures 1-03 and 2-07).

0006 = 504

"NOT" function (Negation)

With the "NOT" function shown in figure 3-02, the CPU of the PLC interrogates a designated signal input point for the "absence" of its signal. For example, if input 0013 is not "ON", then input 511 is turned "ON".

Here, for the purpose of illustration only, output relay 511 is used to show the absence of input 0013. The "NOT" function itself often needs no immediate output, as it may also be embedded within an "OR" logic connection or an "AND" logic connection, or may contribute to an "INHIBITION" function.

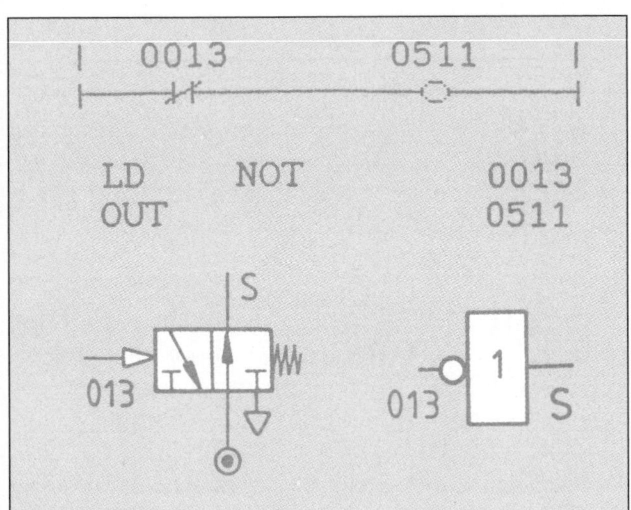

Figure 3-02. Ladder diagram and mnemonic list for a "NOT" function. Output 511 serves the purpose of verifying the logic function.

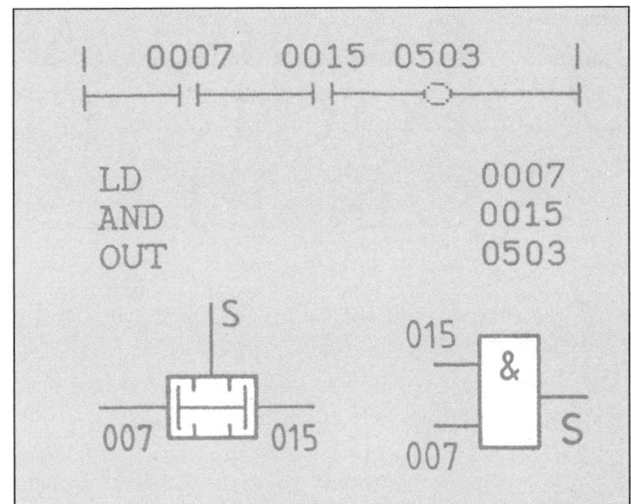

Figure 3-03. Ladder diagram and mnemonic list for the "AND" function. Although the "AND" function in the electronic ladder diagram is shown as a series function, inside the PLC these input signals are "AND" connected by electronic "AND" gates and not in series order!

Conclusion

Output relay 511 is "ON" when input point 0013 receives no signal from its connected sensor or switch. Alternatively, one may say that output 511 is "ON" if the "NOT" contact of bit 0013 is "ON". That is the case when input switch 0013 is not actuated and is not sending a signal to input point 0013 (see also figure 2-07).

$\overline{0013}$ = 511

"AND" function (Conjunction)

With the "AND" function shown in figure 3-03, the CPU interrogates two or more designated input points for the simultaneous presence of their signals. For example, if inputs 0007 and 0015 are "ON" at the same time, then output 503 is turned "ON". The order in which the two inputs are presented in the ladder diagram and the mnemonic list is unimportant as the PLC does not connect them in series in reality, but only by means of "AND" gates.

Here again, for the purpose of illustration only, output relay 503 is used to display the achievement of the "AND" function (or simultaneous presence of signals 0007 and 0015). The "AND" function often requires no direct or immediate output as it may also be connected to and followed by other logic functions, thus terminating in an output further along the logic line (see logic lines in figure 2-06).

Conclusion

Output relay 503 is "ON" when both inputs are "ON" within one scan (0007 and 0015) and therefore receive a signal from their connected switches or sensors.

0007 • 0015 = 503

"OR" function (Disjunction)

With the "OR" function shown in figure 3-04 (inclusive "OR" function), the CPU interrogates two or more designated input points logically connected in parallel for the presence of their signals. For example, if input 0000 is "ON" or input 0008 is "ON", then output 501 is turned "ON".

The order in which the two inputs are listed in the ladder diagram and the mnemonic list is unimportant. Output 501 is "ON" regardless of whether only one or both inputs are "ON" within the same scan. This "OR" function is therefore often also called an inclusive "OR" function. The "OR" function itself often requires no direct or immediate output as it may also be connected to and followed by another logic function, thus terminating in an output further along the logic line (see logic line in figure 2-06).

Conclusion

Output relay 501 is "ON" when input 0000, input 0008 or both receive a signal from their connected limit switches or sensors.

0000 + 0008 = 501

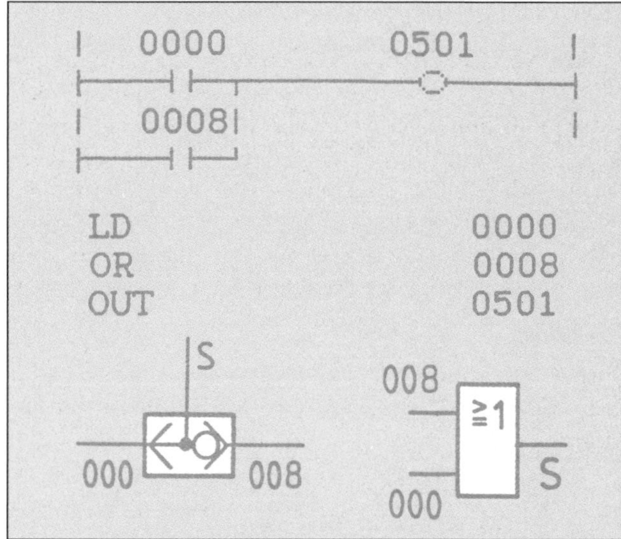

Figure 3-04. Ladder diagram and mnemonic list for an inclusive "OR" function (compare also with exclusive "OR" function shown in figure 3-06).

"INHIBITION" function (Impedance)

With the "INHIBITION" function shown in figure 3-05, the CPU interrogates two designated input points for their simultaneous logic state, one of them for its "presence", the other for its "absence". For example, if input 0004 is "ON" and input 0005 is not "ON", then output 507 is turned "ON".

Conclusion

Output relay 507 is "ON" when a signal is present on input 0004 and no signal is present on input 0005. One may also say that output relay 507 is "ON" when input 0004 is present and input 0005 is absent, or it could be said that input 0005, when present, hinders signal 0004 from passing through the logic gate.

$0004 \bullet \overline{0005} = 507$

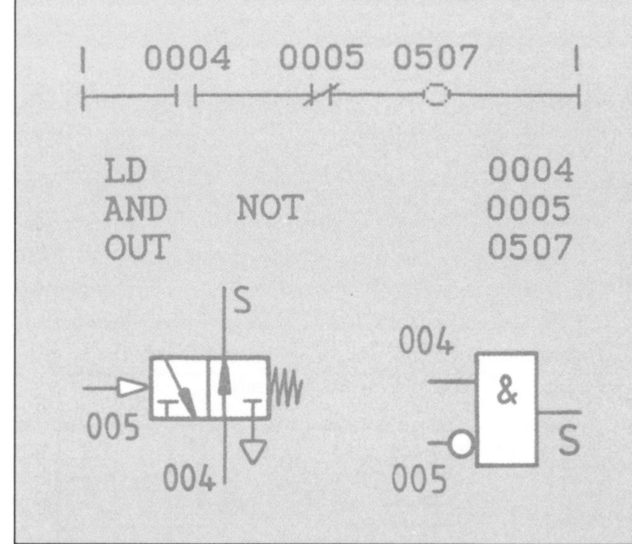

Figure 3-05. Ladder diagram and mnemonic list for the "INHIBITION" function. The order of the contacts on the rung is unimportant, which means that the "NOT" contact could be before the "YES" contact.

The order in which the two signals are listed in the ladder diagram and in the mnemonic list is unimportant. The "INHIBITION" function often requires no direct or immediate output as it may also be connected to and followed by other logic functions further along the logic line (see figure 2-06, addresses 0000 and 0001). The "NOT" function, as an integral part of the "INHIBITION" function, may also be applied to inhibit preceding logic functions, such as "OR" or "AND" (see examples in figure 3-09).

"EXCLUSIVE OR" function (Antivalence)

For the "EXCLUSIVE OR" function, the CPU of the PLC interrogates two inhibition functions (figure 3-06). These inhibition functions consist of each other's contacts (bits), in which the "YES" contact in one of them becomes the "NOT" contact in the other. Hence, the logic function is true only if one or the other input bit is "ON". The logic function is not true, however, if both input bits are "ON" (see also truth table in figure 3-06).

Conclusion

Output 500 is never "ON" if both inputs are simultaneously "ON", but it is always "ON" if only one of the two inputs is "ON".

$$(009 \bullet \overline{003}) + (003 \bullet \overline{009}) = 500$$

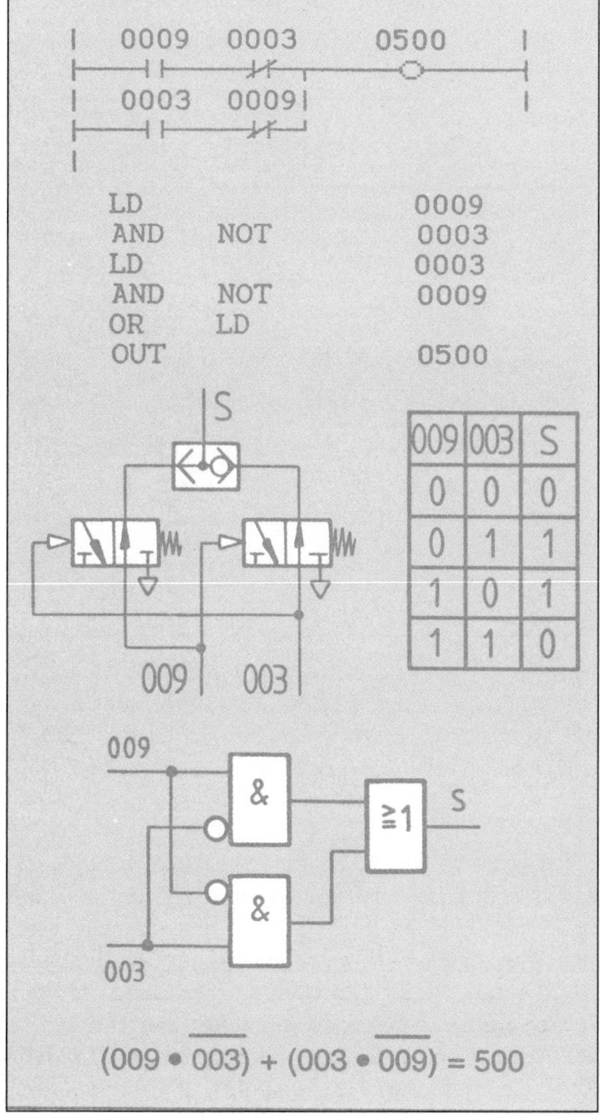

009	003	S
0	0	0
0	1	1
1	0	1
1	1	0

Figure 3-06. "EXCLUSIVE OR" function.

"NAND" function (Exclusion)

With the "NAND" function shown in figure 3-07, a "NOT" function is used to invert the output signal of the preceding "AND" function. This "AND" output is driving internal PLC relay 1000, which is an auxiliary relay. The "NOT" contact of this relay is then used to actuate output 509. Hence, output 509 is "ON" for as long as the "AND" function of inputs 0007 and 0012 is not "ON". Using De Morgan's theorem, this compound logic function can be simplified. This is illustrated at the end of this chapter.

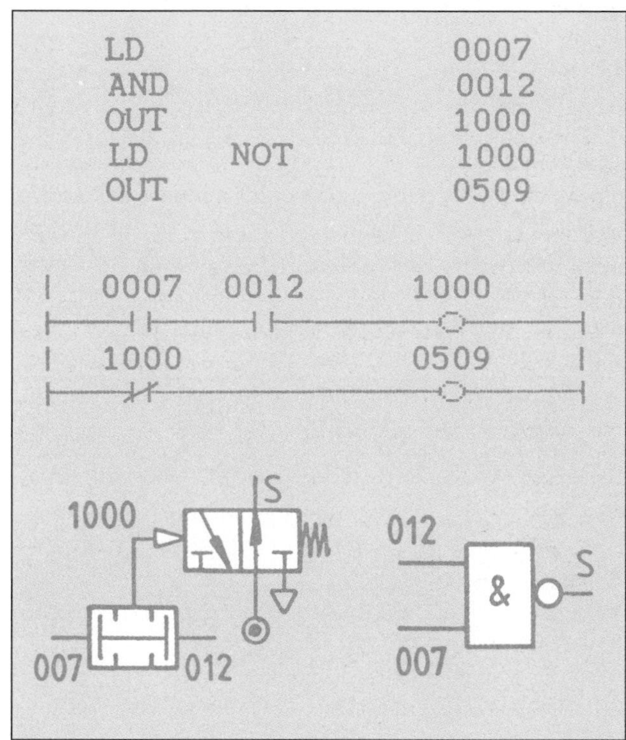

Figure 3-07. "NAND" function.

Conclusion

The output of a "NAND" function is never "ON" if both inputs of the preceding "AND" function are "ON", but it is always "ON" if one or the other or no input at all is "ON".

$$\overline{0007 \bullet 0012} = 509$$

"NOR" function (Rejection)

With the "NOR" function shown in figure 3-08, a "NOT" function is used to invert the output signal of the preceding "OR" function. This "OR" output is driving internal PLC relay 1003. The "NOT" contact of this relay is then used to actuate output 508. Hence, output

508 is "ON" for as long as the "OR" function of inputs 0013 and 0000 is not "ON". Using De Morgan's theorem, this compound logic function can be simplified. This is illustrated at the end of this chapter.

Conclusion

The output of a "NOR" function is never "ON" if one or both of the two "OR" function inputs are "ON", but it is always "ON" if neither of the two inputs is "ON".

$$\overline{0013 + 0000} = 508$$

"AND INHIBIT" concept (Nonjunction)

A single or multiple "AND" function may become inhibited if a "NOT" contact (normally closed contact) is placed after the "AND" function(s). Figure 3-09 (top) shows the ladder diagram, the mnemonic list and the pneumatic circuit equivalent for an "AND INHIBIT" application. The Boolean equation for this incomplete logic line reads as follows:

$$0001 \bullet 0006 \bullet 0004 \bullet \overline{0003} = \dots\dots\dots$$

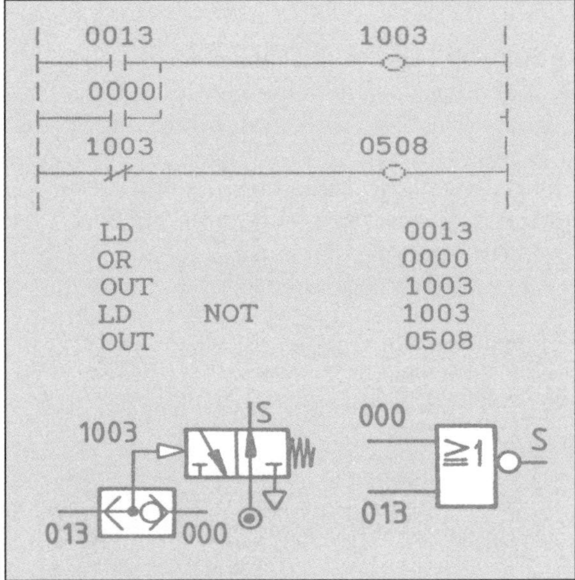

Figure 3-08. "NOR" function.

"OR INHIBIT" concept (Nisjunction)

A single or multiple "OR" function may become inhibited if a "NOT" contact (normally closed contact) is placed after the "OR" function(s). Figure 3-09 (bottom) shows the ladder diagram, the mnemonic list and the pneumatic circuit equivalent for an "OR INHIBIT" application. The Boolean equation for this incomplete logic line reads as follows:

$$(0002 + 0004 + 0502) \bullet \overline{0007} = \dots\dots\dots$$

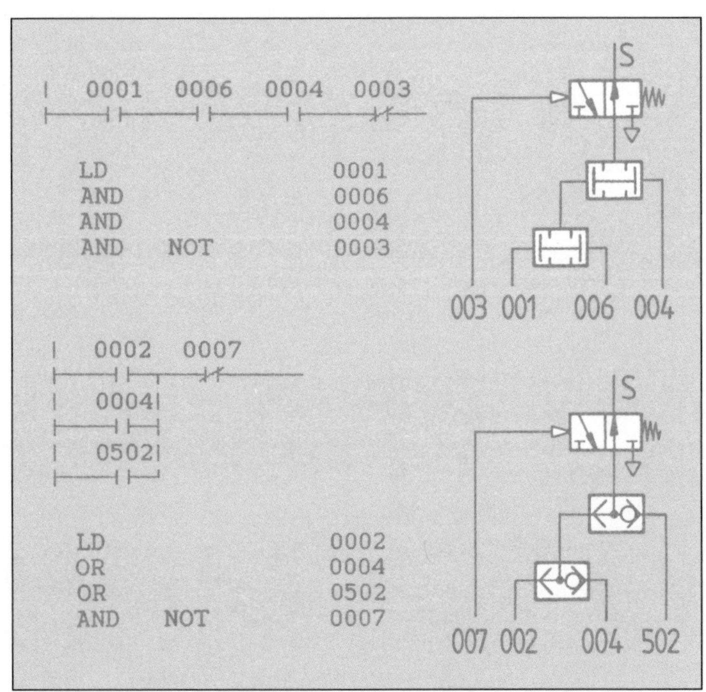

Figure 3-09. Ladder diagram and mnemonic list for logic functions preceding the "NOT" function as part of a compound "INHIBITION" function, called "AND INHIBIT" (top) and "OR INHIBIT" (bottom). Application circuits for "AND INHIBIT" and "OR INHIBIT" are given in figures 8-29 and 9-12.

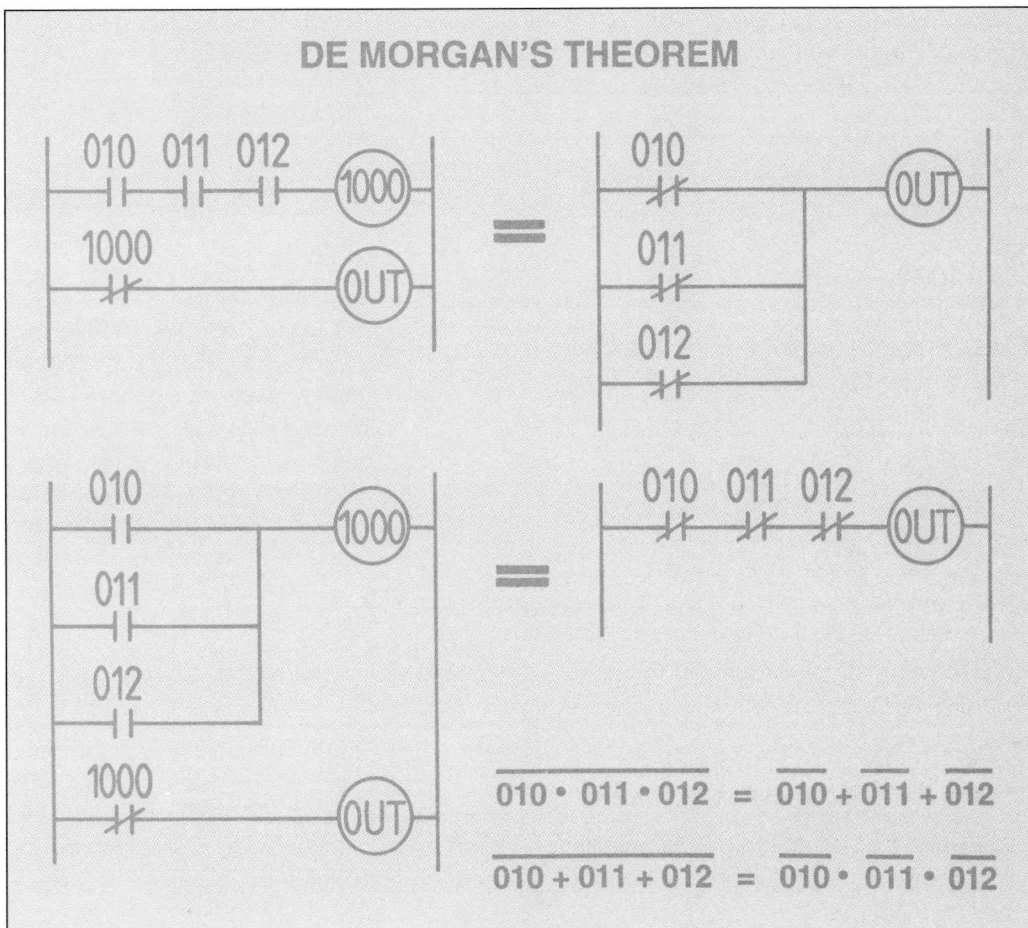

Figure 3-10. De Morgan's theorem.

DE MORGAN'S THEOREM

De Morgan's theorem, sometimes also referred as "the law of duality", is a useful aid to convert "OR" functions into "AND" functions. De Morgan's theorem permits a single "NOT" bar over several input variables (signal contacts or contact bits) to be split into individual "NOT" bars, thus causing the "AND" logic relations to become "OR" logic relations or vice versa (figure 3-10, top). Conversely, individual "NOT" bars can be amalgamated into a single "NOT" bar, thus causing all "AND" logic relations under that "NOT" bar to change to "OR" logic relations (figure 3-10, bottom). Naturally, parentheses or other signs of grouping must be considered.

Example: $\overline{(A + B) \bullet C} = \overline{(A + B)} + \overline{C} = \overline{A} \bullet \overline{B} + \overline{C}$

If De Morgan's theorem is applied to the "NAND" compound logic function shown in figure 3-07, the Boolean switching expression for that function may then be converted as follows:

$$\overline{0007 \bullet 0012} = \overline{0007} + \overline{0012} = 0509$$

Figure 3-11. "NAND" function converted with De Morgan's theorem. Compare figure 3-11 with figure 3-07.

Translated into a ladder diagram, the converted "NAND" function shown in figure 3-07, will then appear as is shown in figure 3-11.

From comparing figure 3-11 with figure 3-07 it becomes obvious that converting this "NAND" function with De Morgan's theorem brings a significant simplification for the programming of a PLC.

If De Morgan's theorem is now also applied to the "NOR" compound logic function shown in figure 3-08, the Boolean switching expression for that function could be converted as follows:

$$\overline{0013 + 0000} = \overline{0013} \bullet \overline{0000} = 0508$$

Translated into a ladder diagram, the "NOR" function shown in figure 3-08 and translated with De Morgan's theorem will then appear as is shown in figure 3-12.

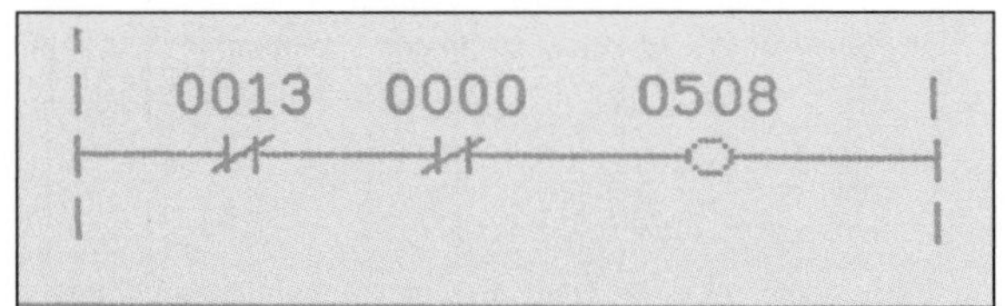

Figure 3-12. "NOR" function converted with De Morgan's theorem. Compare figure 3-12 with figure 3-08.

These three practical applications make it amply clear that De Morgan's theorem is a useful tool for simplifying the programs of PLCs. Further simplification concepts, using Boolean algebra, truth tables and Karnaugh-Veitch maps, are given in Chapter 10.

PROGRAMMING LOGIC FUNCTION ON PLC — CONTROL PROBLEM 1

Convert the four ladder logic switching circuits shown in figure 3-10 (De Morgan's theorem) into signal flowchart symbol circuits.

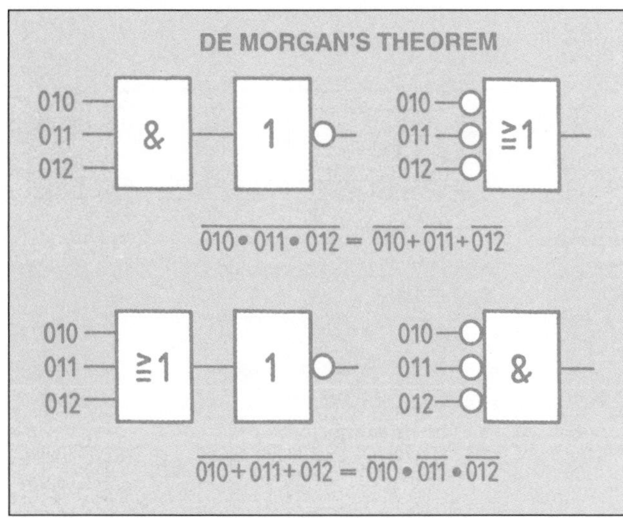

Figure 3-13. Solution to control problem 1.

PROGRAMMING LOGIC FUNCTION ON PLC — CONTROL PROBLEM 2

Convert the two circuits shown in figure 3-09 into signal flowchart symbols.

PROGRAMMING LOGIC FUNCTIONS SUMMARY

The five basic logic functions used in PLC programming are:

YES, NOT, AND, OR, INHIBITION.

When correctly combined, they may produce even the most complex logic networks, including RS flip-flops and binary counters. De Morgan's theorem may be used to convert NOR logic into NAND logic or vice versa. To prove correct compilation of logic functions into a switching circuit, one can use Boolean algebra or truth tables. In Boolean algebra, a dot sign (•) denotes "AND" connection and a plus sign (+) denotes "OR" connection. Logic functions may be illustrated by using pneumatic symbols, system flow chart logic symbols, ladder diagram symbols or Boolean equations. Standardised PLC instructions for the programming of logic functions are:

AND, OR, NOT, AND NOT, OR NOT, LD.

"LD" is used for load connecting a "YES" contact to the left-hand bus bar, and "LD NOT" is used for load connecting a "NOT" contact to the left-hand bus bar.

Standardised instructions for outputs of a logic line on the right-hand bus bar are:

OUT, TIM, CNT, SET, RESET.

"OUT" is used to output a relay. "SET" and "RESET" are used to denote the set and reset instruction for RS flip-flops on some PLC brands.

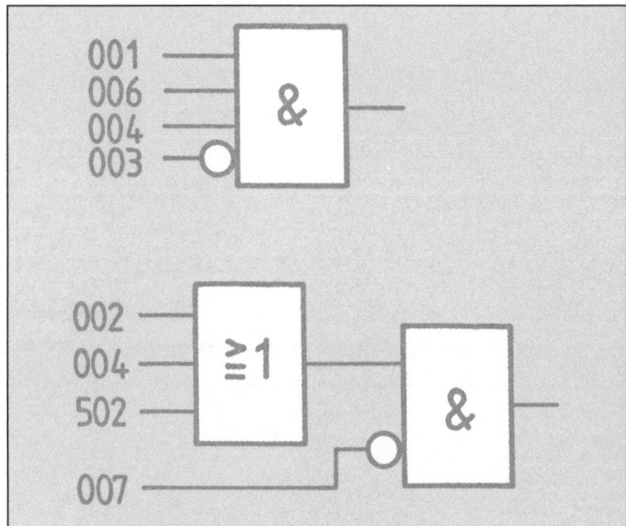

Figure 3-14. Solution to control problem 2.

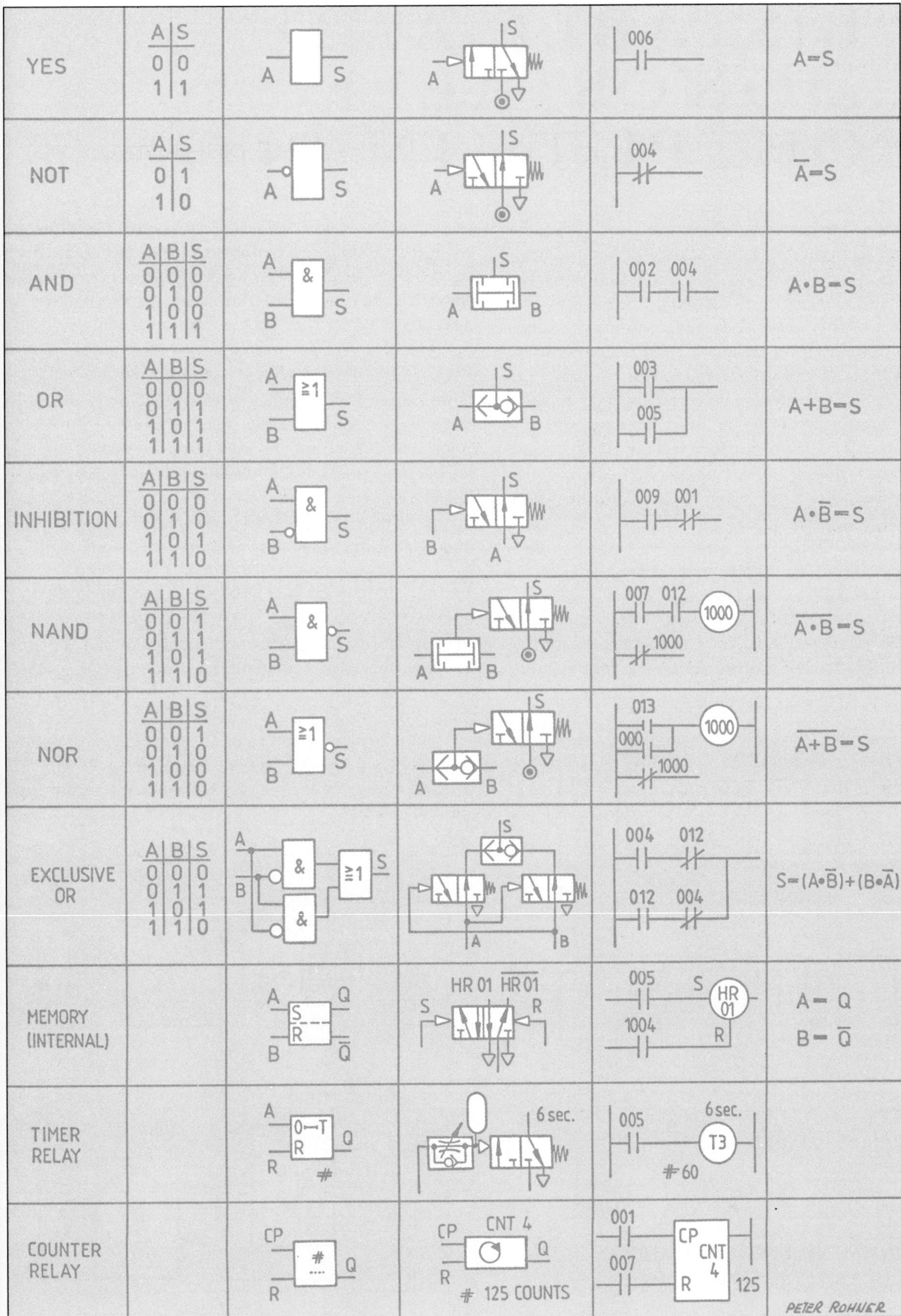

Figure 3-15. Basic logic functions for PLC.

4 BISTABLE AND MONOSTABLE OUTPUTS IN PLC

PLCs have output bits (relays) that may be used for external output connection activation—to drive solenoids, lamps, motor starters, heaters, buzzers etc. They may also be used for internal output processing or for both. When used for external output activation, these bits activate an output relay, usually a miniature mechanical relay, which may be used for voltages ranging from 0 to 240V AC or DC. The maximum current is approximately 0.5A for transistorised outputs or 2A for miniature mechanical relays (check the installation manual for the PLC in use to obtain exact details).

When used as internal relays, the contacts driven by these relays are logically linked in "AND", "OR", "NOT", "YES", "INHIBITION" and "EXCLUSIVE OR" connection to create basic logic output enabling functions (figures 3-07, 3-08 and 3-15).

If looked at from a behaviour characteristics viewpoint, however, these internal and external output relays can be categorised into "bistable" and "monostable" relays (bits).

BISTABLE PLC OUTPUT RELAYS

Bistable means to be stable in two states and therefore to have memory characteristics. That is, when turned "ON", the relay moves to the "ON" state and remains in that state, even if its "SET" signal disappears. When turned "OFF", it moves to the "OFF" state and again stays in that state, whether its "RESET" signal disappears or stays "ON". Such bistable relays are frequently called flip-flops. Because of this bistability they can be used to record information (figures 4-01 top and 4-02 top).

MONOSTABLE PLC OUTPUT RELAYS

If the output bit has only one stable state, it is said to be "monostable", or sometimes "astable", or one calls it a "monoflop" to contrast it from the bistable flip-flop. The relay bit will return to the "OFF" state when its enabling signal (set signal) disappears. The four valves shown in figures 4-01 and 4-02 will help to explain and demonstrate the concepts of monostability and bistability (monoflop and flip-flop).

RELAYS (BITS) WITH COMPLEMENTARY OUTPUT

Relays, whether bistable or monostable, may be programmed and used for their "ON" state only (Q), for their "OFF" state only ($\overline{Q}$), or for both their "ON" state and their "OFF" state (figures 4-01 and 4-02). If both states are used, the relay is said to have complementary outputs ($\overline{Q}$ is "ON" when Q is absent; alternatively, Q is "ON" when $\overline{Q}$ is absent). Hence, one of the two relay outputs is always "ON".

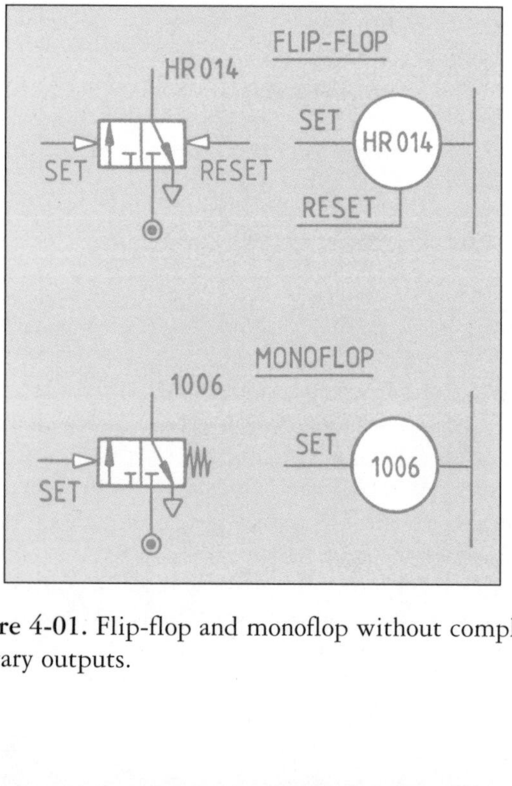

Figure 4-01. Flip-flop and monoflop without complementary outputs.

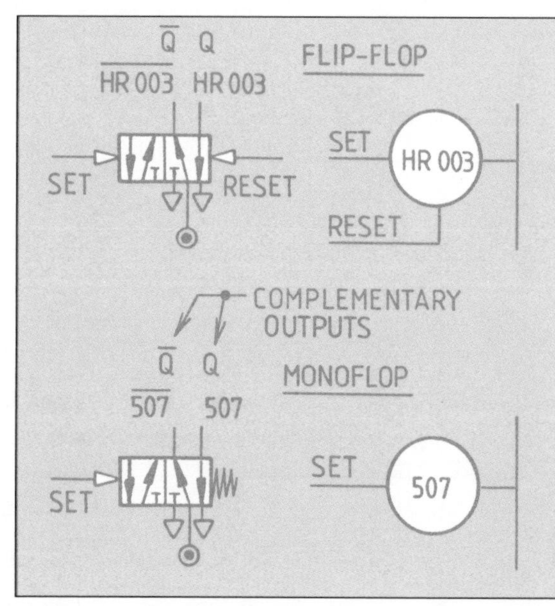

Figure 4-02. Flip-flop and monoflop with complementary outputs (Q and $\overline{Q}$).

RS FLIP-FLOP

Of all the commonly known flip-flops, the RS flip-flop is probably the most widely used in PLC programs. Industry and PLC manufacturers use a variety of names for this flip-flop (figure 4-03). The most often used names are Keep relay, Latching relay, Flag, Haftmerker (German), Holding relay, Retentive relay and RS flip-flop (see also explanation of Holding relay, Latching relay and Retentive relay, below).

The RS flip-flop has two inputs, designated by the letters R and S. The R input **resets** the relay into its "OFF" state, whereas the S input **sets** the relay into its "ON" state. The RS flip-flop may be used with or without complementary outputs. In figures 4-03 and 4-04 the complementary output bits of the RS flip-flop (HR 006) are used to enable outputs 0501 and 0500 (see addresses 0004 to 0006).

Where an RS flip-flop needs to be memory-retentive during PLC supply power failure, special programming is required to maintain the flip-flop state. To guarantee such logic state retention, PLC manufacturers offer PLCs with an in-built back-up battery for some memory sections. In Omron PLCs, the RS flip-flop programming instruction needs to be combined with instruction FUN 11 ("KEEP") to make the relay memory-retentive during power failure. It must be noted that the latching circuit shown in figure 4-05, although regarded as a relay with RS flip-flop characteristics, is not memory-retentive during power failure (see also description of latching relay, below). For applications of RS flip-flops, see the circuits given in figures 4-10, 5-11 and 5-13. The RS flip-flop does not, in itself, need outputs to function correctly, but to enable the reader to test the function of this flip-flop, outputs 501 and 500 have been added in figure 4-03 to show the flip-flop's "ON" and "OFF" states.

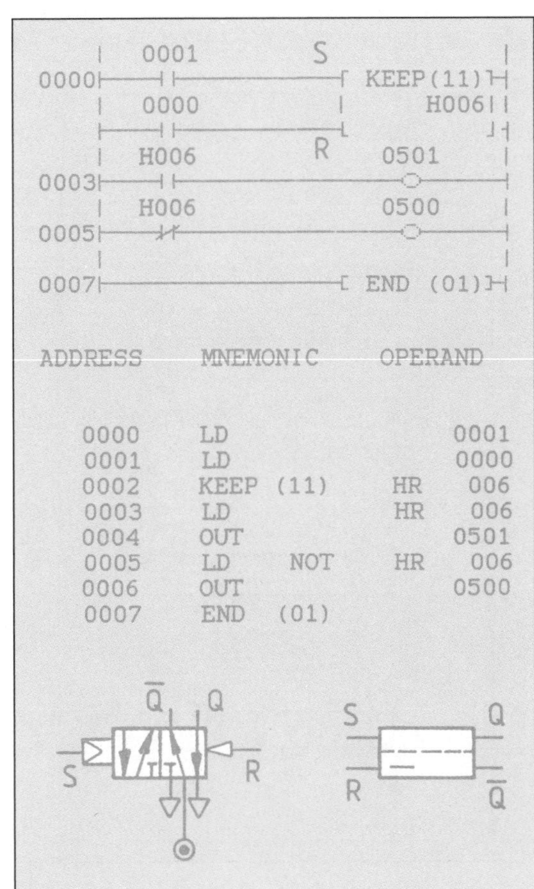

ADDRESS	MNEMONIC	OPERAND
0000	LD	0001
0001	LD	0000
0002	KEEP (11)	HR 006
0003	LD	HR 006
0004	OUT	0501
0005	LD NOT	HR 006
0006	OUT	0500
0007	END (01)	

Figure 4-03. RS flip-flop with reset dominance.

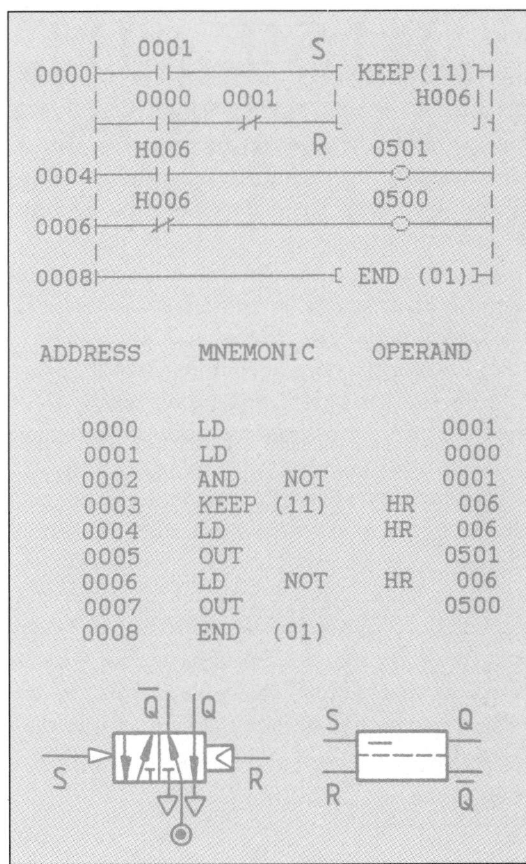

ADDRESS	MNEMONIC	OPERAND
0000	LD	0001
0001	LD	0000
0002	AND NOT	0001
0003	KEEP (11)	HR 006
0004	LD	HR 006
0005	OUT	0501
0006	LD NOT	HR 006
0007	OUT	0500
0008	END (01)	

Figure 4-04. RS flip-flop with set dominance.

KEEP FUNCTION (FUN 11)

The Omron C28K contains 190 holding relays, which are also called keep relays (figure 4-11). Only keep relays have true memory capacity, which makes them memory-retentive during power failure. With Omron PLCs, the keep function (FUN 11) may also be assigned to output relays or to internal auxiliary relays to convert them into set–reset type relays with separate set and reset input instructions to enable signals (see figures 8-17 A and 9-12, address 0265 in section 4). This does not, however, make the output relay or the internal auxiliary relay memory-retentive during power failure; it merely converts it from a monoflop to a flip-flop.

SET- OR RESET-DOMINANT FLIP-FLOPS

Although RS type flip-flops should never be given a set and a reset signal at the same time, when such an undesirable situation happens, the flip-flop needs to react in a predictable manner. If for such a situation the flip-flop is to assume the reset state, it is said to be "reset-dominant". This means that resetting the flip-flop has preference over setting it. Conversely, if the flip-flop needs to take on the set state, it is then said to be "set-dominant". This dominance is shown in the logic symbol with a short horizontal line next to either the reset or set input line (figures 4-03 and 4-04; ISO system flow chart symbols are shown). Flip-flops in PLCs may be programmed for set dominance by using the set signal to inhibit the reset signal. If reset dominance is essential, one simply programs the reset signal after the set signal, which means that when the PLC scans the program from top to bottom (address 000 to end instruction), it encounters the reset last (after the set), and therefore resets the flip-flop. Hence, normal flip-flop programming gives the relay reset dominance, but if for exceptional circumstances set dominance is required, one simply inserts the set contact as an inhibitor contact into the reset line (figure 4-04).

LATCHING RELAY

The latching relay, often also called self-holding relay or seal-in relay, consists of a monoflop and an "OR" function in its set-enabling line, whereby a normally open contact of the monoflop is used for the latching operation in the "OR" rung. In figure 4-05, relay 506, being a monoflop, is used for the latching relay, and its contact, programmed as OR 0506, forms the latch. Input bit 0006 sets the relay to "ON" and input bit 0000 resets the relay to "OFF" by breaking the latch. Latching a relay is a common practice to create an RS flip-flop, but one needs to remember that such a latched relay is definitely not memory-retentive during power failure. This means that it does not return to the "ON" state it had before power failure when power is restored to the PLC! For this reason, most PLCs also possess a number of keep relays, which maintain switching state with the aid of a back-up battery if power to the PLC should fail (figures 4-03 and 4-04).

Caution! In some PLCs, the keep relay, being memory-retentive, may be called a latching relay. Therefore, always find out the true memory-retentive characteristics of such relays, by studying the manual for the PLC in use.

T FLIP-FLOP (TRIGGER FLIP-FLOP)

The T flip-flop has only one input signal (trigger signal) and it may be used with or without complementary outputs (figure 4-06). The T flip-flop changes state each time a trigger signal appears and remains in its last state in the absence of further trigger signals. Hence if the first trigger signal changes it to the "ON" state, the second trigger signal reverses it to the "OFF" state (see truth table in figure 4-07). If used with complementary outputs (Q and $\overline{Q}$, if the first signal turns output Q "ON", the second trigger signal turns Q "OFF" and turns $\overline{Q}$ "ON". Thus, for counting, scaling or dividing circuits, the T flip-flop can be used to divide. This is demonstrated in figure 4-07, where six trigger signals produce three output signals (Q). For this reason, the T flip-flop is also called binary reducer or binary divider. Other names given to this flip-flop are bistable multivibrator, Eccles-Jordan circuit or trigger circuit. It must, however, be noted that this flip-flop has two stable states, as is shown in the truth table of figure 4-07. This figure shows the construction of a T flip-flop using a shift register, whereas in figure 4-06 the T flip-flop is built from two keep relays. An application for the use of a T flip-flop is given in figure 6-07.

The T flip-flop in figure 4-06 is more economical from a programming point of view, as it uses only two keep relay bits. The T flip-flop in figure 4-07 uses an entire word for its shift register. Shift register concepts and the programming of shift registers are explained and illustrated in Chapter 7.

The most economical version of a T flip-flop is shown in figure 5-09, where the T flip-flop consists of a differentiate-up function allocated to relay 1000. The first appearance of a bit 1000 pulse sets HR 800 to "ON", and the next appearance of a bit 1000 pulse resets HR 800 to "OFF".

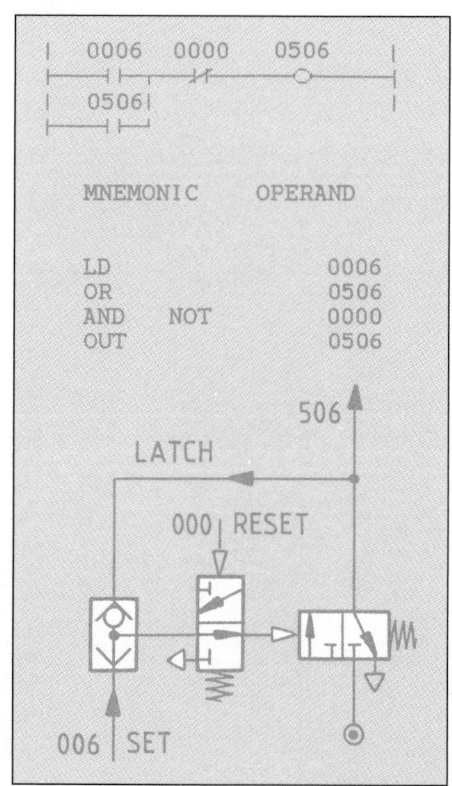

Figure 4-05. Latching relay (not memory-retentive during or after power failure).

JK FLIP-FLOP

The JK flip-flop is controlled by pulse signal 015. J is the set signal 001 and K may be regarded as the reset signal 1201 (figure 4-08). If both inputs are disabled before a clock pulse appears, the flip-flop does not change state even when a clock pulse 015 appears (the flip-flop is stable). If the J input is enabled and the K input is disabled, the flip-flop will assume the 1 logic condition when a clock pulse arrives. Conversely, if the K input is enabled and the J input is disabled, the flip-flop will assume the 0 logic condition when a clock pulse arrives. If both the J and K inputs are enabled when a clock pulse arrives, the flip-flop will complement (toggle), which means the flip-flop assumes the opposite state (see truth table in figure 4-08). Thus it may be said that with both inputs being in the "ON" state (enabled), the JK flip-flop behaves like a T flip-flop, but in any other circumstance it may be compared to a clock-pulse-controlled RS flip-flop. The JK flip-flop shown here is in fact built from a shift register. Shift registers are extensively explained and illustrated in Chapter 7.

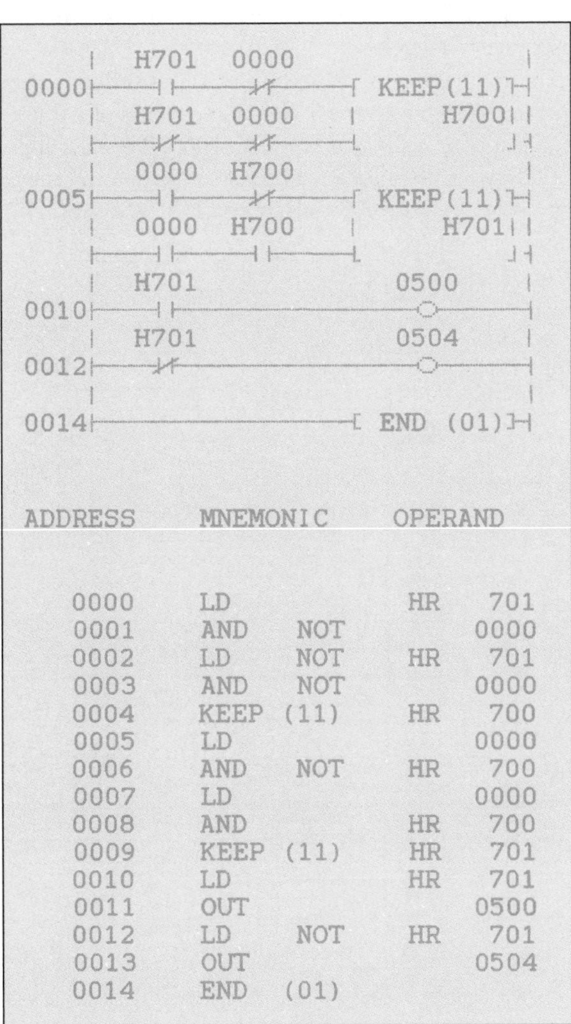

Figure 4-06. T flip-flop (Trigger flip-flop) built from two holding relays (more economical version).

D FLIP-FLOP

The D flip-flop has a (clock) pulse input 015 and a "set" signal input 001. Its output (Q = HR 300) is determined by the input that appeared one pulse earlier. Hence, the output logic state (1 or 0) is always delayed by one pulse. This gives the flip-flop its name D for delay. For

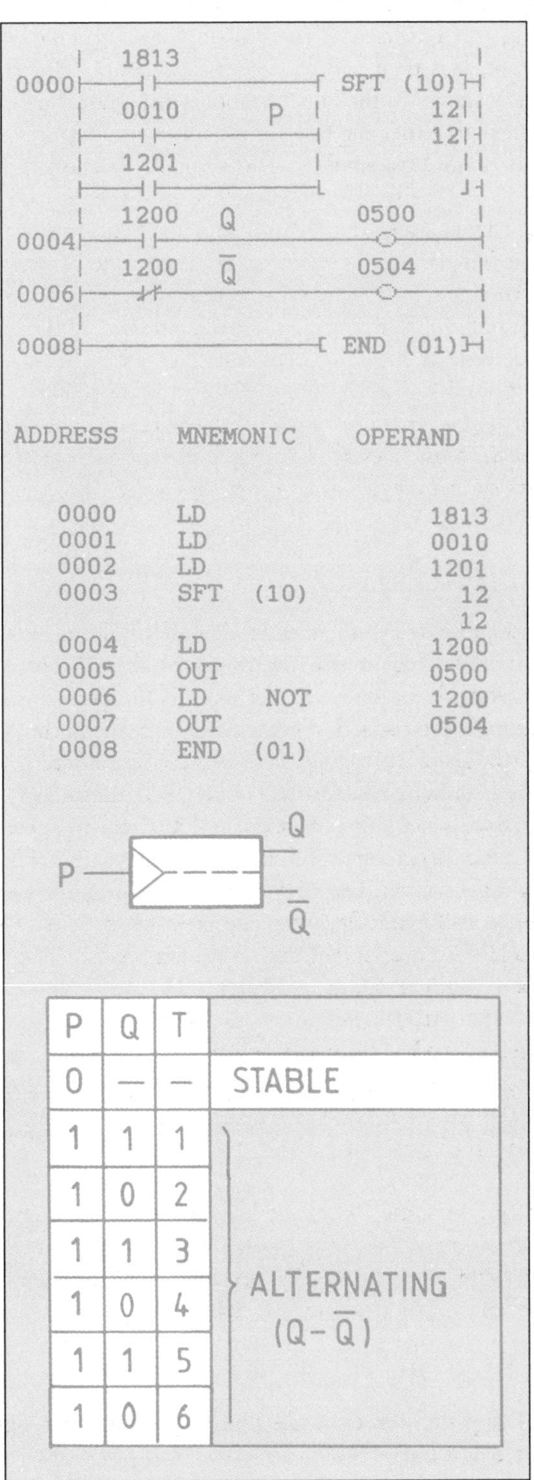

Figure 4-07. T flip-flop built from a shift register (less economical version).

example, if during the previous pulse the set signal was logic 1, the output Q after the next pulse will be logic 1. With no pulse and no set signal present, the flip-flop remains stable. With a set signal given but no pulse present, the flip-flop is also stable (see truth table in figure 4-09).

MISCELLANEOUS INTERNAL OUTPUTS

Apart from internal auxiliary relay outputs, which include holding relays and output relays not specifically allocated for output connections, a PLC also has timer and counter outputs, and an output category called special flags. Special flags may include count pulse bits, always-on bits, always-off bits and other, more-or-less exotic flags. A PLC's relay repertoire varies from manufacturer to manufacturer, but they all have some commonality. For an Omron C20, its external outputs range from word 05 to word 09 (Omron sometimes uses "channel" for "word", but "word" is widely used with other PLC manufacturers). For each of these words, only the first 12 bits (bit 00 to bit 11) may be used. Internal auxiliary relays range from word 10 to the first half of word 18 (see figure 4-11). In these words, all 16 bits may be used. If the memory-retentive relay section is used, the available relays range from word 00 to word 09. When programming them, one needs to place the acronym HR before the operand number (see figure 4-09) and FUN 11 into the mnemonic column. In each of these words, all 16 bits may be used. Thus, if one uses for example relay 502 as an externally used output relay, this would then constitute word 05 and bit 02. Together they make relay 502. The 0 in front of the 5 need not be programmed in as the PLC ladder circuit, when printed out, shows it automatically (see figure 4-10, address 0017). Relay 502 in figure 4-10 is used to actuate solenoid 502 on the pneumatic valve.

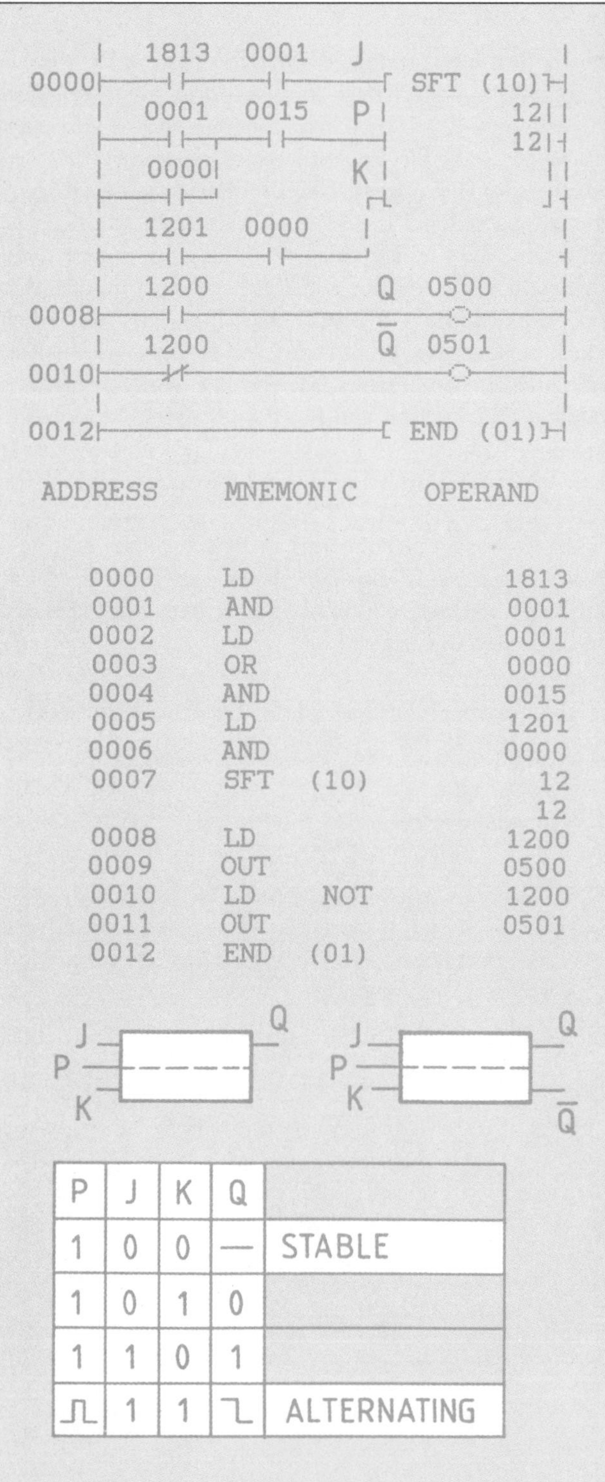

ADDRESS	MNEMONIC		OPERAND
0000	LD		1813
0001	AND		0001
0002	LD		0001
0003	OR		0000
0004	AND		0015
0005	LD		1201
0006	AND		0000
0007	SFT	(10)	12
			12
0008	LD		1200
0009	OUT		0500
0010	LD	NOT	1200
0011	OUT		0501
0012	END	(01)	

P	J	K	Q	
1	0	0	—	STABLE
1	0	1	0	
1	1	0	1	
⎍	1	1	⎤	ALTERNATING

Figure 4-08. JK flip-flop with complementary outputs.

FLIP-FLOP APPLICATION CIRCUIT — CONTROL PROBLEM 1

A potato hopper/washer is to be refurbished from entirely pneumatic control to electropneumatics with PLC control. The machine's two linear pneumatic actuators remain, and its two directional control valves are to be replaced by solenoid-actuated valves. All other switching and control devices will be electrical; either as part of the PLC's program (HR 000, HR 001) or as switches and sensors producing signals to the PLC input points (figure 4-10). Actuator Ⓐ vibrates the potatoes during the washing process. Actuator Ⓑ moves the guard lock pin into locking position to prevent the guard from being opened during machine operation. For input and output relay number allocation, including the three lamps, the solenoids and switches, see the pneumatic circuit in figure 4-10.

A design improvement needs to be incorporated so that a drop of pneumatic system pressure during machine operation below 700 kPa (pressure switch 005) signals the machine to stop ("OFF" signal). For this use signal inversion of input 005 ($\overline{005}$).

To signal machine start ("ON" signal or set signal for

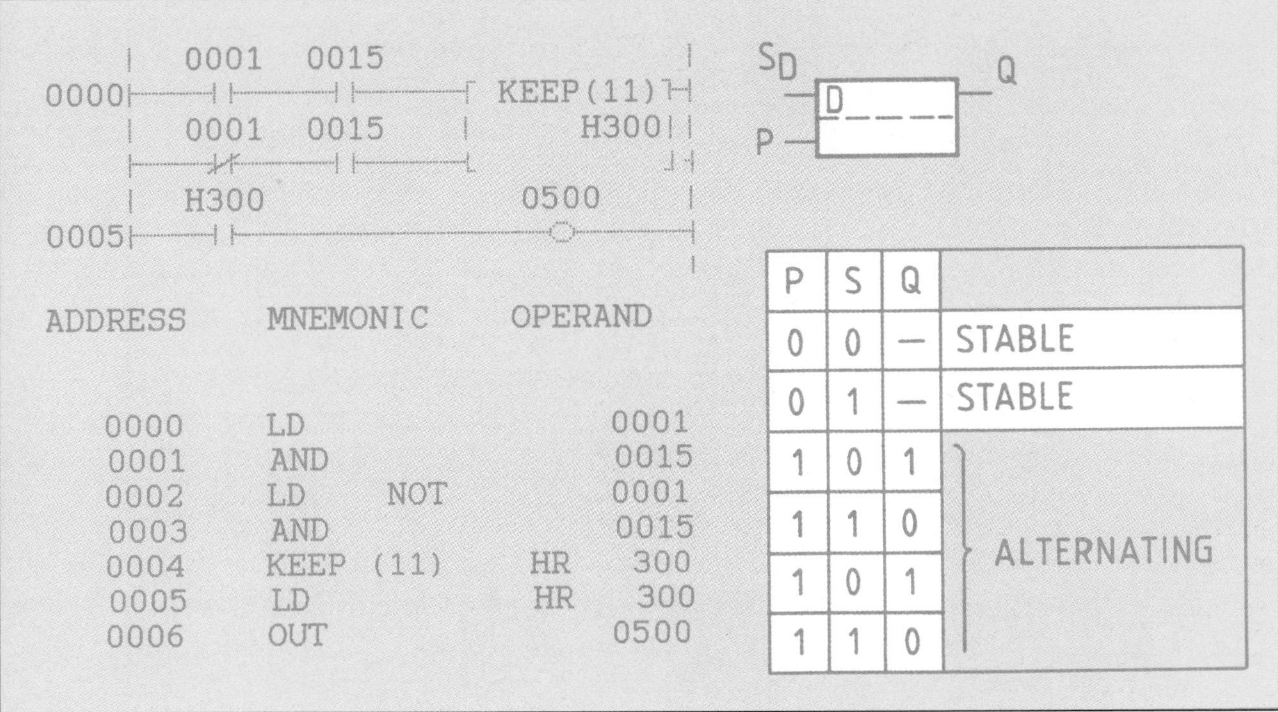

Figure 4-09. D flip-flop (delay flip-flop).

HR 000), the guard (004) must be closed, the start switch (000) needs to be actuated, and the potato supply conveyor must be in motion (006). For adjustment and trial run purposes, the hopper/washer may be started by maintenance personnel without the guard being closed but only with a special key-operated start switch (003), and the conveyor, for safety reasons, must not be running (006)! For both starting conditions, the hopper actuator Ⓐ needs to be retracted (001), as shown in the pneumatic circuit, and compressed air supply pressure must be at least 700 kPa (pressure switch 005). The guard pin actuator Ⓑ is permitted to extend, although the guard remains open when the machine is started with key switch 003.

Normal machine stopping is signalled with stop switch 013. Emergency machine operation interrupt (stopping) is signalled with stop switch 014. Emergency stop signalling is stored by holding relay HR 001. This holding relay can be reset (unlocked) only with key switch 015. The three operator selection modes—run, reset or emergency stop—must be announced with indication lamps 509, 510 and 511.

BISTABLE AND MONOSTABLE OUTPUTS SUMMARY

PLCs have output relays that are exclusively used for internal outputs, some relays that may be used for internal as well as external outputs. PLC relays may have monostable characteristics or bistable characteristics. When bistable, they are called "flip-flops"; when monostable they are called "monoflops". Flip-flops require a distinct "SET" instruction to turn "ON" and a distinct "RESET" instruction to turn "OFF". Flip-flops, when memory-retentive during power loss to the PLC, need back-up from a battery. Such flip-flops are known as "KEEP RELAYS" or "HOLDING RELAYS". Bits activated by output relays may also be used as input bits to form logic circuits the same way as normally closed or normally open switches. Where output relays are used as inputs, one may use the bit as an N/C bit or as an N/O bit.

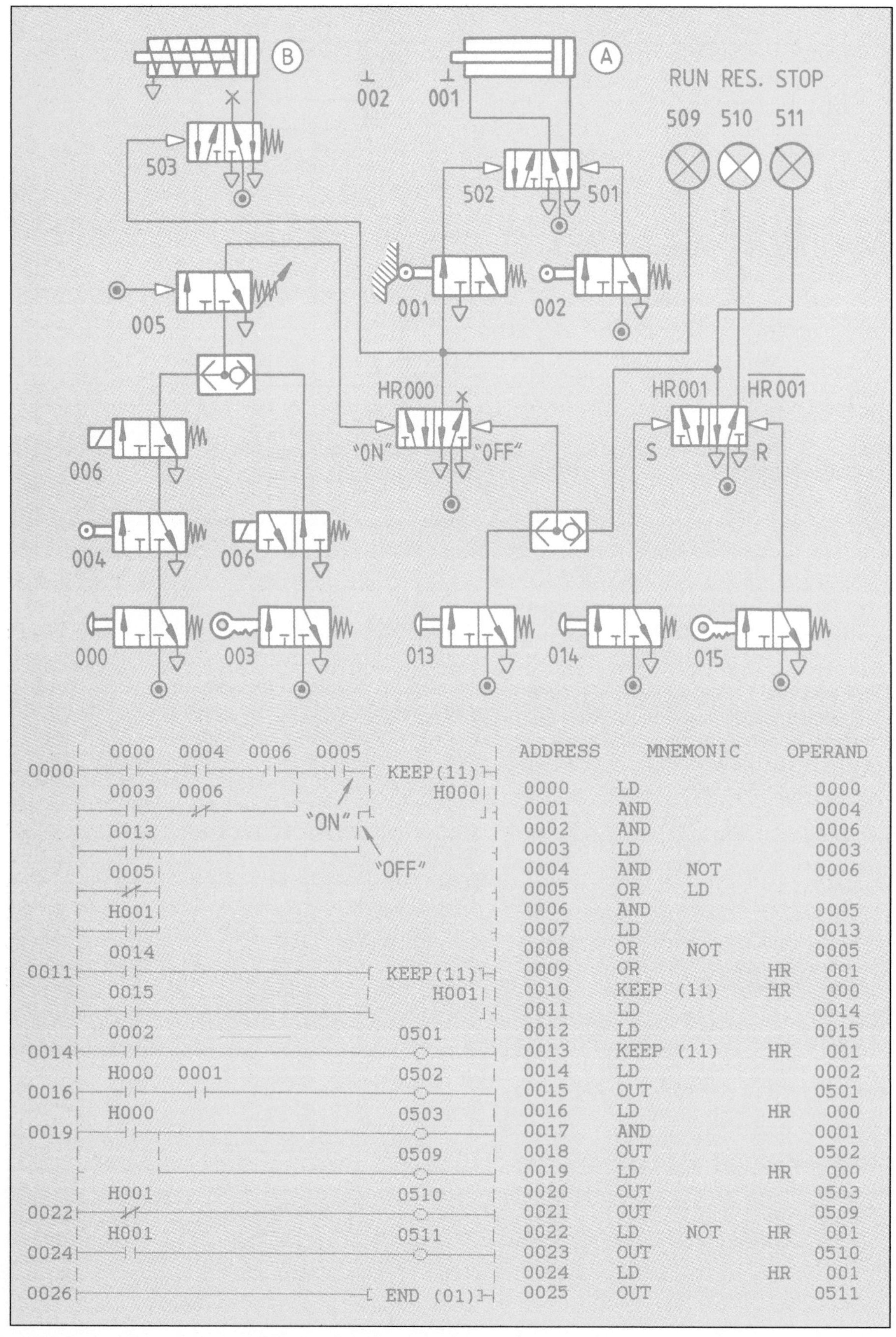

ADDRESS	MNEMONIC		OPERAND	
0000	LD			0000
0001	AND			0004
0002	AND			0006
0003	LD			0003
0004	AND	NOT		0006
0005	OR	LD		
0006	AND			0005
0007	LD			0013
0008	OR	NOT		0005
0009	OR		HR	001
0010	KEEP	(11)	HR	000
0011	LD			0014
0012	LD			0015
0013	KEEP	(11)	HR	001
0014	LD			0002
0015	OUT			0501
0016	LD		HR	000
0017	AND			0001
0018	OUT			0502
0019	LD		HR	000
0020	OUT			0503
0021	OUT			0509
0022	LD	NOT	HR	001
0023	OUT			0510
0024	LD		HR	001
0025	OUT			0511

Figure 4-10. Potato hopper/washer — control problem 1.

Figure 4-11. Assignment of input/output words (channels) and bit (relay) numbers for an Omron C20 programmable electronic controller. Word and bit numbers are similar for Omron C28K.

Name	No. of points	Relay number									
Input relay	80	0000 to 0415									
		00CH		01CH		02CH		03CH		04CH	
		00	08	00	08	00	08	00	08	00	08
		01	09	01	09	01	09	01	09	01	09
		02	10	02	10	02	10	02	10	02	10
		03	11	03	11	03	11	03	11	03	11
		04	12	04	12	04	12	04	12	04	12
		05	13	05	13	05	13	05	13	05	13
		06	14	06	14	06	14	06	14	06	14
		07	15	07	15	07	15	07	15	07	15
Output relay	60	0500 to 0915									
		05CH		06CH		07CH		08CH		09CH	
		00	08	00	08	00	08	00	08	00	08
		01	09	01	09	01	09	01	09	01	09
		02	10	02	10	02	10	02	10	02	10
		03	11	03	11	03	11	03	11	03	11
		04	12	04	12	04	12	04	12	04	12
		05	13	05	13	05	13	05	13	05	13
		06	14	06	14	06	14	06	14	06	14
		07	15	07	15	07	15	07	15	07	15
Internal auxiliary relay	136	1000 to 1807									
		10CH		11CH		12CH		13CH		14CH	
		00	08	00	08	00	08	00	08	00	08
		01	09	01	09	01	09	01	09	01	09
		02	10	02	10	02	10	02	10	02	10
		03	11	03	11	03	11	03	11	03	11
		04	12	04	12	04	12	04	12	04	12
		05	13	05	13	05	13	05	13	05	13
		06	14	06	14	06	14	06	14	06	14
		07	15	07	15	07	15	07	15	07	15
		15CH		16CH		17CH		18CH			
		00	08	00	08	00	08	00			
		01	09	01	09	01	09	01			
		02	10	02	10	02	10	02			
		03	11	03	11	03	11	03			
		04	12	04	12	04	12	04			
		05	13	05	13	05	13	05			
		06	14	06	14	06	14	06			
		07	15	07	15	07	15	07			
Holding relay (retentive relay)	160	HR000 to 915									
		00CH		01CH		02CH		03CH		04CH	
		00	08	00	08	00	08	00	08	00	08
		01	09	01	09	01	09	01	09	01	09
		02	10	02	10	02	10	02	10	02	10
		03	11	03	11	03	11	03	11	03	11
		04	12	04	12	04	12	04	12	04	12
		05	13	05	13	05	13	05	13	05	13
		06	14	06	14	06	14	06	14	06	14
		07	15	07	15	07	15	07	15	07	15
		05CH		06CH		07CH		08CH		09CH	
		00	08	00	08	00	08	00	08	00	08
		01	09	01	09	01	09	01	09	01	09
		02	10	02	10	02	10	02	10	02	10
		03	11	03	11	03	11	03	11	03	11
		04	12	04	12	04	12	04	12	04	12
		05	13	05	13	05	13	05	13	05	13
		06	14	06	14	06	14	06	14	06	14
		07	15	07	15	07	15	07	15	07	15

Name	No. of points	Timer/counter number					
Timer/counter	48	TIM/CNT00 to 47					
		00	08	16	24	32	40
		01	09	17	25	33	41
		02	10	18	26	34	42
		03	11	19	27	35	43
		04	12	20	28	36	44
		05	13	21	29	37	45
		06	14	22	30	38	46
		07	15	23	31	39	47

5 TIMERS IN PLC

Timers in PLCs delay output signals, terminate signals, cause machine processes to start or stop, or advance machine sequence steps at given time intervals. Timers control speed of products being processed. They may be used to determine filling levels in containers for filling or packaging operations, or they may control curing time in gluing or bonding processes, for example. Whatever the industrial application may be, timers are definitely an extremely useful element within the PLC controller.

TIMER CONCEPT

Timer operation is much easier to understand if one compares it to the traditional electromechanical timer, which consists of a motor-driven clock and a multitude of contacts, being either opened or closed when the clock has decremented (or lapsed) to zero. The timer's clock mechanism and the contacts operated by the clock mechanism are given the same number or label. This practice is also used for PLC timers. The PLC timer is connected to an internal quartz pulse generator controlled by the PLC's microprocessor. Some PLCs have several pulse generators or pulsing flags that pulse at different pulse rates. The Omron PLC C28K has three quartz pulse generators. Pulse generator 1 operates bit 1900 (flag 1900) with a 0.1 second pulse. Pulse generator 2 operates bit 1901 with a 0.2 second pulse. Pulse generator 3 operates bit 1902 with a 1.0 second pulse. For each of these three pulsing bits, the pulse "ON" duty time is half the rated pulse time. This means that for bit 1902, the "PULSE-ON" time is 0.5 second and its "PULSE-OFF" time is also 0.5 second (figure 5-01). These pulse bits (flags) may be integrated into a circuit like any other bit (or contact), as, for example, LD 1900, OR 1900, AND 1900, or in its inverted form, where necessary, as LD NOT 1900, OR NOT 1900, AND NOT 1900. For an integration application of these pulsing bits see figures 5-04, 5-06 and 10-17.

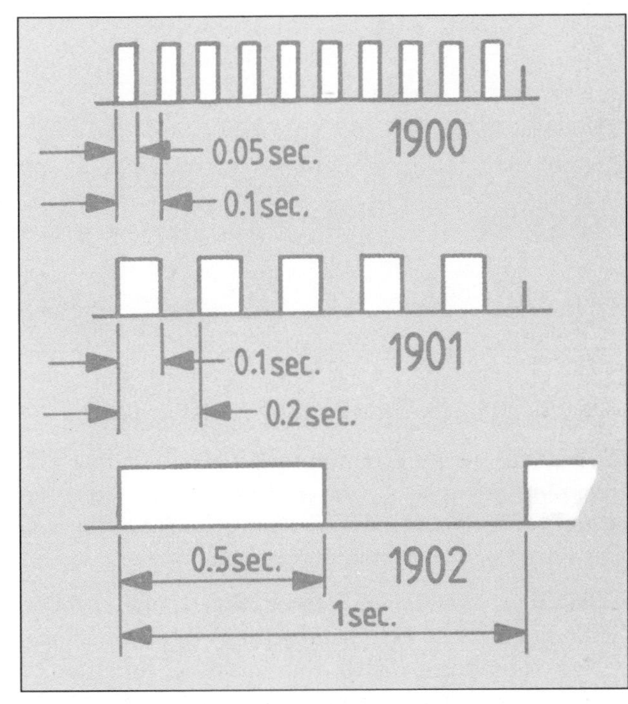

Figure 5-01. Pulses generated by an Omron C28K PLC. Bit 1902 pulses with a pulse length of 1.0 second. Its "on-duty" is 0.5 second and its "off-duty" is 0.5 second.

The pulsing rate at which the timer decrements is called the time base. On some PLC products the programmer may determine and select the time base, as it is not preselected by the PLC manufacturer. The timers in an Omron C28K PLC are, however, programmed with the 0.1 second pulse time base or pulse rate. This time base cannot be altered, since it is factory-determined. So for a timer to be set for, say, 39 seconds, its set value (SV) would be #390. With some PLC manufacturers, this set value (SV) may be called the K-value or preset value.

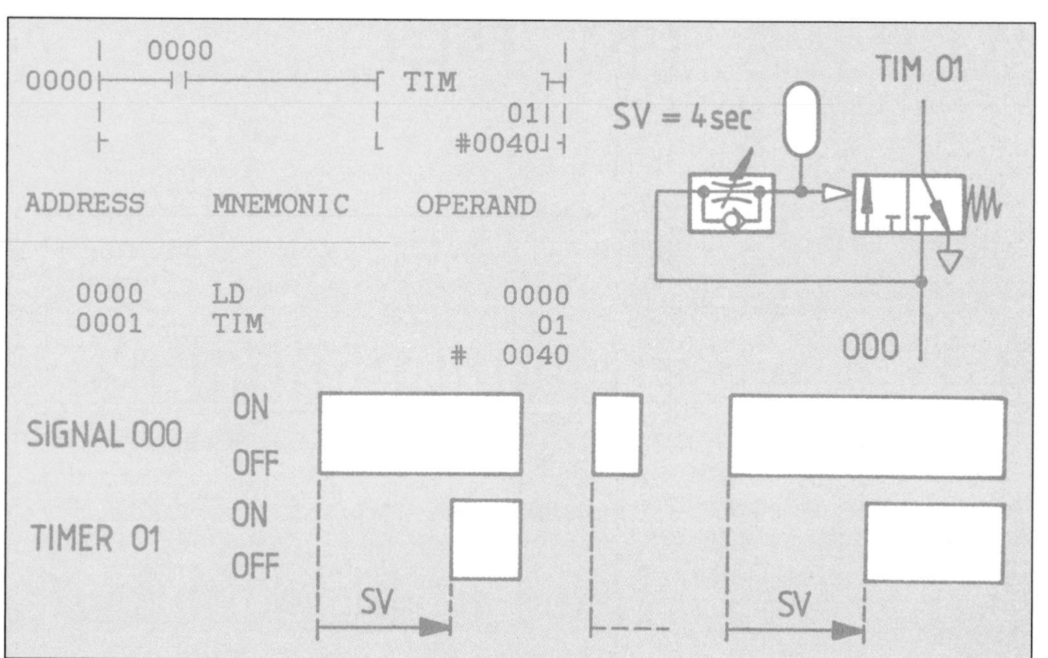

Figure 5-02. "ON" delay timer. The timer is programmed with an enabling signal (000), the timer flag number (TIM 01) and the set value (#40). An equivalent, normally closed pneumatic timer and a timer function bar chart are also given to illustrate the operation and limitation of this timer.

TIMER PROGRAMMING INSTRUCTIONS

To illustrate the programming procedures, a timer programming example is given in figure 5-02. From this example it becomes clear that a timer circuit requires at least three essential programming instructions.

- The first programming instruction is the enabling condition (some PLC manufacturers call it the execution condition). When this enabling condition is "ON", the timer starts to accumulate time (decrement from SV to 0000). The enabling condition may be a single contact, as shown in figure 5-02, which uses input 000, or it may be a complex logic circuit consisting of several logic functions or function blocks (figure 7-03).
- The second programming instruction nominates the timer flag, for example TIM 01. The first part of this instruction is a mnemonic instruction (TIM); the second is an operand bit (01). Together they become the timer flag. Some PLC manufacturers call this flag the "Done bit" because when the timer has "timed out" it has *done* its timing job. In an Omron C28K PLC, for example, there are 48 such timers ranging from TIM 00 to TIM 47. They may be used as timers or counters (see Chapter 6 and figure 1-12).
- The third programming instruction determines the timer's set value (SV). In the example circuit, the timer has been given a set value of #40 (40 × 0.1 second = 4 seconds). For most PLC timers, the set value ranges from 0000 clock pulses to 9999 clock pulses. Hence, based on a 0.1 second time base, the maximum possible set value becomes 999.9 seconds (0.1 second short of 1000 seconds). If the set value must exceed this time, one may use cascaded timers or a combination of timer and counter (figures 5-06 and 6-03). When programming the set value, the decimal point need not be entered. Thus, a 38.3 second delay is simply entered as #0383. (The # symbol need not be entered either as it automatically appears on the screen of the programming console.) The timer set value may be programmed as a decimal constant (#) or as a word address in the PLC's data memory area in hexadecimal or BCD form.

Note! A timer will never time-out if the "ON" time length of the enabling signal is less than the timer's preset value (see bar chart of enabling signal 000 in figure 5-02). If an enabling signal by nature is only a short pulse, and is less than the timer set value, the enabling signal will then require storage by using a holding relay (RS flip-flop) to store its temporary presence (figure 5-03). To make the timer ready for a new timing task, the holding relay must be reset as soon as the timer has performed its duty. One may even use the timer bit (flag) to reset the enabling signal holding relay (see figure 9-12, addresses 0228 to 0231 and 0278 to 0279).

Like an electromechanical timer, PLC timers may drive a large number of N/O and N/C contacts. These timer contacts are often also referred to as timer bits. The contacts driven by the timer carry the same number as their timer flag. To understand this, see figure 5-04, where timer 24, when timed out, activates auxiliary relay 1003 and output relay 504 with its timer

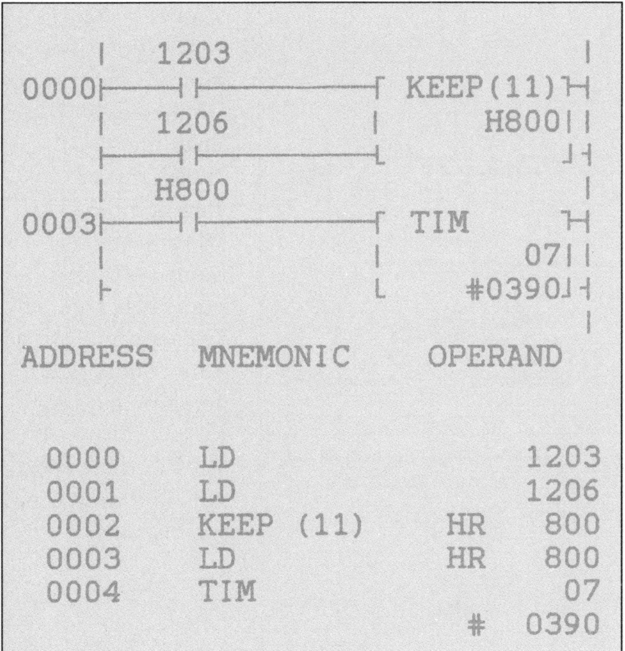

ADDRESS	MNEMONIC	OPERAND
0000	LD	1203
0001	LD	1206
0002	KEEP (11)	HR 800
0003	LD	HR 800
0004	TIM	07
		# 0390

Figure 5-03. A short timer-enabling signal or a pulse-like timer-enabling signal is stored in a holding relay to enable and drive timer 07 with a set value of 39 seconds. The holding relay must be reset as soon as the timer has performed its duty.

bits. It also sets holding relay HR 309 and resets latching relay 500. These actions are simultaneous.

"ON"-DELAY AND "OFF"-DELAY TIMERS

Timers are generally grouped into "ON"-delay timers and "OFF"-delay timers. An "ON"-delay timer does not produce an output signal (or turn "ON") until its set value (SV) is expired, which means the accumulated time becomes equal to the set value (figure 5-02). An "OFF"-delay timer, however, produces an output signal as soon as its enabling signal is "ON", and it terminates its timer signal (timer flag) when the set value is expired (figure 5-05). For pneumatic timers, an "ON"-delay timer is called a "normally closed" timer, whereas the pneumatic "OFF"-delay timer is known as a "normally open" timer (compare figures 5-02 and 5-05).

The electronic PLC "OFF"-delay timer shown in figure 5-04 is in reality also an "ON"-delay timer, similar to the one shown in figure 5-02, but its timer contact (bit), driving output relay 500, is a normally closed contact. Upon actuation of the enabling signal 000, output relay 500 turns instantly "ON", and 0.8 second later, when the timer flag TIM 01 turns "ON", output 500 is cancelled. This circuit may, for example, be used to activate a lamp. When the timer has timed out, the lamp will turn off.

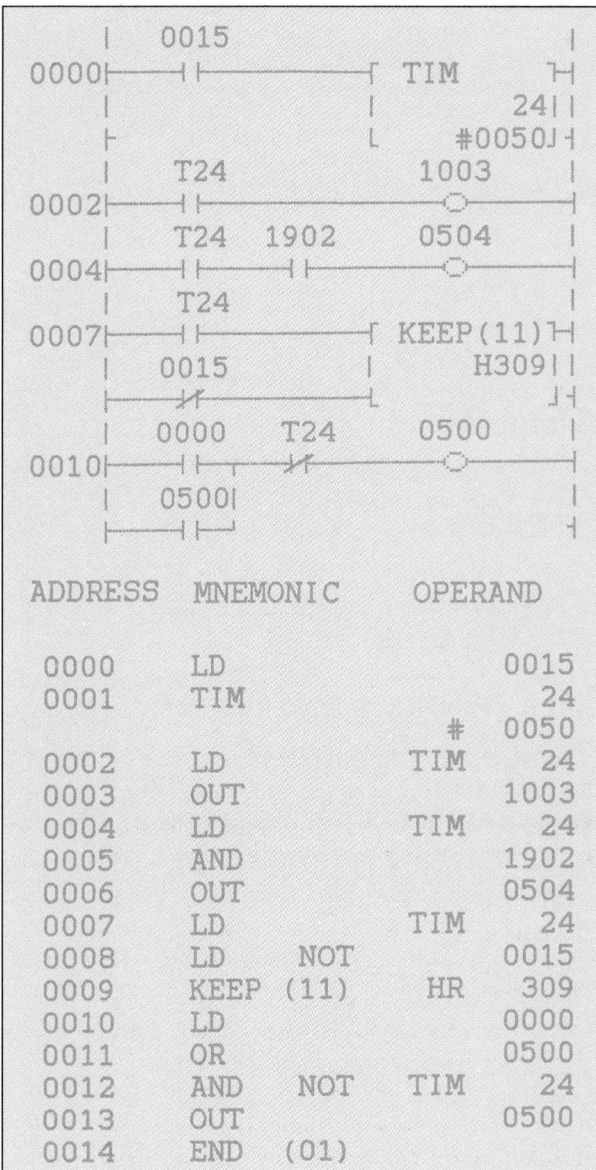

ADDRESS	MNEMONIC	OPERAND
0000	LD	0015
0001	TIM	24
		# 0050
0002	LD	TIM 24
0003	OUT	1003
0004	LD	TIM 24
0005	AND	1902
0006	OUT	0504
0007	LD	TIM 24
0008	LD NOT	0015
0009	KEEP (11)	HR 309
0010	LD	0000
0011	OR	0500
0012	AND NOT	TIM 24
0013	OUT	0500
0014	END (01)	

Figure 5-04. Timers drive numerous contacts. When activated, timer flag TIM 24 activates internal auxiliary relay 1003 and output relay 504 with its timer bits. At the same time it sets holding relay HR 309 and cancels latching relay 500.

TIMERS AND PLC POWER FAILURE

Some timers are not memory-retentive, which means that any time elapsed when power is interrupted is lost. The timer thus resets to SV and starts afresh when power to the PLC is restored. Therefore, if accumulated time at the point of PLC power interruption must be recorded, the timer must be a retentive timer. For such a case, one must use a counter to create the timer function. An example for this is given in figure 6-02.

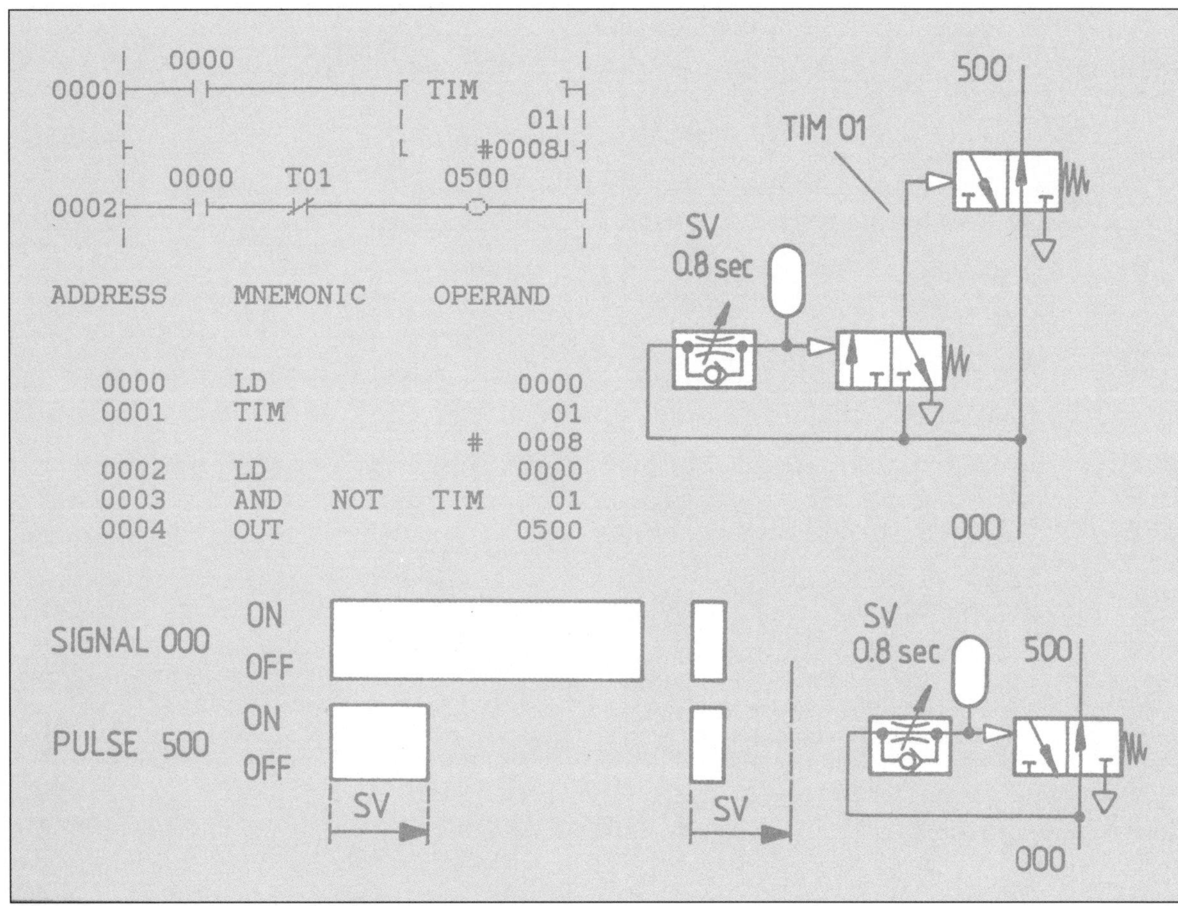

Figure 5-05. "OFF"-delay timer. Equivalent pneumatic timers and a function bar chart are provided to illustrate the operation and limits of this timer.

CASCADING TIMERS FOR EXTRA-LONG TIME

Cascading timers are commonly used to create timing functions that exceed the capacity of a single timer. Cascading two timers to achieve an extended timer function is shown in figure 5-06. With such an arrangement one could reach a maximum timing length of 999.9 seconds plus 999.9 seconds, or 1999.8 seconds, or 33.33 minutes. The timer shown in figure 5-06 is programmed for 26 minutes. For longer timing spans, the timer-counter combination as shown in figure 6-03 is recommended.

HIGH-SPEED TIMER

The high-speed timer has the same attributes as a normal timer, but its clock drives the timer with a pulse rate of 0.01 second instead of 0.1 second. Timer flags (relays) range from 00 to 47, which means they share the same timer/counter register area with the normal timers and counters (see figure 1-12). For this reason, one must never give a counter or timer the same flag number. (Example: One must not give flag number TIM 23 to a timer and to a counter such as CNT 23.)

The set value of the high-speed timer ranges from 00.00 to 99.99 seconds. Example: A set value for say 13.08 seconds would be programmed as SV #1308. Although set values 00.00 and 00.01 may be set when programming, SV 00.00 will disable the timer whenever the timer flag turns "ON", and set value 00.01 is, according to Omron, not reliably scanned! A scan time greater than 10 ms will adversely affect the accuracy of the high-speed timer.

If a high-speed timer needs to be memory-retentive, one must use a counter with a pulsing rate of 0.02 second (see "Retentive Timers" in Chapter 6 and figure 6-02). Omron C20 and C28K PLCs do not have a 0.01 clock pulse bit that may be readily used for this purpose. Omron C200H PLCs, however, do have a pulsing bit with a pulse rate of 0.02 second (bit 25401). A high-speed timer is programmed by using the FUN 15 (TIMH) instruction instead of the TIM instruction key as for normal timers. This automatically changes its clock rate from 0.1 second to 0.01 second. After entering the FUN 15 instruction and pressing the "ENTER" key, the set value # is entered in 0.01 second units, and the ENTER key is pressed again.

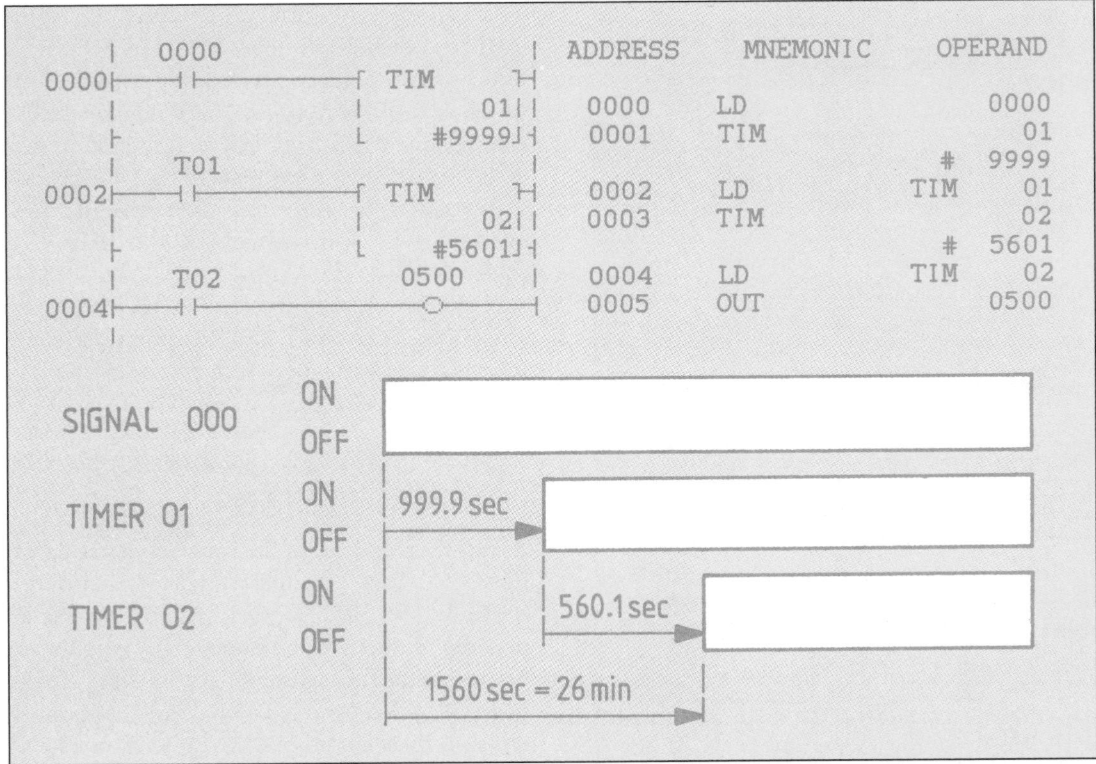

Figure 5-06. Cascaded timers. Cascading timers means timer 2 is added to the maximum time span of timer 1. The set time span for these two timers (SV) is 1560 seconds, or 26 minutes.

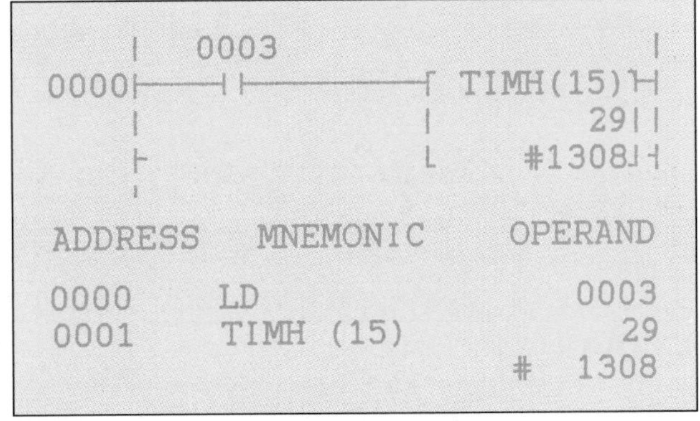

Figure 5-07. High-speed timer (FUN 15) with a decrementing rate of 0.01 second units.

DIFFERENTIATION FUNCTION (EDGE-PULSE TIMER)

Impulse or short-pulse electronic building blocks within a PLC's programming repertoire are absolutely essential for modern control technology programming. Such building blocks convert persisting, unwanted and long-term signals, which may cause problems, into a short pulse of one scan length. The differentiate functions, although not strictly being timer functions, are deliberately mentioned in this chapter since they are often used for the same purpose as the "OFF"-delay timer. Both are used to modify or shape an undesirably long signal into a short pulse. Components used for such signal conversion or pulse-shaping modification are:

- impulse timers or so-called "OFF"-delay timers (figure 5-05)
- differentiate-up pulse to create a scan-length pulse at the leading edge of an enabling signal (figure 5-08)
- differentiate-down pulse to create a scan-length pulse at the trailing edge of an enabling signal (figure 5-08).

The differentiate-up function, sometimes called positive edge function, leading edge function or edge pulse function, is an instruction given to an operand bit (relay) to turn "ON" for only one scan time when its enabling signal goes from "OFF" to "ON". This enabling signal could be a single contact from an input or it may be the execution condition from a complex logic circuit, as is shown in figure 7-13, addresses 9 to 13. The differentiate function can be allocated to auxiliary, output or holding relays. In figure 5-08 the enabling signal is input 001 and the differentiation instruction is given to internal auxiliary relay 1000 (bit 1000). For Omron PLCs, the differentiate-up function is function 13 (DIFU (13)).

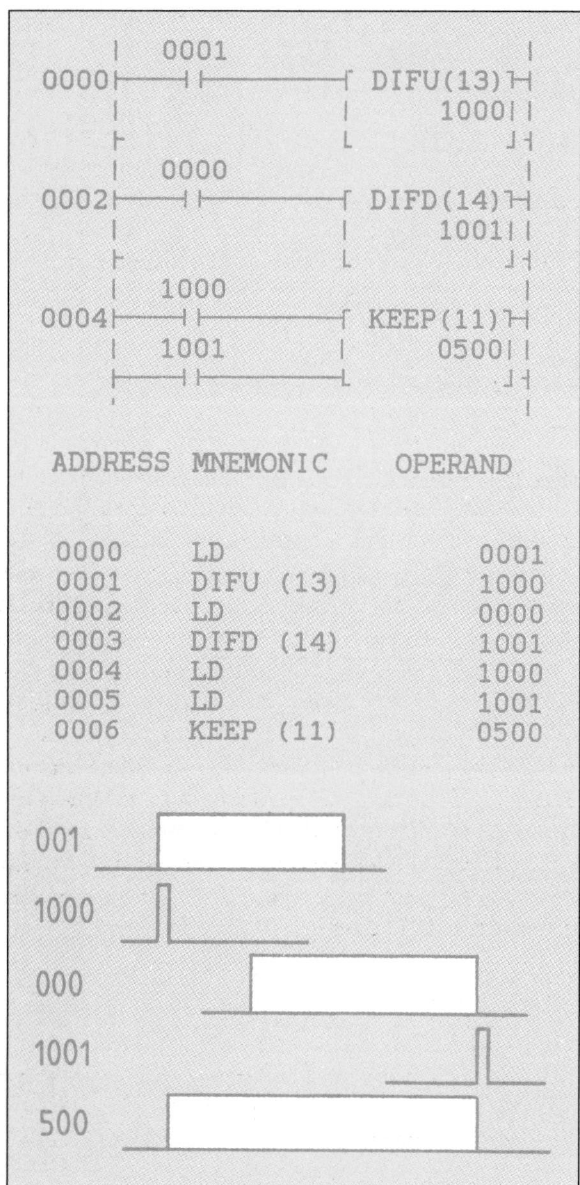

Figure 5-08. A hypothetical circuit combination of DIFU (13) and DIFD (14).

The differentiate-down function, sometimes called negative edge or trailing edge function, is used to modify a long enabling signal into a short scan-time-length pulse when the enabling signal goes from "ON" to "OFF". For Omron PLCs, this is function 14 (DIFD (14)).

Note: In neither differentiate-up nor differentiate-down can the scan pulse be repeated unless the enabling signal or execution condition has disappeared first. When it then reappears it then creates the next pulse. In figure 5-08 input 001 creates scan pulse 1000 with its leading edge. This pulse sets KEEP 500 (flip-flop 500). Input 000 with its trailing-edge scan pulse bit 1001 resets KEEP 500 (DIFD (14)).

In figure 5-09, the differentiate-up instruction is used to turn holding relay 800 (HR 800) into a T flip-flop, whereby every second input signal of bit 000 causes HR 800 to go "ON". To demonstrate the operation of this T flip-flop, HR 800 is used to drive output 501. For the operation principle of T flip-flops using a shift register, see the truth table in figure 4-07. For the operation of a T flip-flop with two "KEEP" relays, see figure 4-06.

Differentiate-up functions are frequently used in PLC arithmetic to ensure that a computation is not repeated several times with every scan and thus give a false result (see figures 10-22, 10-23, 10-25 and 10-34).

TIMER APPLICATION CIRCUIT — CONTROL PROBLEM 1

A practical and versatile timer application is given in figure 5-10. A pneumatic control circuit needs to be converted to PLC control. After conversion, all that remains of the previous pneumatic circuit are the two linear pneumatic actuators (cylinders). The two pneumatically-pilot-operated directional control valves are replaced by solenoid-operated pneumatic control valves. The limit valves become limit switches and the three timers plus the memory type RS flip-flop are part of the PLC control system.

In rest position, actuator Ⓐ is normally extended and actuator Ⓑ is normally retracted. The "ON-OFF" detent

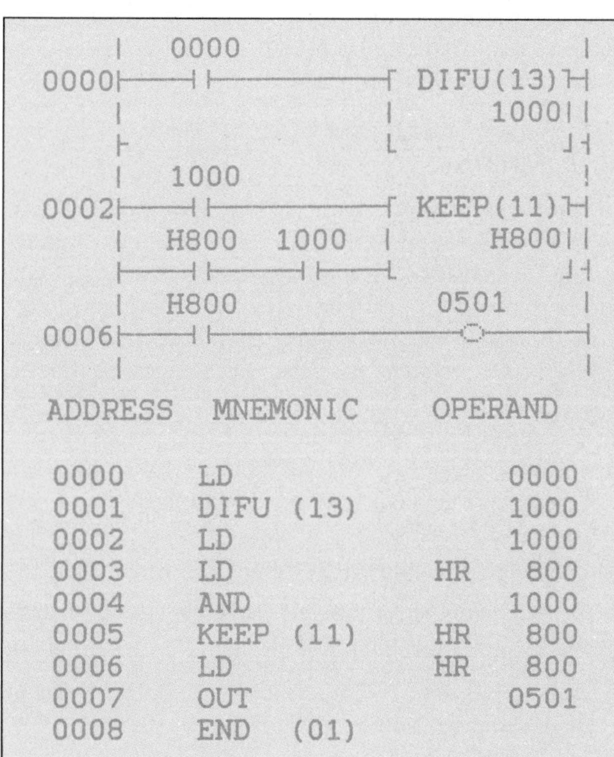

Figure 5-09. T flip-flop, using DIFU (13) to create the trigger behaviour of this relay.

type start valve must be replaced by two push-button switches and an RS flip-flop (holding relay) to store the start command (008 to start, 009 to stop). Upon actuation of the start switch, actuator Ⓑ begins to reciprocate between its two limit switches 004 and 003. Its reciprocation movements are automatically terminated by the "OFF"-delay timer TIM 03. This timer has a set value of 15 seconds (#150). Actuator Ⓐ does not immediately respond to the given start signal until 10 seconds have elapsed. This signal delay is controlled by timer 01 (TIM 01). Timer 01 is an "ON"-delay timer with a set value of 10 seconds. To avoid an opposing signal created by the memory type "ON-OFF" valve (now HR 000), which constantly produces pneumatic pilot signal A0 at the pneumatic directional control valve, an impulse valve is placed into signal line A0. This impulse valve, being a normally open timer by construction, is replaced by a 0.8 second "OFF"-delay timer (TIM 02). The holding relay HR 000 (RS flip-flop) becomes the stored enabling signal for all three timers (see addresses 0003, 0011 and 0013). For PLC input/output assignment numbers, see the pneumatic circuit in figure 5-10.

Timer 02 in this control problem is used to turn the persistent and long start signal into a short 0.8 second pulse to prevent it from opposing solenoid actuating signal 502 (A1). This signal modification could, on first sight, also be accomplished by using the differentiate-up instruction for the pulsing of output signal 501, whereby the edge pulse function (FUN 13 DIFU) is given to relay 501 (figure 5-11). The logic line ending in relay 501 would then read as is shown in figure 5-11.

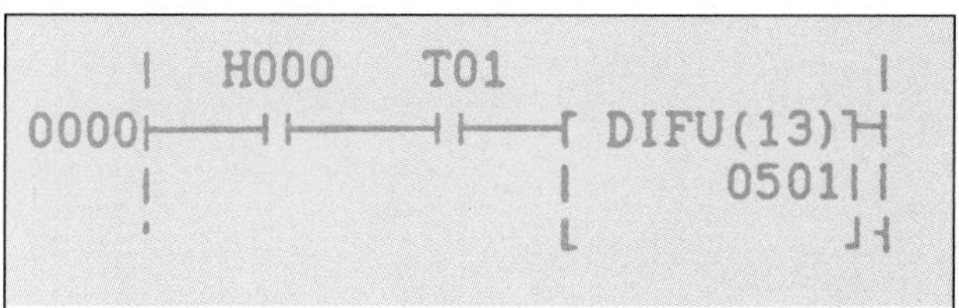

Figure 5-11. Edge pulse function for output signal 501.

Such a modification would, however, not work!

Note! The differentiate function pulse is never longer than one scan time! This scan time is normally a few milliseconds, depending on program length. To energise the solenoid of a pneumatic or hydraulic valve, this extremely short scan pulse could, if applied to an output relay for the solenoid valve actuation, be *too* short. As a result, the valve spool would not respond at all, or it may not move completely into its new flow position. This in turn could cause a malfunction of the fluid power circuit. It is therefore safer to use an "OFF"-delay timer for such critical pulse-shaping functions. With an "OFF"-delay timer, the pulse length can be determined with the timer's set value (SV), and can be tailor-made to the minimum time required to shift the valve spool. The minimum time is approximately 0.5 second.

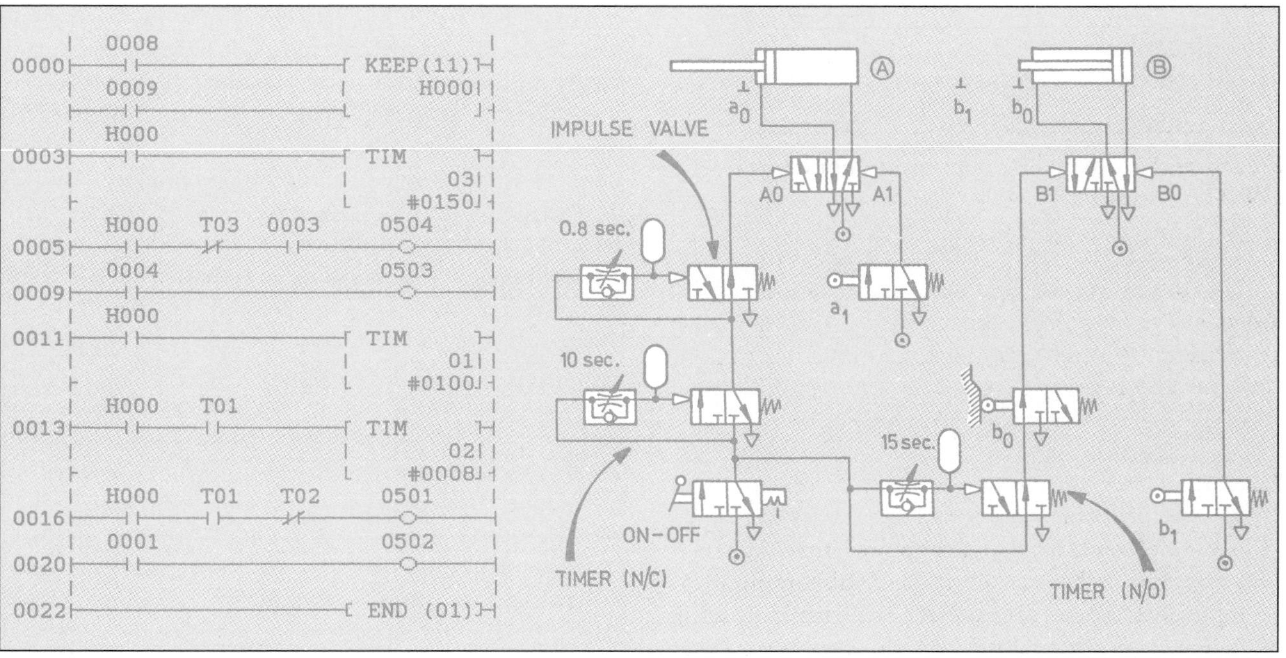

Figure 5-10. Timer application circuit — control problem 1. This circuit uses an "ON"-delay timer (TIM 01), an "OFF"-delay timer (TIM 03) and an impulse timer to prevent an opposing solenoid actuation signal (TIM 02).

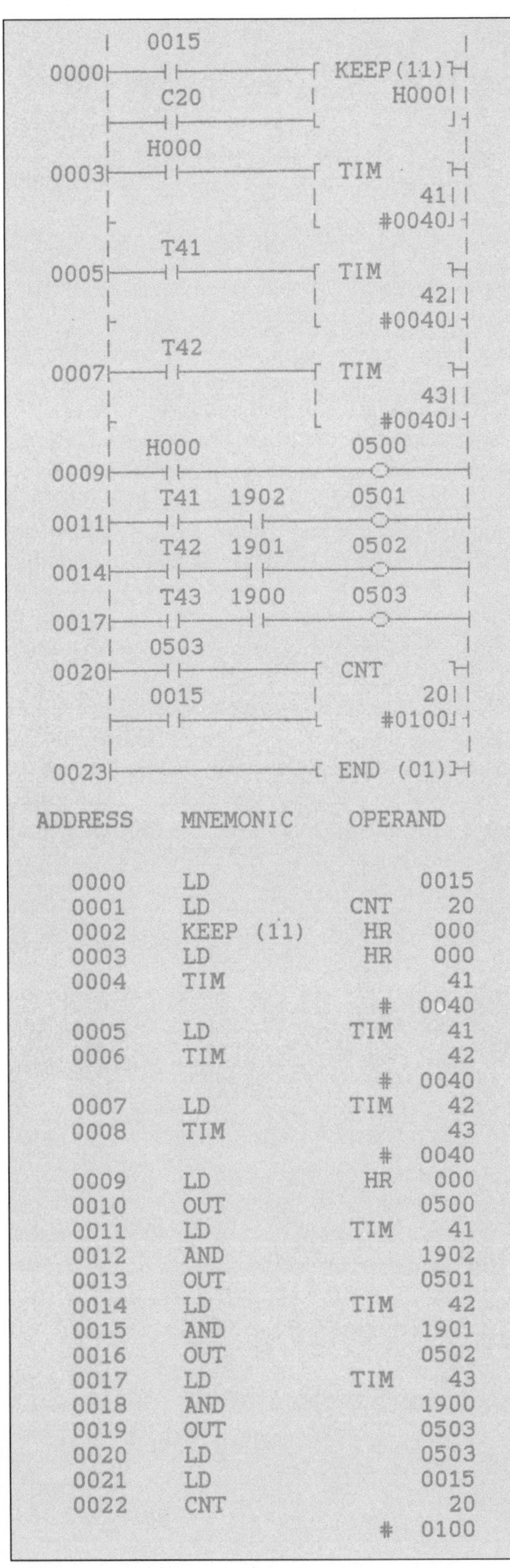

ADDRESS	MNEMONIC	OPERAND
0000	LD	0015
0001	LD	CNT 20
0002	KEEP (11)	HR 000
0003	LD	HR 000
0004	TIM	41
		# 0040
0005	LD	TIM 41
0006	TIM	42
		# 0040
0007	LD	TIM 42
0008	TIM	43
		# 0040
0009	LD	HR 000
0010	OUT	0500
0011	LD	TIM 41
0012	AND	1902
0013	OUT	0501
0014	LD	TIM 42
0015	AND	1901
0016	OUT	0502
0017	LD	TIM 43
0018	AND	1900
0019	OUT	0503
0020	LD	0503
0021	LD	0015
0022	CNT	20
		# 0100

Figure 5-12. Machine malfunction alarm system as described in timer application circuit — control problem 2.

TIMER APPLICATION CIRCUIT— CONTROL PROBLEM 2

A PLC-controlled alarm system to announce machine malfunction is activated when momentary alarm signal (input) 015 is recorded. An alarm lamp (output) 500 is then to be activated and it must remain "ON" until the alarm system is turned "OFF". Four seconds after the alarm has been raised, flicker lamp 501 must turn "ON". Its flicker or blinking rate is one pulse per second (0.5 second "ON" and 0.5 second "OFF"). Four seconds thereafter, alarm flicker lamp 502 must turn "ON". Its flicker rate is five pulses per second. A further four seconds later, alarm flicker lamp 503 turns "ON" with a flicker or blinking rate of ten pulses per second. By this time, all four alarm lamps are blinking with their respective pulsing rates. Alarm lamp 500 does not blink but it is also "ON". After 100 blinking pulses of alarm lamp 503, the entire alarm system is automatically turned "OFF" (figure 5-12).

To record the momentary alarm signal 015, use holding relay (KEEP relay) HR 000. For the counting of flicker pulses of lamp 503, use counter 20 (CNT 20). For the three timers, use TIM 41, TIM 42 and TIM 43. The flicker pulses are created by the pulse clock bits 1900, 1901 and 1902 (see figure 5-01 for pulse bits).

TIMER APPLICATION CIRCUIT — CONTROL PROBLEM 3

The washing basket of a vegetable-rinsing machine is powered by a pneumatic, double-acting, linear actuator (figure 5-13). This machine has so far been completely pneumatically controlled. It now has to be converted to PLC control with only the pneumatic actuator and its speed control devices remaining. Apart from the 30-second reciprocating pattern, the machine needs to be redesigned for the inclusion of a second reciprocating pattern of 70 seconds. The two patterns are selectable by two push-button switches on the control panel. If emergency stopping is signalled by the operator stop button, or automatically if the guard is opened before normal termination of reciprocation movements, the pneumatic actuator must extend immediately, regardless of its motion direction or interim position. At the same time, a warning lamp must announce the stopping action. This warning lamp must flash with a 0.5 second "ON" and "OFF" rate (1-second pulses) and it must not stop flashing until being reset by a key-type reset switch. In addition, a machine "RUN" lamp must be included, which remains "ON" for as long as the machine is reciprocating. Include also lamps indicating the respective reciprocating time selection (a lamp for timer 1 and a

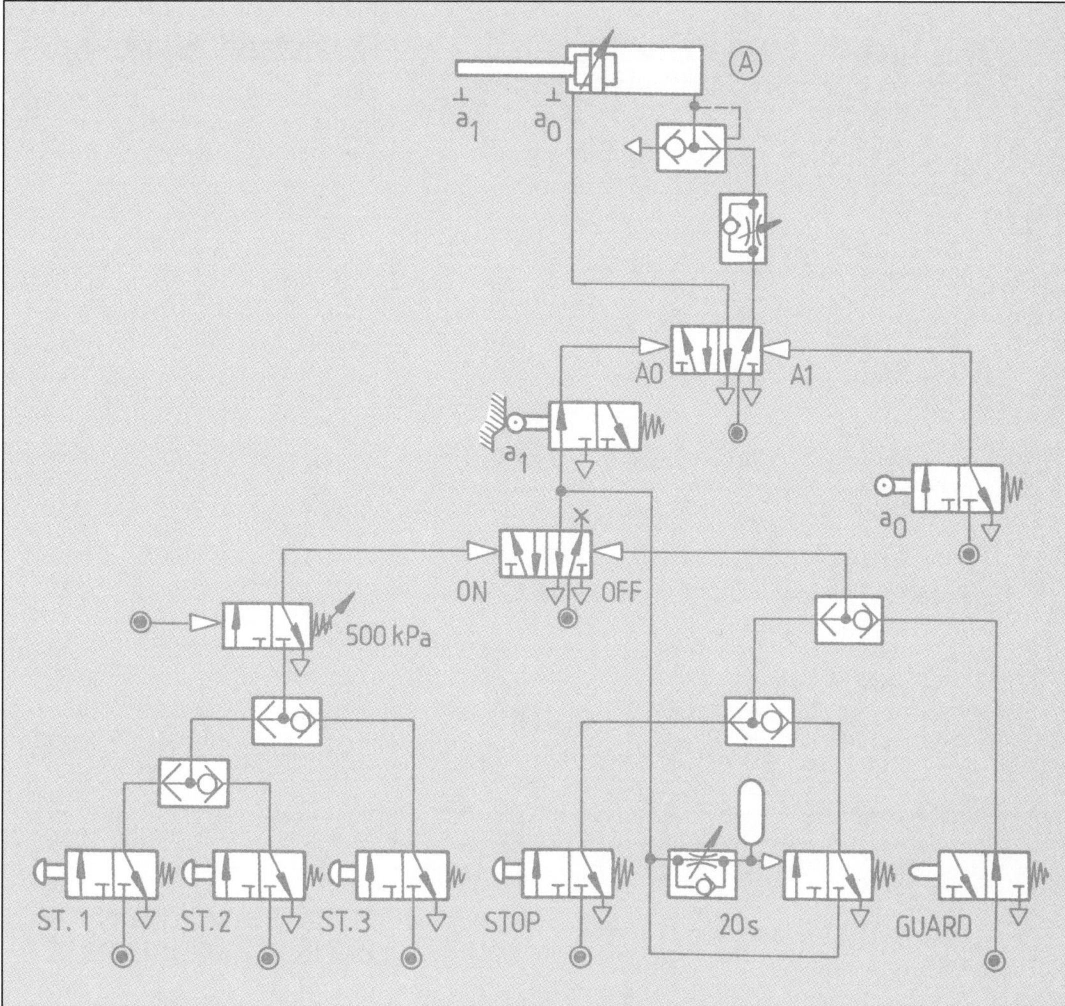

Figure 5-13. Vegetable-rinsing machine. Timer application — control problem 3. The machine control must be altered to achieve a selectable termination of reciprocation motions by timer 1 (30 seconds) or timer 2 (70 seconds). For safety purposes, the stop and guard switches must be normally closed switches that signal the machine to stop in case of wire breaking between machine and PLC.

lamp for timer 2).

Design the ladder diagram and its corresponding mnemonic list and draw an input/output assignment list of all the required input/output connections to and from the PLC.

INPUT-OUTPUT ASSIGNMENT LIST FOR CONTROL PROBLEM 3 (VEGETABLE-RINSING MACHINE)

PLC inputs:

001 = Ⓐ retracted switch
002 = Ⓐ extended switch
003 = pressure switch
004 = start 1 switch
005 = start 2 switch
006 = start 3 switch
007 = stop switch
008 = guard switch
009 = reset key switch
014 = select TIM 01 switch
015 = select TIM 02 switch

PLC outputs:

501 = retract Ⓐ solenoid
502 = extend Ⓐ solenoid
508 = TIM 01 selected lamp
509 = TIM 02 selected lamp
510 = machine RUN lamp
511 = machine STOP lamp
HR 000 = ON/OFF relay
HR 001 = timer select relay
HR 002 = EM stop recording relay
TIM 01 = 30 second reciprocation
TIM 02 = 70 second reciprocation

TIMER SUMMARY

The timer is actuated when its enabling condition goes "ON" and it is reset to SV when the enabling condition goes "OFF". When being enabled, the timer measures time in units of 0.1 second from SV. For a timer to be given the chance to time out, its enabling signal must be longer than its SV. Timer accuracy for most PLCs is approximately ±0.1 second. Timers and counters must never be given the same flag number.

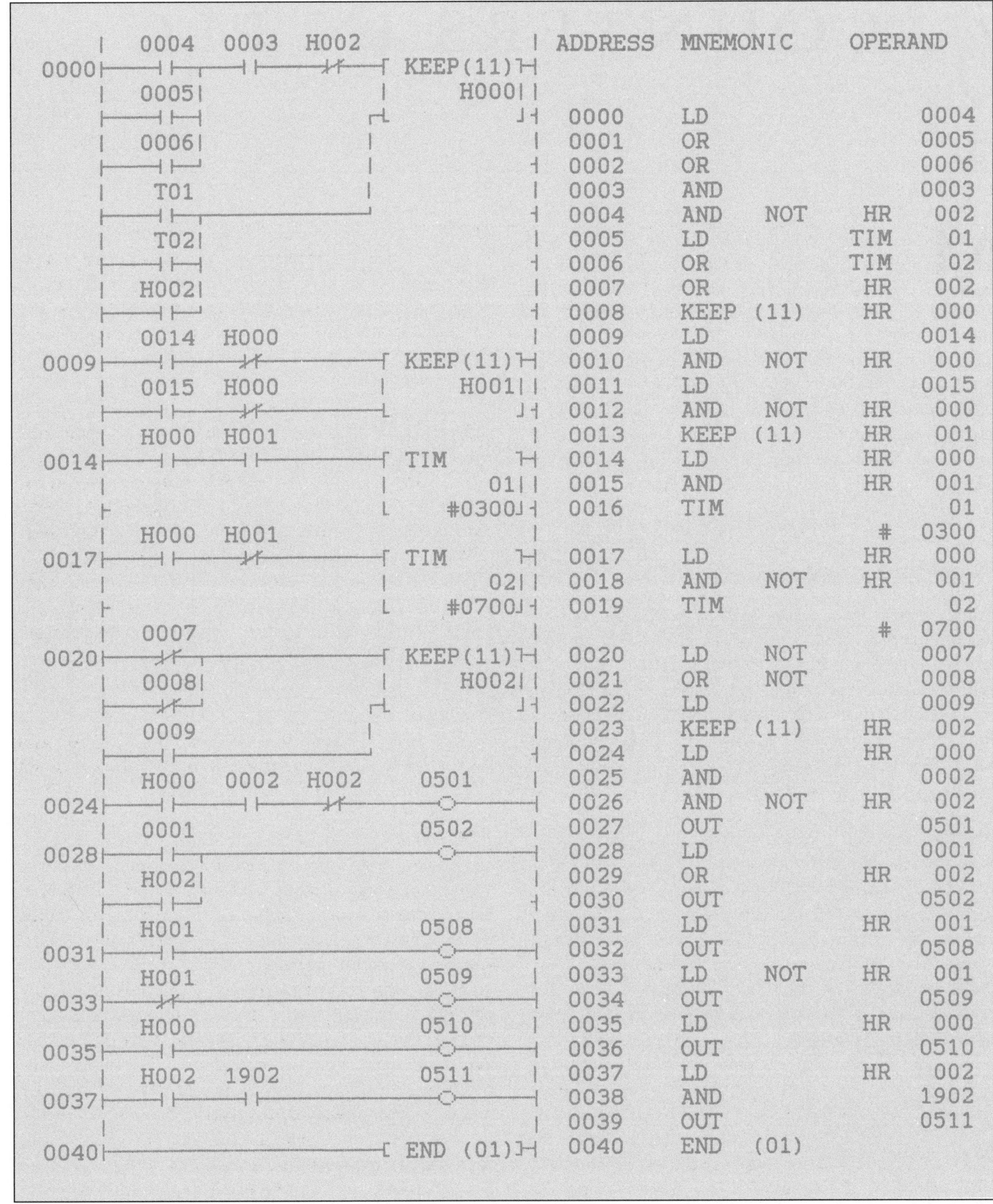

ADDRESS	MNEMONIC	OPERAND
0000	LD	0004
0001	OR	0005
0002	OR	0006
0003	AND	0003
0004	AND NOT	HR 002
0005	LD	TIM 01
0006	OR	TIM 02
0007	OR	HR 002
0008	KEEP (11)	HR 000
0009	LD	0014
0010	AND NOT	HR 000
0011	LD	0015
0012	AND NOT	HR 000
0013	KEEP (11)	HR 001
0014	LD	HR 000
0015	AND	HR 001
0016	TIM	01
		# 0300
0017	LD	HR 000
0018	AND NOT	HR 001
0019	TIM	02
		# 0700
0020	LD NOT	0007
0021	OR NOT	0008
0022	LD	0009
0023	KEEP (11)	HR 002
0024	LD	HR 000
0025	AND	0002
0026	AND NOT	HR 002
0027	OUT	0501
0028	LD	0001
0029	OR	HR 002
0030	OUT	0502
0031	LD	HR 001
0032	OUT	0508
0033	LD NOT	HR 001
0034	OUT	0509
0035	LD	HR 000
0036	OUT	0510
0037	LD	HR 002
0038	AND	1902
0039	OUT	0511
0040	END (01)	

Figure 5-14. PLC ladder diagram circuit for vegetable rinsing machine of control problem 3.

6 COUNTERS IN PLC

Most modern PLCs are equipped with in-built counter flags (counter relays). These counter flags operate and are programmed similarly to timer flags. A counter may be used to count the total number of products being produced by a manufacturing machine in a given time or it may group components into batches for packaging or wrapping. It may serve to keep track of the r.p.m. of a revolving shaft (r.p.m. counter) to maintain and regulate r.p.m. consistency. In short, counters are an inseparable part of PLC controllers and their use in industrial automation control tasks.

COUNTER PROGRAMMING

Counters by construction may be "down-counters" (decrementing counters) or "up-counters" (incrementing counters). Some PLC manufacturers may even offer both "up and down circular" counters. A down-counter counts down from its set value (SV) to 0000. Each count pulse arriving at the counter decrements the counter from its present value (PV) by 1. When the counter is decremented to 0000 present value (PV = 0000), its flag will turn "ON" and it will stay "ON" until the reset signal returns the counter to the SV. Note the two terms used with counters: set value (SV) and present value (PV). The present value is the accumulated value at a given point in the counting process. The set value is the predetermined count value to which the counter is preselected before the counting process starts.

The operation of a decrementing counter is illustrated in figure 6-01. The counter in that illustration is CNT 04 (counter flag 04). Its SV is 0013 (#0013), and its enabling signal is input bit 000. The counter decrements whenever a pulse from bit 008 enters the counter. The enabling signal and the pulse signal are "AND" connected. When the counter has reached PV 0000 (after 13 pulses have been entered into the counter), its flag CNT 04 turns "ON" and remains "ON" until reset signal 015 causes the counter to reset to SV 0013. At that time the counter is again ready to resume a new counting process, when its enabling signal is "ON" and pulses from bit 0008 enter the counter. The bar chart in figure 6-01 shows the operation of a down-counter. To demonstrate the turning "ON" of the counter flag CNT 04, a normally open contact driven by counter 04 is used to turn output relay 500 "ON". For practical purposes, this output may be used to activate a lamp, a solenoid valve or a buzzer, or any other complex circuit or subcircuit (see figures 6-12, 8-24, 9-17 and 10-31).

Note: A counter will never activate the counting completion flag if recorded count pulses (PV) are less than the counter's SV or if a counter is caused to reset before the counter has reached zero. As with timers, a counter may also drive numerous N/O and N/C contacts (counter bits). The contacts (bits) driven by the counter flag carry the same number as the counter flag (see figure 6-10, where the counter CNT 02 drives a normally closed contact in address 002 and a normally open contact in address 007).

Figure 6-01 shows that the contacts 000, 008 and 015 are added to the counter box in the ladder diagram as operating conditions. These contacts are no direct part of the actual counter, they merely cause the counter to function. By programming a counter into a circuit, one may therefore select any contacts from the entire range of available PLC contacts (bits). Such bits, used to enable, reset or pulse the counter, may be taken from internal relays, holding relays, timer relays, counter relays or output relays.

In the case of a down-counter, if the pulsing contact changes from logic 0 to 1, the counter will decrement by one unit from its PV to the new (lower) PV. This decrementation takes place at the leading edge of the count pulse. Whether this count pulse is "ON" briefly or for an hour makes no difference: the decrementing value is still only 1. This means that the counter sees only the electrical change at the PLC input point go from 0 to 1, caused by the pulse-generating switch or sensor.

Unlike timers, counters are memory-retentive during PLC power supply failure. This means the counter's PV is held during power interruption. This attribute is used to advantage if timers require PV retention during power failure. Figure 6-02 shows such a memory-retentive timer built from a counter.

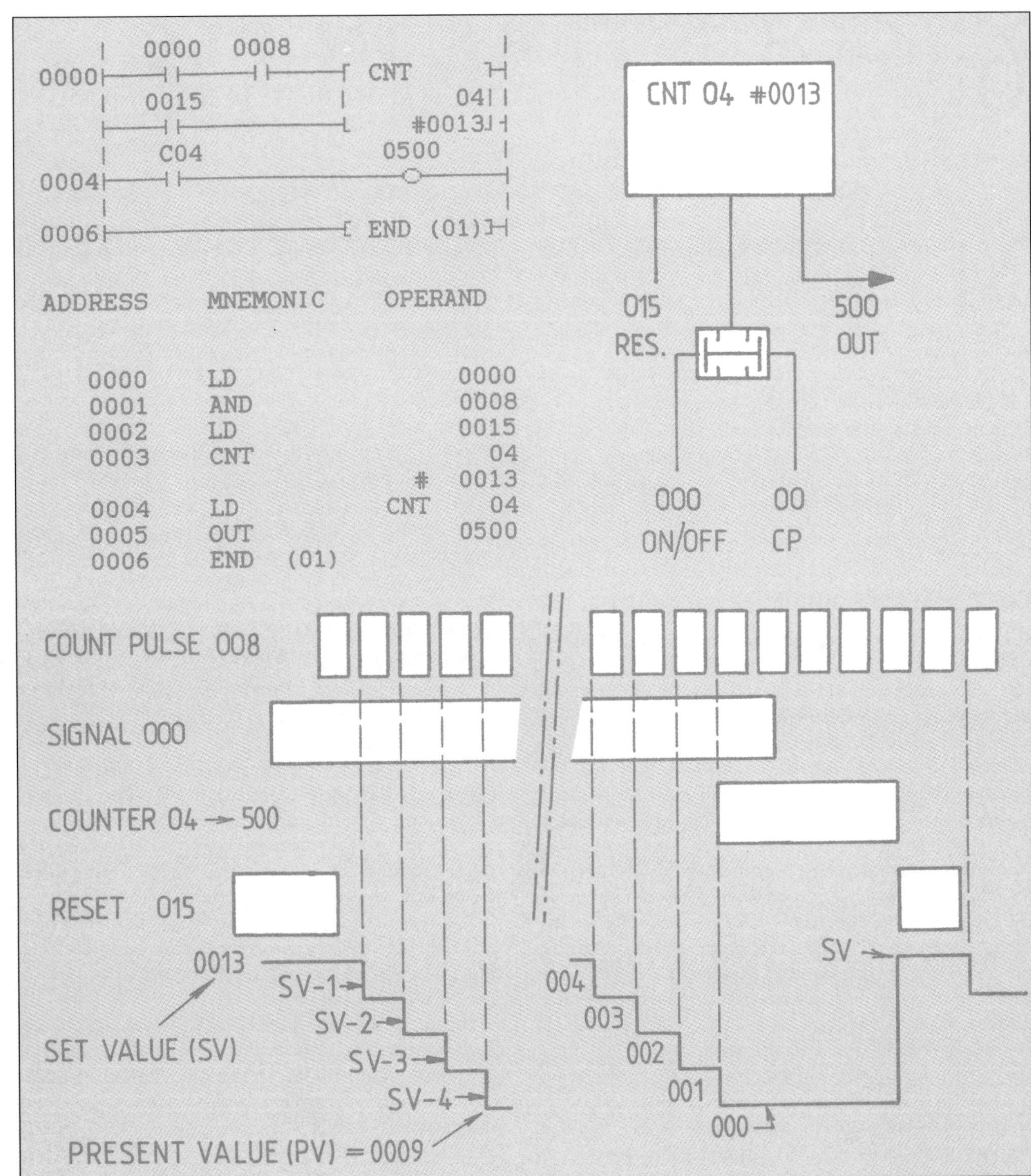

Figure 6-01. This down-counter's flag CNT 04 is turned "ON" when 13 count pulses of bit 008, enabled by input bit 000, have reduced the counter's present value (PV) from its set value (SV) to 0000. The counter flag CNT 04 remains "ON" until reset bit 015 is "ON". This reset signal (015) causes the counter to return to its SV 0013 to make it ready for a new counting task.

RETENTIVE TIMERS (COUNTER TIMERS)

Counters in most PLCs are retentive, whereas timers in some PLCs do not retain the PV or so-called accumulated time. Therefore, where accumulated time of the enabled timer must be retained when power failure to the PLC interrupts the timing process, one needs to use counters instead of timers to program a so-called retentive timer. The enabling signal causing the timer process to start must then be "AND" connected to a clock pulse contact (figure 6-02). Thus, the counter is counting precise clock pulses from the PLC's internal clock bits (bits 1900, 1901 or 1902). When the counter is decremented to zero, it turns its output flag "ON". This output flag signal may then be reset when the counting/timing task is completed. The retentive timer (counter) shown in figure 6-02 resets automatically when the enabling signal 007 disappears.

Figure 6-02 shows a retentive timer that uses a counter (CNT 04) to render an output signal (counter flag C 04) when 35 clock pulses of clock bit 1900 have decremented the counter from SV to 0000. Of course, one may use other clock pulses such as 1901 (0.2 second pulses) or 1902 (1.0 second pulses) to pulse the counter (timer), but a greater accuracy is obtained by using the shorter pulses of bits 1900 or 1901. The counter resets as soon as the enabling signal 007 is cancelled.

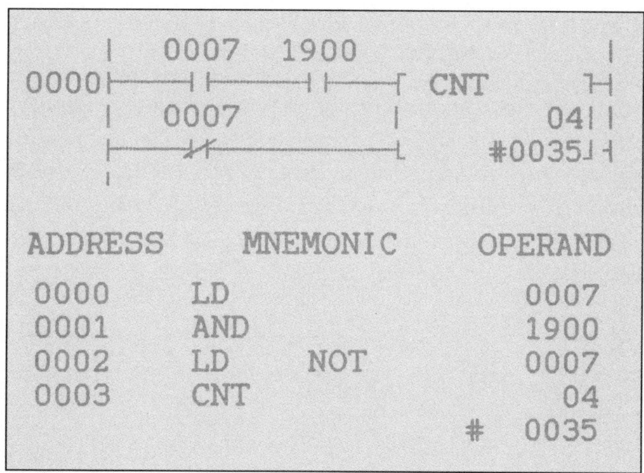

ADDRESS	MNEMONIC		OPERAND
0000	LD		0007
0001	AND		1900
0002	LD	NOT	0007
0003	CNT		04
			# 0035

Figure 6-02. A counter (CNT 04) is used as a retentive timer to turn "ON" when 3.5 seconds have elapsed. It automatically resets when its enabling signal 007 disappears.

EXTRA-LONG TIMER WITH COUNTER COMBINED

To accomplish extra-long timing processes, a timer may be combined with a counter. With only one timer, a time delay task of 999.9 seconds maximum is achievable (SV #9999). By combining a single timer with a counter, as is shown in figure 6-03, one can achieve a maximum time delay of 9 998 000.1 seconds, or approximately 3.8 months.

LADDER DIAGRAM

```
     | 0000   T01    C02         |
0000 |--| |----|/|----|/|----[ TIM      ]-
     |                       |     01   |
     |                       |  #6000   |
     | T01                              |
0004 |--| |-------------------[ CNT      ]-
     | 0015                  |     02   |
     |--| |------------------|  #1000   |-
     | C02                     0500     |
0007 |--| |-----------------------( )----|-
     |                                  |
0009 |-------------------------[ END (01)]-
```

ADDRESS	MNEMONIC		OPERAND	
0000	LD			0000
0001	AND	NOT	TIM	01
0002	AND	NOT	CNT	02
0003	TIM			01
			#	6000
0004	LD		TIM	01
0005	LD			0015
0006	CNT			02
			#	1000
0007	LD		CNT	02
0008	OUT			0500
0009	END	(01)		

Figure 6-03. Non-retentive extra-long timer using a combination of a timer and a counter. The set value for this timer is 600 000 seconds. Output 0500 turns "ON" after 600000 seconds if the timer has been enabled for at least the same time and the counter has not been reset during this time.

EXTRA-LONG RETENTIVE TIMER

When timing processes exceed the capacity of a single retentive timer (counter), one may use cascaded retentive timers (figure 6-04). Two such retentive timers cascaded into a timer package can achieve a set value of up to 3.8 months' timing span (9999 seconds × 9999 nesting cycles). For more detail, see next section, "Cascaded Counters".

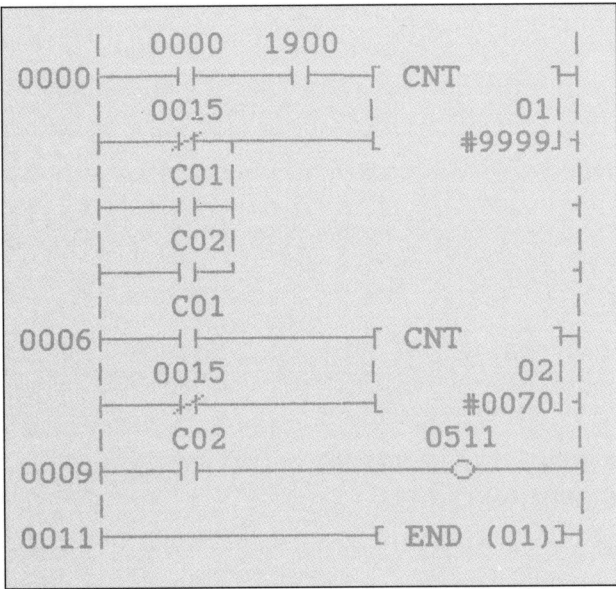

Figure 6-04. Extra-long retentive timer using two counters.

CASCADED COUNTERS

To accomplish a counting task exceeding the maximum counting capacity of a single counter (9999 counts maximum), one may cascade several counters into a counter package. This practice is also called nesting. Figure 6-05 shows such a cascaded counter or counter package. Whenever counter 01 has decremented to 0000 PV and its counter flag turns "ON" (CNT 01), it sends a count pulse to counter 02. Counter 01 then resets itself at the same time and counting restarts at counter 01. When counter 02 has reached zero (0000), which means it has decremented from SV 10 to 0, the counter package has counted a total of 9999 × 10 pulses, which amounts to 99 990 pulses. At that time, counter 02 resets counter 01 and counting of further pulses stops until both counters are reset by reset signal 015 (when input 015 is logic 0). Input 000 is the enabling signal and input 008 is the count pulse input for counter 01. A contact of counter 01 becomes the count pulse input for counter 02.

```
     | 0000   0008                   |
0000 |--| |----| |-----[ CNT        |-|
     | 0015              |      011 |
     |--|/|--------------|    #9999 |
     | C011                         |
     |--| |--|                      |
     | C021                         |
     |--| |--|                      |
     | C01                          |
0006 |--| |-----------[ CNT        |-|
     | 0015              |      021 |
     |--|/|--------------|    #0010 |
     | C02                  0500    |
0009 |--| |--------------------( )--|
     |                              |
0011 |-------------------[ END (01) |
```

ADDRESS	MNEMONIC		OPERAND	
0000	LD			0000
0001	AND			0008
0002	LD	NOT		0015
0003	OR		CNT	01
0004	OR		CNT	02
0005	CNT			01
			#	9999
0006	LD		CNT	01
0007	LD	NOT		0015
0008	CNT			02
			#	0010
0009	LD		CNT	02
0010	OUT			0500
0011	END	(01)		

Figure 6-05. Cascaded counters to achieve extra-long counting tasks.

In figure 6-05, the counter flag CNT 02 is used to drive output relay 500, which announces the counting task as being completed. Cascaded counters, like all Omron counters, are memory-retentive, which means the accumulated value (or present value) is retained during PLC power failure.

REVERSIBLE COUNTER

A reversible counter, also called an up-down circular counter, is used to count between 0000 and SV according to changes in two enabling conditions. The reversible counter, therefore, requires at least three operating conditions. The first of these three operating inputs is the enabled pulse input to cause incrementing (count up). The second operating input is the enabled pulse input to cause decrementing (count down). The third is the reset signal input. The incrementing pulse input is usually "AND" connected to the enabling signal, and the decrementing input is usually "AND" connected to its separate enabling input (figures 6-06 and 6-12). When decrementing from 0000, PV is set to SV and the completion flag is turned "ON" until the PV is decremented again. When incrementing beyond SV, the PV now being SV is set to 0000, and the completion flag is turned "ON".

Because of these attributes, the reversible counter is sometimes also called a ring counter, as it is counting in circles (0000 → SV; then automatic reset; then again 0000 → SV; etc; figure 6-08). With the reversible counter the PV will not be incremented or decremented while the reset signal is "ON". Counting will begin again when the reset signal goes off.

If enabled pulse changes have occurred at both the down- and up-counting inputs, PV will not be changed (counter is blocked).

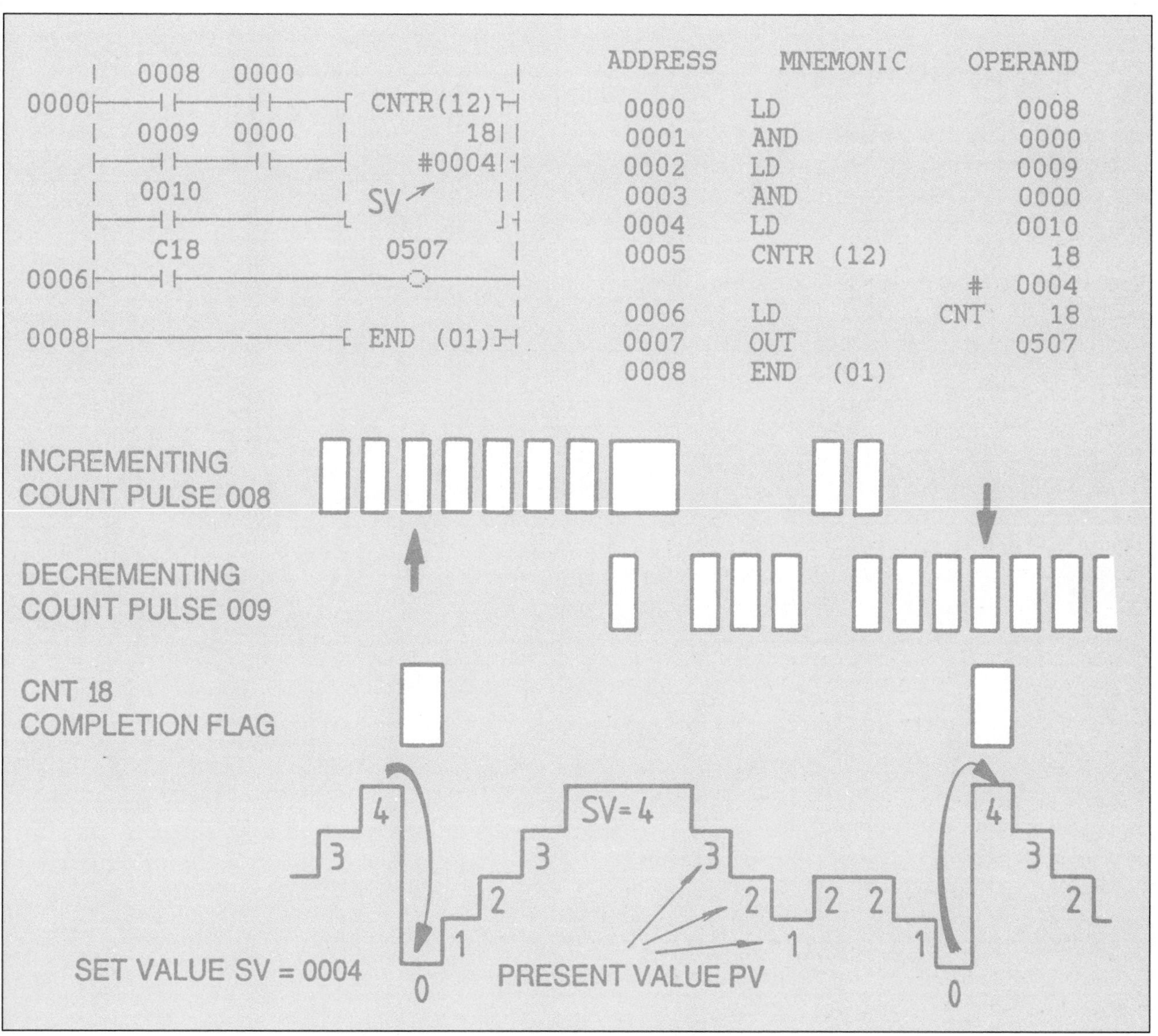

Figure 6-06. Reversible counter, also called up-down counter or ring counter.

COUNTER APPLICATION CIRCUIT — CONTROL PROBLEM 1

A conveyor system to divert bottles to two parallel exit conveyors is to be designed and programmed. Separation is accomplished with a diverter blade moved by a pneumatic, double-acting, linear actuator. For logistic purposes, the conveyor needs to be stopped automatically after a total of 1300 bottles have been counted. The multiplied SVs of counters 01 and 02 accomplish this task. The diverter blade passes four bottles, and then diverts the next four bottles to the parallel conveyor. To achieve this toggling motion, a T flip-flop is applied to command the pneumatic actuator to extend or retract alternately with every incoming count pulse of counter 01. Switch 000 enables the conveyor system, and switch 006, being the bottle detector sensor signal, creates the count pulses. Switch 014 is used to reset the diverter counter only, and switch 015 resets the entire conveyor system (figure 6-07).

COUNTER APPLICATION CIRCUIT — CONTROL PROBLEM 2

An extra-long (extended) retentive timer is to be designed and programmed. A chemical reaction process for a glue used in the manufacturing of motor car components must be automatically stopped after exactly 13 days, 6 hours, 28 minutes and 10 seconds. This set value (SV) amounts to a total of:

13 days	=	1123 200 sec.
6 hours	=	21 600 sec.
28 min.	=	1680 sec.
10 sec.	=	10 sec.
Total	=	1 146 490 sec.

9999 × 114	=	1 139 886 sec. (C01 × C02)
remainder	=	6604 sec. (C03)

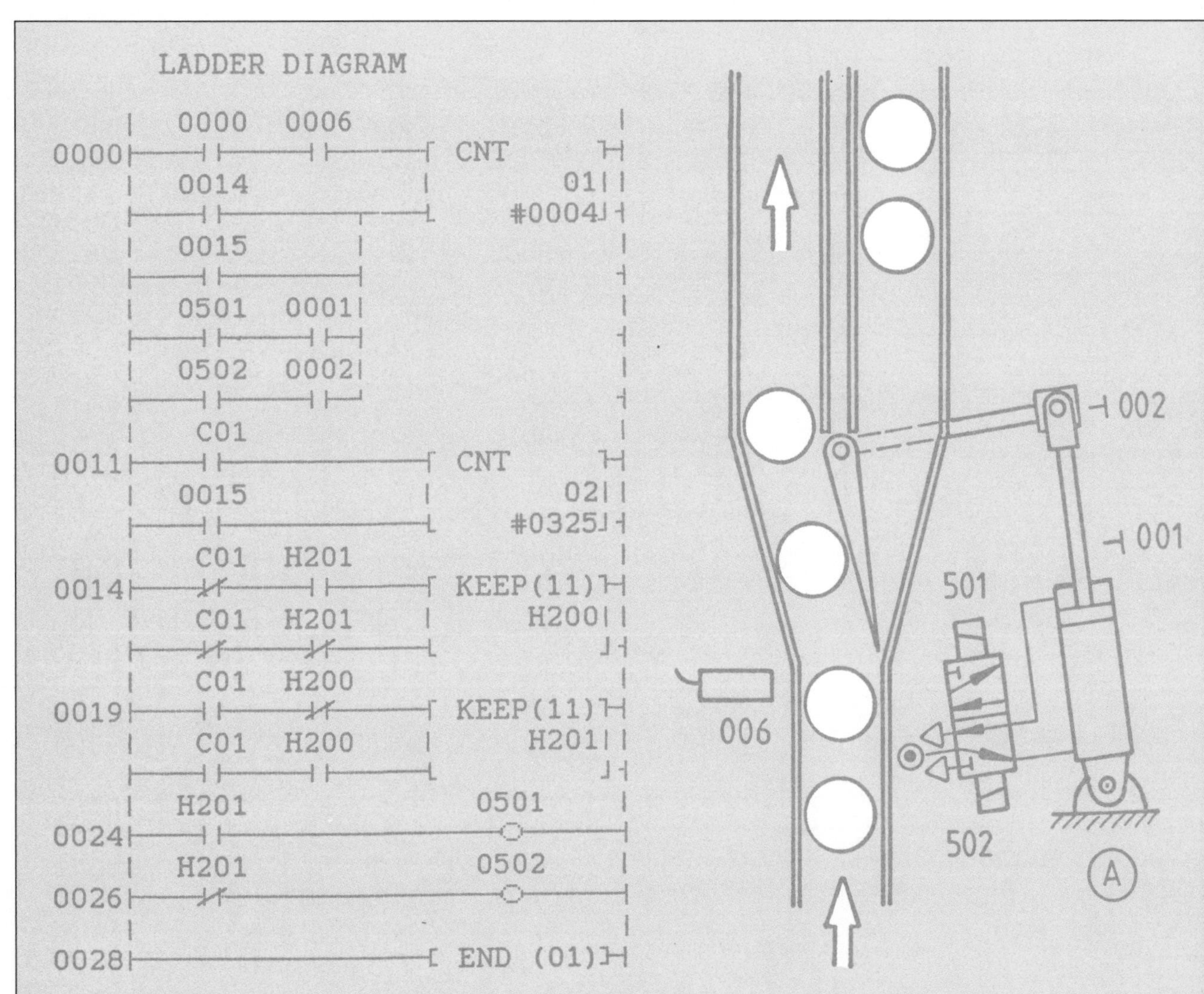

Figure 6-07. Counter application — control problem 1. Bottle sorter conveyor circuit.

Counter 01 (#9999) multiplied by counter 02 (#0114) in a cascading counting arrangement amounts to 1399 886 pulses with 6604 counts remaining. The remainder, since it is less than 9999, must be counted by the remainder counter CNT 03. When counting is completed, counter 03 sets holding relay HR 000, and holding relay HR 000 in turn activates output 508. Output 508 announces the timing process as being completed. Output 509, when activated by the enabling signal input 0000, activates lamp 509, which is "ON" during the entire timing process. Input 015 resets counters 01, 02 and 03, and input 014 resets the entire timing circuit. The enabling signal, input 000, must be a toggle type switch (detent) that remains "ON" for as long as the timing circuit is in operation (figure 6-08).

COUNTER APPLICATION CIRCUIT — CONTROL PROBLEM 3

A time-dependent bottle-filling process is controlled with a PLC. Bottles are transported and placed under the filling valve on a conveyor belt. The filling valve is operated with a pneumatic, double-acting, linear actuator (figure 6-09). Retraction of the actuator (signal 501) opens the filling valve. Possible power failure to the PLC must not upset the filling quantity. The timer controlling the filling quantity therefore needs to be a retentive timer. As the bottle presence signal 000 may be a persistent signal, the differentiate-up function (FUN 13) has been used to shape it into a pulse. This enabling pulse in "AND" connection to the clock pulse bit 1902 decrements the timer every second by one unit. Counter flag CNT 06 resets the holding relay HR 800, and this in turn cancels solenoid

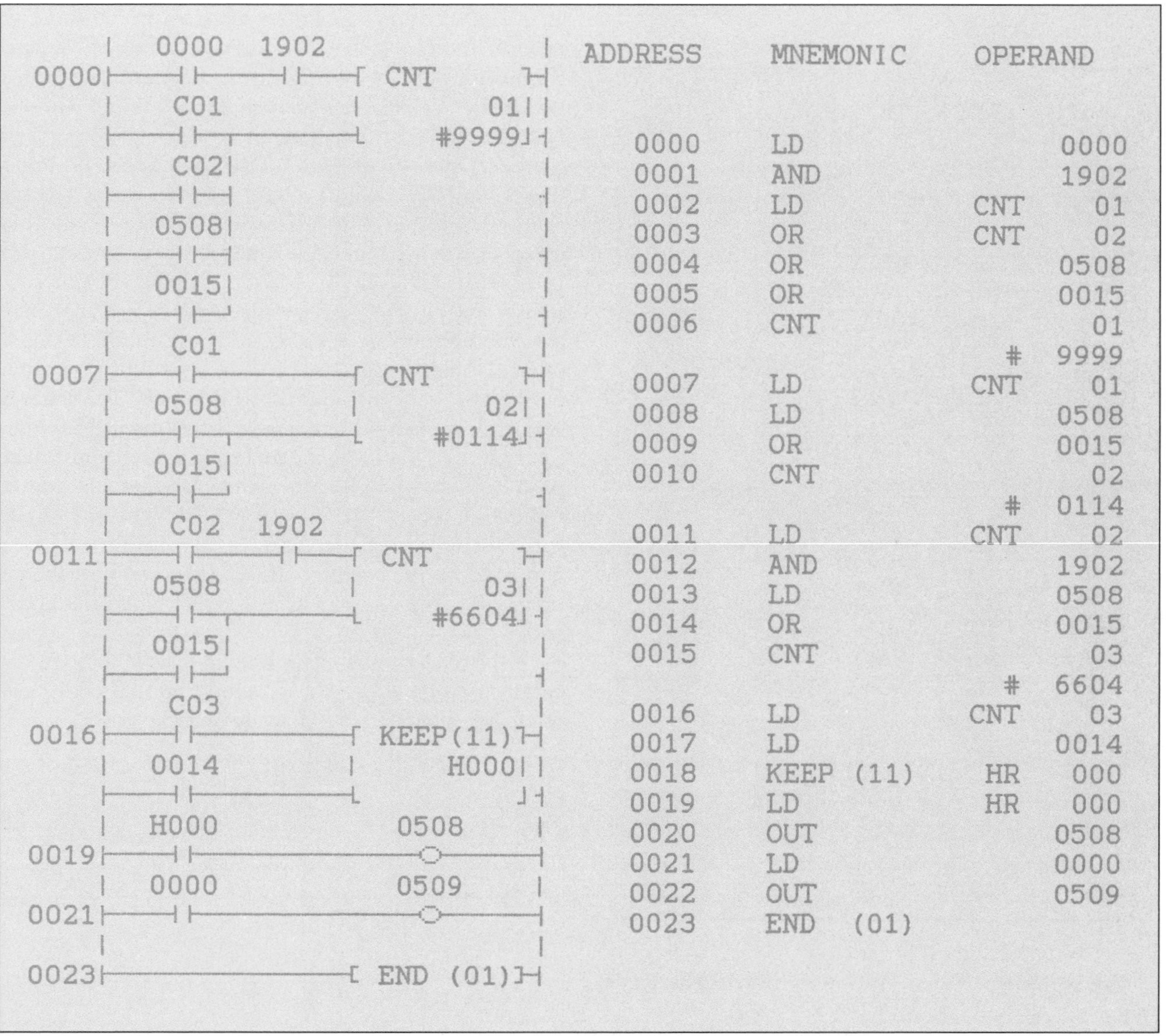

ADDRESS	MNEMONIC	OPERAND
0000	LD	0000
0001	AND	1902
0002	LD	CNT 01
0003	OR	CNT 02
0004	OR	0508
0005	OR	0015
0006	CNT	01
		# 9999
0007	LD	CNT 01
0008	LD	0508
0009	OR	0015
0010	CNT	02
		# 0114
0011	LD	CNT 02
0012	AND	1902
0013	LD	0508
0014	OR	0015
0015	CNT	03
		# 6604
0016	LD	CNT 03
0017	LD	0014
0018	KEEP (11)	HR 000
0019	LD	HR 000
0020	OUT	0508
0021	LD	0000
0022	OUT	0509
0023	END (01)	

Figure 6-08. Ladder diagram circuit for counter application — control problem 2.

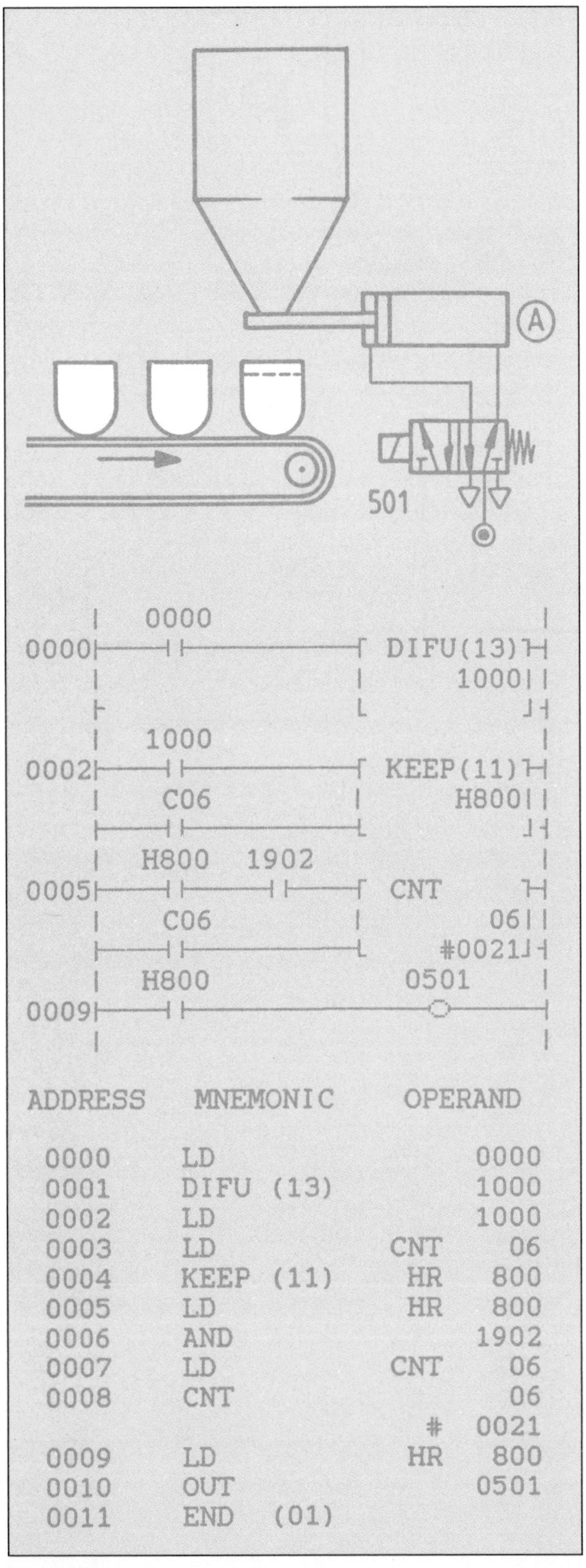

Figure 6-09. Counter application — control problem 3. Filling machine with retentive timer.

signal 501, which causes the spring on the pneumatic valve to shift the valve spool and extend the pneumatic actuator. This then closes the filling valve, and the filling cycle for one bottle is completed.

COUNTER APPLICATION CIRCUIT — CONTROL PROBLEM 4

A reversible counter is used to keep a count of motor cars being parked in an underground car park. The entrance to the car park is controlled by an automatic boom gate and a ticket machine (figures 6-10 to 6-12). To enter the car park, the driver automatically activates switch 03 at the ticket machine. Removal of the printed ticket activates switch 007 momentarily, which causes the boom to lift (HR 000 → output 102). When proceeding through the boom area, the entering car moves past switches 005 and 006 (figures 6-11 and 6-12). The combination 005 • 006 resets the holding relay. If the car is no longer directly under the boom, combination 005 • 006 signals the boom to lower, thus preventing a second car from slipping through without being accounted for or without its driver having taken a ticket. Once the car park is full (137 cars) and counter flag 47 is "ON", holding relay HR 000 cannot be set again (inhibition through contact C 47). Thus, the boom does not lift and the ticket printer (output 103) is disabled. For every car passing through the exit gate (sensor switch 004), a car waiting at the entrance may pass through the boom and ticketing station. The boom can be manually signalled to lift (push-button switch 008) and lower (push-button switch 009). The counter may be incremented by pressing adjustment push-button switch 013 or decremented by pressing adjustment push-button switch 014. Switch 015 resets the counter to 0000. Car park "FULL" is signalled by lamp 109; car park "FREE" is signalled by flashing lamp 110. A green lamp 111 on the boom is turned "ON" when the boom lifts, and a red lamp is turned "ON" when the boom lowers.

If the total number of cars having passed through the entrance needs also to be accounted for, an additional totalising counter could be incorporated.

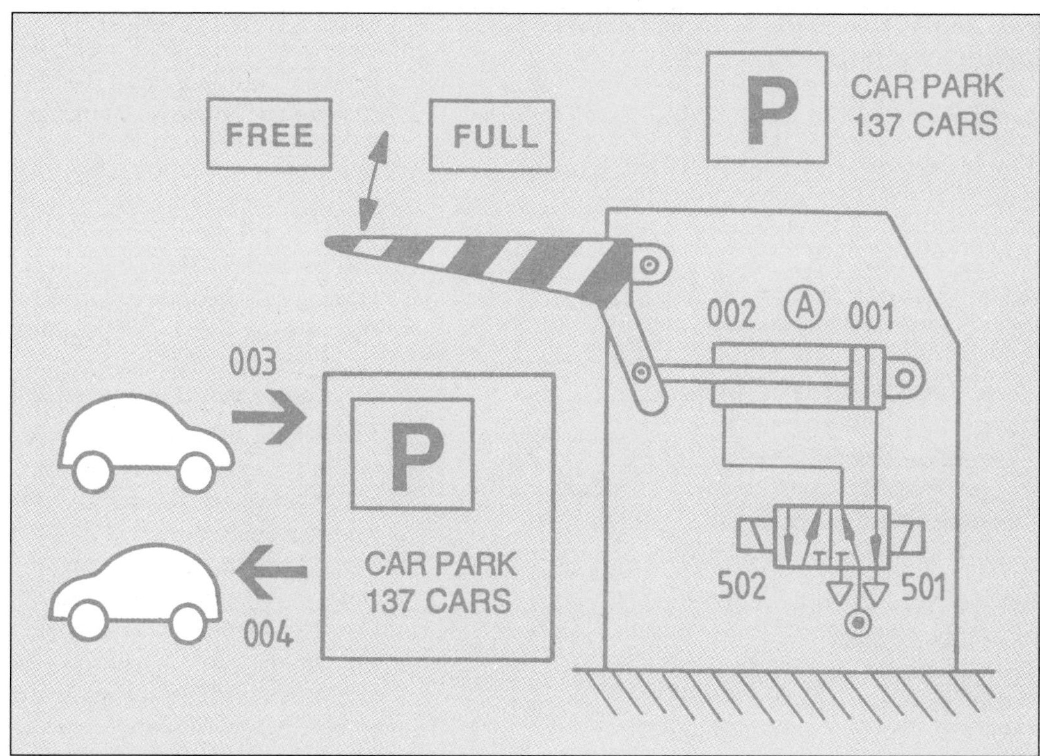

Figure 6-10. Control problem 4 with reversible counter for car park boom control.

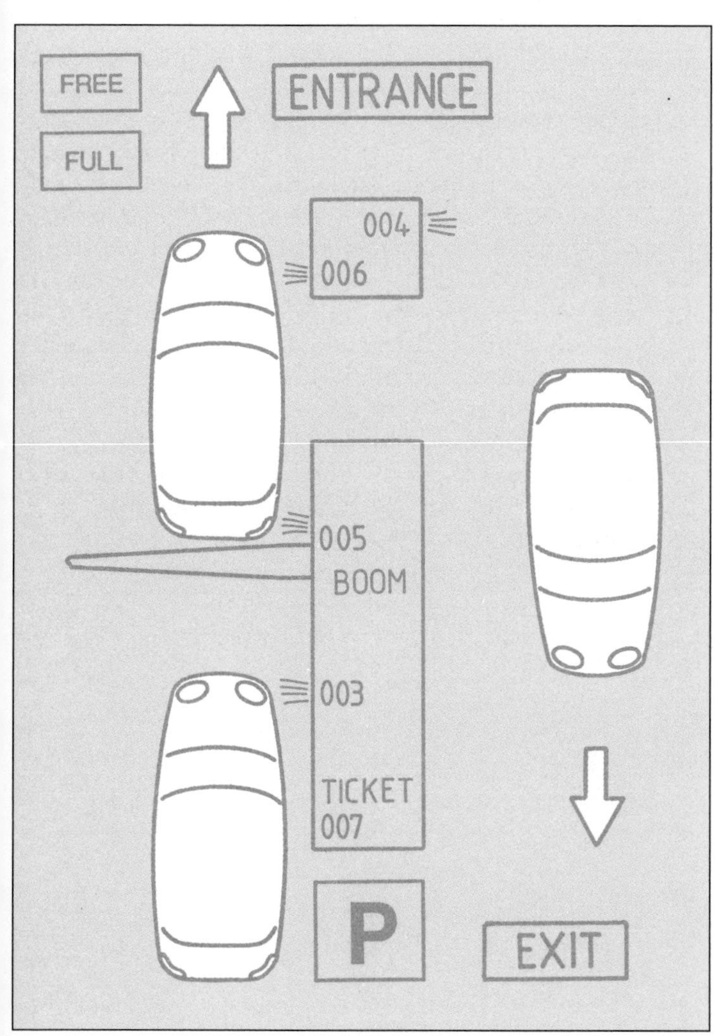

Figure 6-11. Control problem 4. Plan view of car park entrance and exit with sensors and switches for boom control, ticketing machine and counter operation.

PLC INPUT/OUTPUT ASSIGNMENT LIST (OMRON C28K)

PLC inputs:

000 = enable boom and ticketing system
001 = actuator Ⓐ retracted
002 = actuator Ⓐ extended
003 = car at entrance
004 = car exiting park garage
005 = car directly under boom
006 = car driving into car park
007 = ticket taken from machine
008 = manually raise boom
009 = manually lower boom
013 = increment counter
014 = decrement counter
015 = reset counter to zero (0000)

PLC outputs:

0100 = boom lamp green
0101 = boom-lowering solenoid
0102 = boom-lifting solenoid
0103 = ticket printer starting signal
0109 = car park "FULL" lamp
0110 = car park "FREE" lamp
0111 = boom lamp red
HR 000 = lift boom combination recorded
CNTR 47 = reversible counter #136

For Omron C28K PLCs output words are of the numbers 01, 03 and 05. For a non-expanded C28K PLC, the output word is 01. For this reason, outputs in control problem 3 range from 0100 to 0111. Omron C20 PLCs do not include a reversible counter.

COUNTER SUMMARY

Counters with Omron PLCs are memory-retentive. When power to the PLC fails, their accumulated value (or present value) is retained. When being enabled, and count pulses enter the counter, and the counter is reset, decrementing of the counter from set value towards 0000 starts. Any interim value is called present value (PV). (On some PLCs this value may be called the accumulated value.) For a counter to be given the chance to reach 0000 and turn its flag "ON", entered count pulses must be equal to or more than its set value. Counters and timers must never be given the same flag number (see figure 1-12).

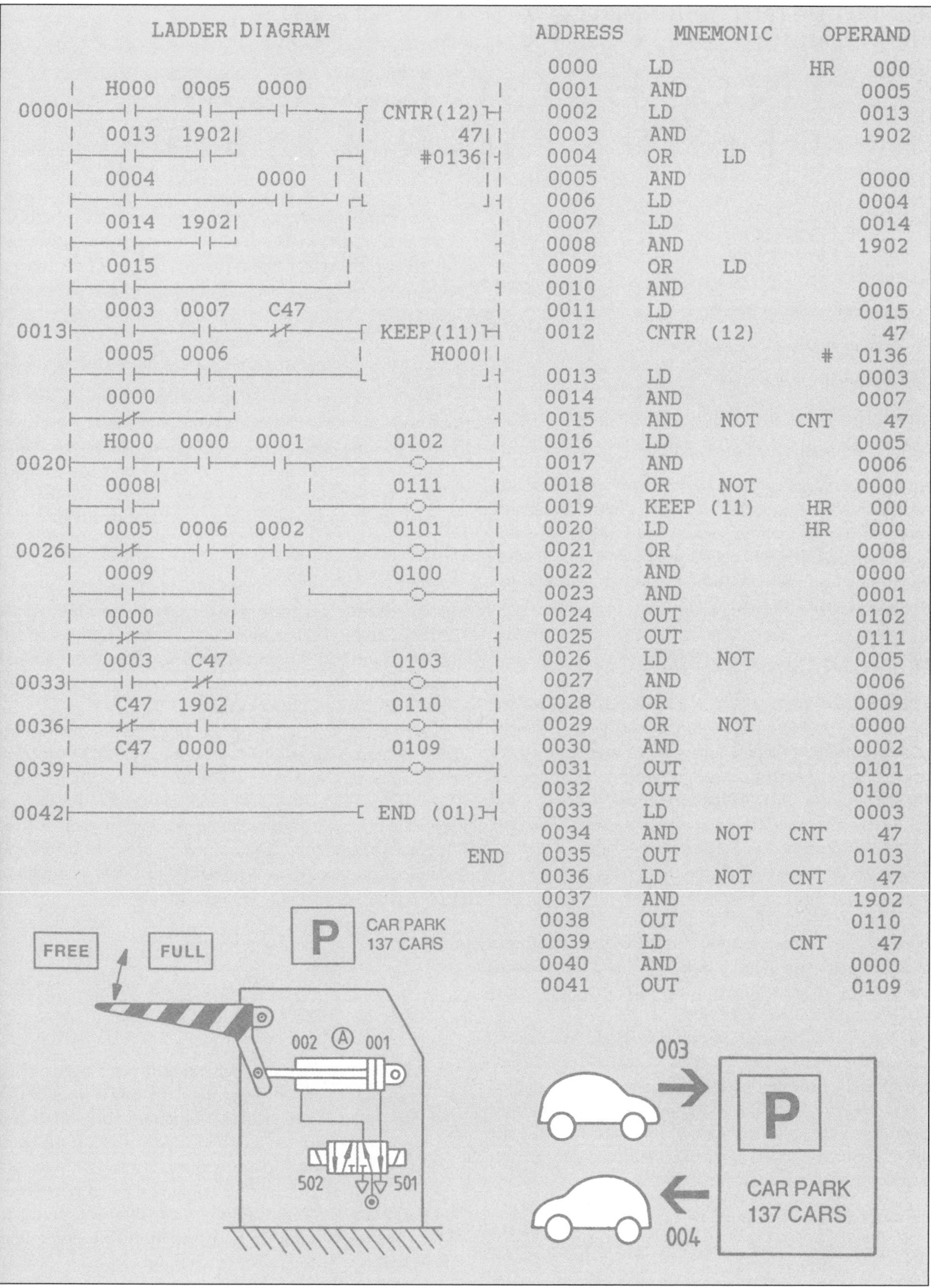

ADDRESS	MNEMONIC	OPERAND
0000	LD	HR 000
0001	AND	0005
0002	LD	0013
0003	AND	1902
0004	OR LD	
0005	AND	0000
0006	LD	0004
0007	LD	0014
0008	AND	1902
0009	OR LD	
0010	AND	0000
0011	LD	0015
0012	CNTR (12)	47
		# 0136
0013	LD	0003
0014	AND	0007
0015	AND NOT	CNT 47
0016	LD	0005
0017	AND	0006
0018	OR NOT	0000
0019	KEEP (11)	HR 000
0020	LD	HR 000
0021	OR	0008
0022	AND	0000
0023	AND	0001
0024	OUT	0102
0025	OUT	0111
0026	LD NOT	0005
0027	AND	0006
0028	OR	0009
0029	OR NOT	0000
0030	AND	0002
0031	OUT	0101
0032	OUT	0100
0033	LD	0003
0034	AND NOT	CNT 47
0035	OUT	0103
0036	LD NOT	CNT 47
0037	AND	1902
0038	OUT	0110
0039	LD	CNT 47
0040	AND	0000
0041	OUT	0109

Figure 6-12. Ladder diagram for control problem 4.

7 COMPOUND BUILDING BLOCKS FOR PLC PROGRAMMING

Industrial automation circuits for programmable logic controllers may be subdivided into the following five categories of circuit characteristics:

- compound building blocks
- pure sequential control circuits
- fringe condition modules
- computation and data acquisition control circuits
- pure combinational control circuits.

Pure combinational controls are very rare and will therefore seldom be encountered by PLC circuit designers. Sequential control circuits, however, are frequently used, but in most industrial PLC circuits they would be complemented with fringe condition modules and sometimes with data acquisition control circuits.

PLC PROGRAMMING APPROACH

As these five circuit types are of great significance for the programming of PLCs and require a completely different design approach, a separate chapter is provided for each of the first four. In most control applications, the boundaries of these five circuit types are rather indistinct and often even the most routine programming "expert" would be hard pressed to define a control's precise nature. Compound building blocks, the first of the five types of circuit design listed above, are seldom found on their own. They are by nature building blocks and thus are embedded and integrated with sequential or combinational circuits. Compound building blocks consist of timers, counters, basic logic functions, flip-flops, shift registers or combinations of these.

The circuit design section of this book therefore starts with these building blocks. Some control applications may use these building blocks just as they are presented in this chapter, but there are numerous applications where they have to be adapted and modified to suit the control application they serve. The building blocks presented here are:

- A before B circuit (figure 7-01)
- one-shot signal (short version figure 7-02)
- variable flicker signal blocks (figures 7-03 and 7-04)
- ON/OFF delay block (figure 7-05)
- basic shift register (figures 7-06 and 7-07)
- multiple word shift register (figure 7-08)
- simultaneous event signalling block (figure 7-09)
- automatic actuator reversal circuit block (figure 7-10).

Each and every one of these circuit building blocks has been "discovered" in the same way as the wheel was invented—by necessity! And since they are often absolutely essential to make a system work, or to design a more economical circuit diagram, I had no choice but to pass them on to the diligent reader of this book.

A BEFORE B CIRCUIT

In an assembly machine, where two products need to be attached to each other, and the assembling process must not start until both products are present but one product must arrive before the other, this circuit becomes an invaluable building block (figure 7-01).

If product "b" (input 015) is not present when product "a" arrives (input 014), the logic "AND" function consisting of a • $\overline{b}$ is established at the "AND" valve (014 • $\overline{015}$). This "AND" function creates the latch x (contact of relay 1001). When product b then arrives (input 015), the top "AND" valve switches to "ON" and produces output signal y (500). Thus, 500 is equal to 015 • latch, or in Boolean expression form:

1001	=	$\overline{015}$ • 014	= first stage
1001	=	014 • 1001	= transition stage (latch)
500	=	015 • 1001	= final stage

A close investigation shows that input a needs to be ahead of input $\overline{b}$ for at least the time taken to establish the latch and hold the "AND" function while input $\overline{b}$ is disappearing and input b is being established. If the transition from b to b complementing is faster than the latch creation, output y (500) will not appear until b has been removed and a enters before b with sufficient time lag. This time lag is probably in the small millisecond range. (For complementing outputs see figure 4-02.)

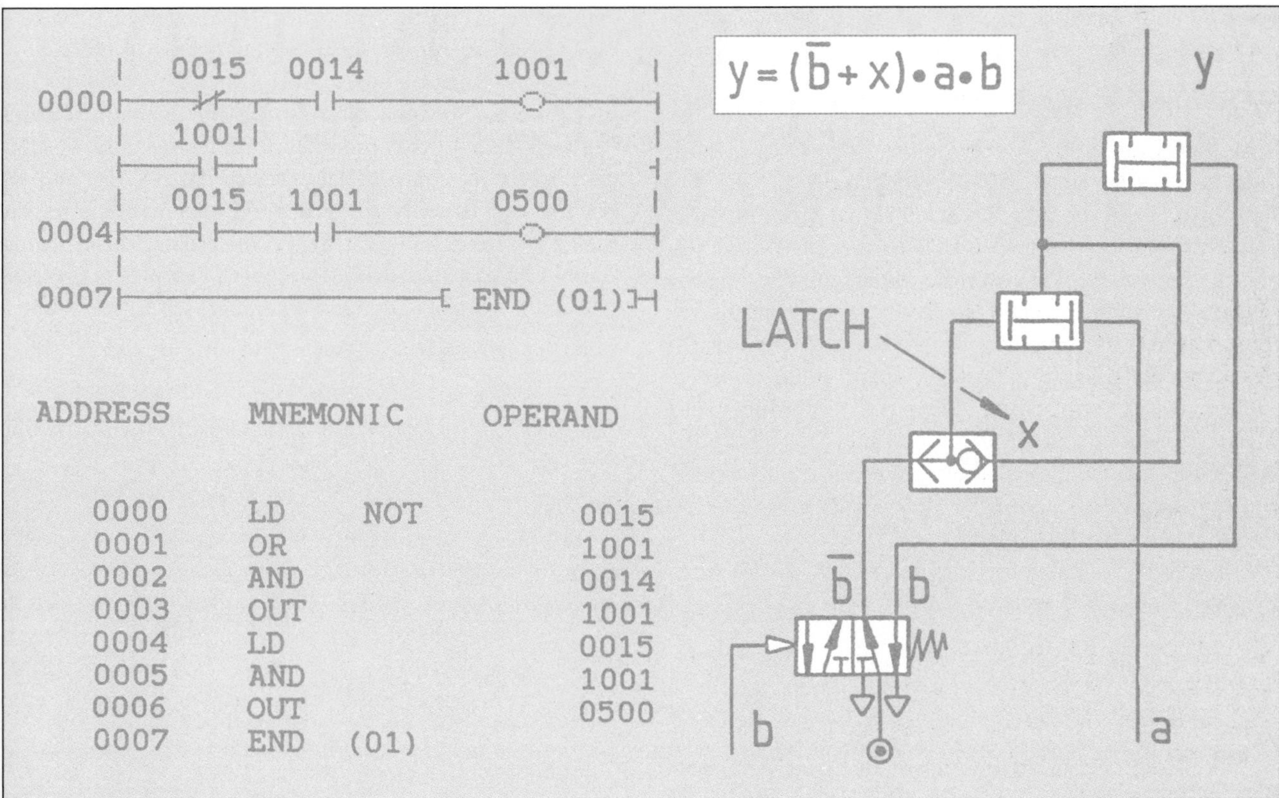

Figure 7-01. A before B circuit.

ONE-SHOT SIGNAL

As has been explained and illustrated in Chapter 5, an off-delay timer will not time out if its enabling signal is shorter than the set value (SV) of the timer. Thus, if a pulse of say 1.2 seconds had to be created by the timer, its enabling signal had to be present for at least 1.2 seconds. A "one-shot signal" circuit, however, has the output characteristics of an off-delay timer, but its enabling signal may be as short as a scan length, or as long as the program designer wishes it to be. In other words, the enabling signal, when recognised as being "ON", triggers the "one shot", but its length has absolutely no bearing on the one-shot output. To see this unique feature, see figure 7-02. One may select any output relay for the

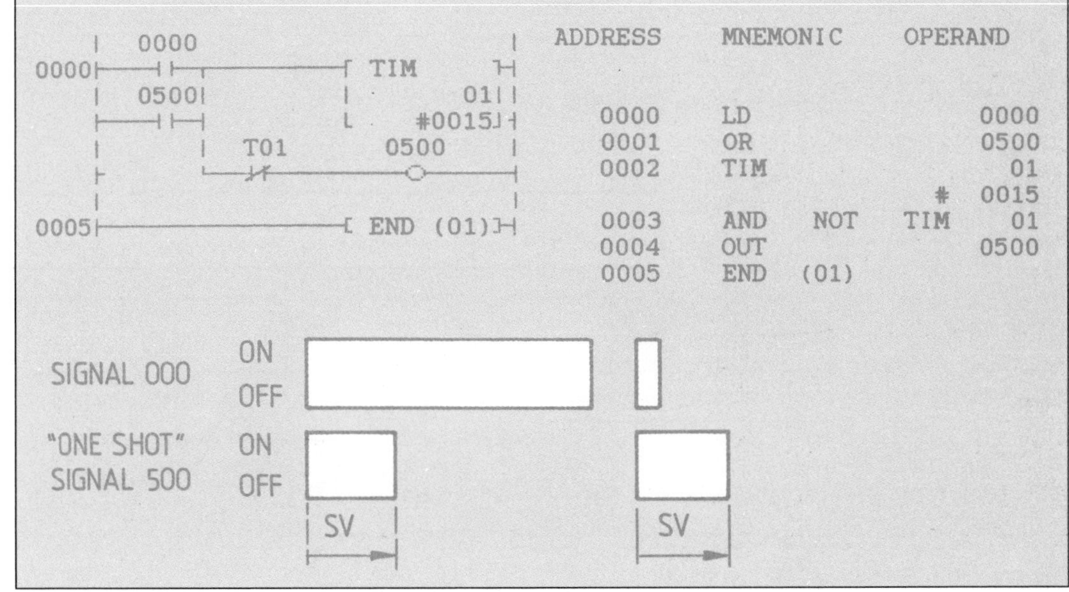

Figure 7-02. "One-shot signal". The "ON" time length of the enabling signal (000) has no bearing on the occurrence of the one-shot output signal (500).

one-shot output (HR, auxiliary relays or output relays; see figure 1-14).

VARIABLE FLICKER SIGNAL BLOCKS

Some PLC brands do not provide the "luxury" of in-built pulse flags as described in Chapter 5 and figure 5-01. For such a situation, the variable flicker signal block provides a helpful solution (figure 7-03). This variable flicker signal block gives the PLC user the opportunity to create a pulsing bit with ultimate flexibility, which means the frequency rate of the pulse, the "ON-duty" time of the pulse and the pulse's "OFF-duty" span can be varied to suit any control application. The pulse "OFF-duty" span is determined by the set value of timer 01. The pulse "ON-duty" time is governed by the set value of timer 02, and both set values added (SV 1 + SV 2) gives the frequency rate of the flicker pulse.

SV 1 = 1.5 sec.

SV 2 = 3.0 sec.

Frequency 4.5 sec.

The flicker signal circuit shown in figure 7-03 produces the first pulse 1.5 seconds after being enabled by input 000. This implies that the very first pulse is delayed by set value 1 of timer 01. However, if the first pulse needs to appear simultaneously with the enabling signal (called "pulse onset control"), the circuit provided in figure 7-03 then needs some minor modifications (figure 7-04). In this modified circuit, the pulse's "ON-duty" time is controlled by timer 01 (SV 1), and its "OFF-duty" time by timer 02 (SV 2). Furthermore, the flicker bit needs to be "AND" connected to the enabling signal and its timer-driving contact (TIM 01) is inverted (it becomes a normally closed contact; see figure 7-03).

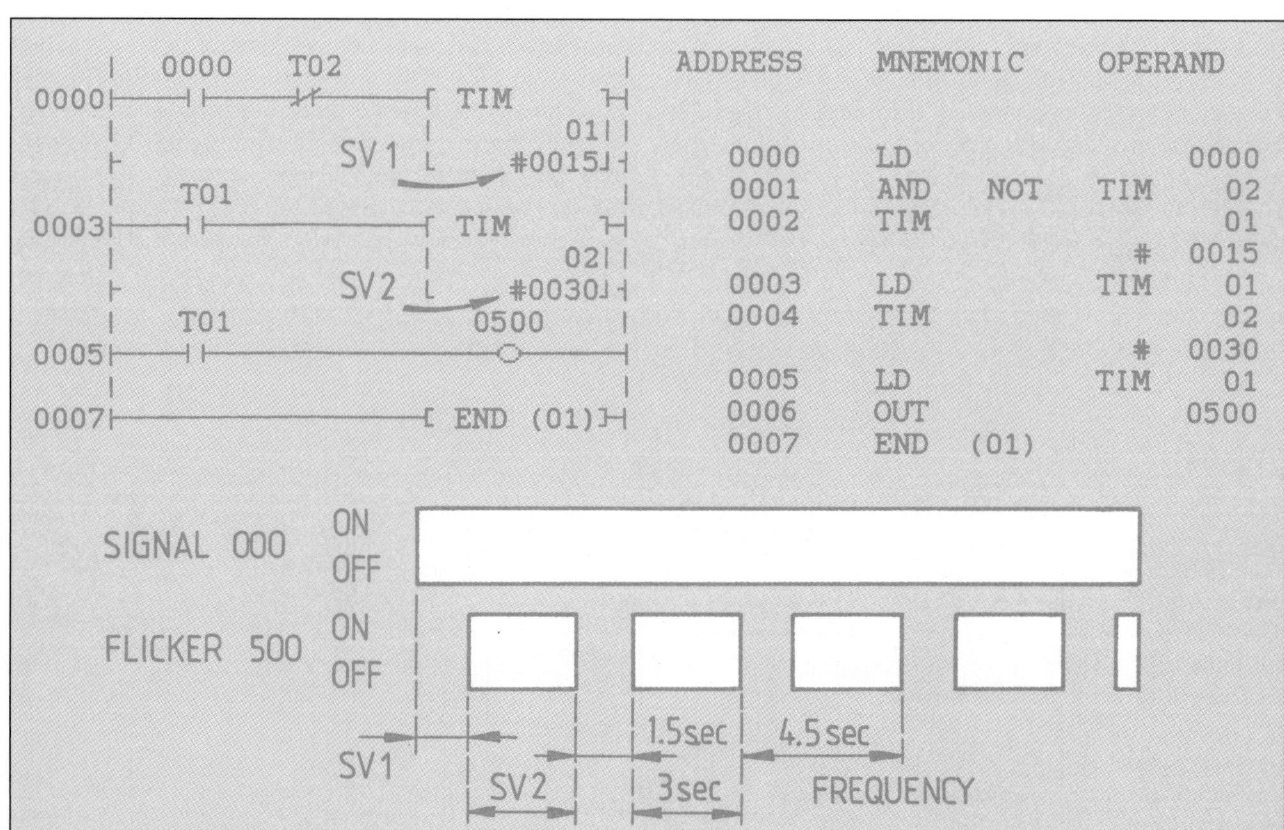

Figure 7-03. Variable flicker signal block.

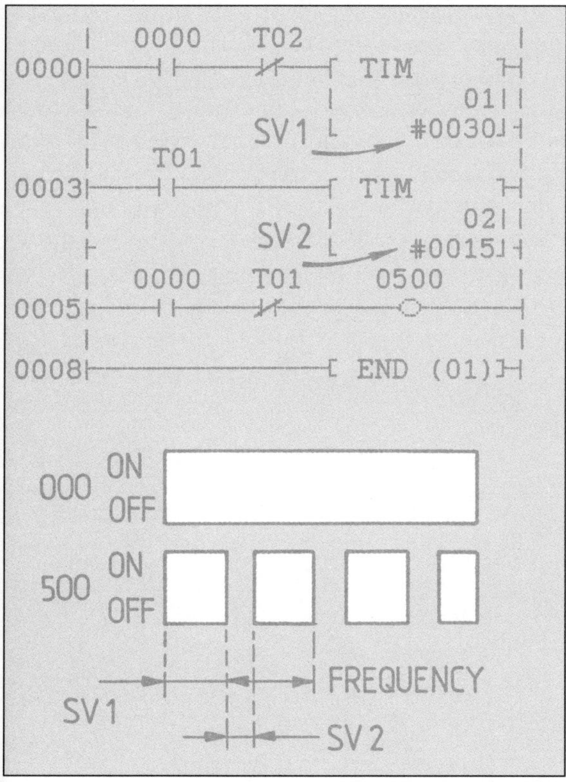

Figure 7-04. Variable flicker signal with "pulse onset" control.

ON/OFF DELAY BLOCK

In an industrial control application where a prewarming process, for example, needs to be "ON" for a certain time before the actual machine sequence can be started, and a machine clearing process needs also to be "ON" for a certain time before the machine is permitted to stop when production finishes, the ON/OFF delay block is a useful subcircuit (figure 7-05).

The interim time, beginning at the time-out point of timer 01 (4 seconds) and ending at the termination of the enabling signal (000), may be variable or fixed and has no influence on the "ON delay" (TIM 01) or the "OFF delay" (TIM 02). Example: Containers are packed into cardboard boxes. The boxes are formed (folded) and glued within the packaging machine. Hence, the packaging machine has a multiple function: it forms and glues the boxes, it then packs the formed boxes with filled containers and it closes and glues them for shipping.

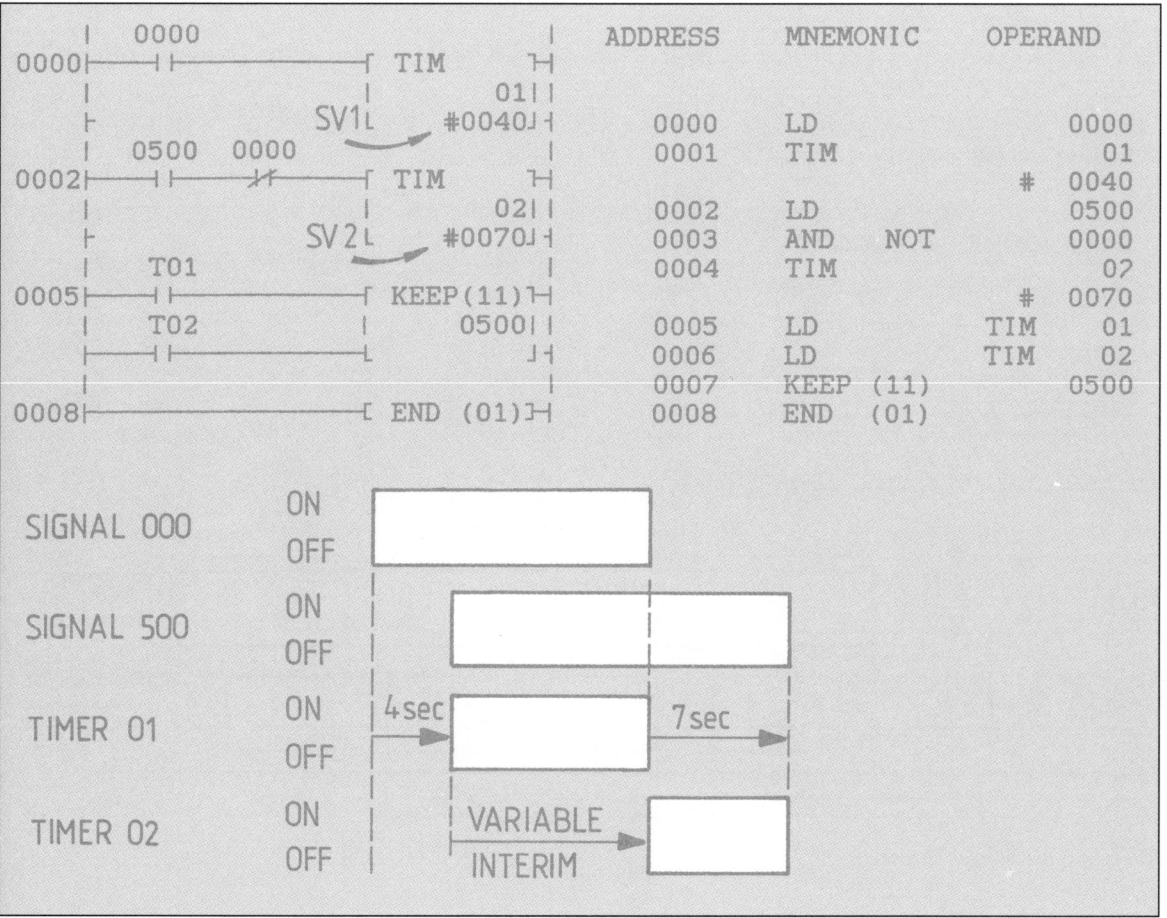

ADDRESS	MNEMONIC	OPERAND
0000	LD	0000
0001	TIM	01
		# 0040
0002	LD	0500
0003	AND NOT	0000
0004	TIM	02
		# 0070
0005	LD	TIM 01
0006	LD	TIM 02
0007	KEEP (11)	0500
0008	END (01)	

Figure 7-05. ON/OFF delay block.

In the morning, before the sequential box-forming and packaging process is permitted to start, the glue resin must be warmed up for 12 minutes. This is controlled with the "ON delay" timer TIM 01 (#7200). In the evening, before the machine is permitted to be stopped, it needs to clear out partly processed boxes. This clearing process takes approximately 47 seconds. Thus, the start signal 000 is switched "ON" in the morning, but production does not start until 12 minutes later (prewarming time), signalled by the production "START" command output 500. Upon cancellation of the start signal 000, the machine keeps automatically running for a further 47 seconds, controlled by the "OFF delay" timer TIM 02 (#470), which, when timed-out, cancels the production "START" command output 500. By now the machine is completely cleared and stopped and ready for a new production run.

BASIC SHIFT REGISTER CONCEPT

Although principally data manipulation instructions, shift instructions or shift registers are used to shift bits, digits or complete words. The latter two are explained in Chapter 10. Bit shifting means that the value (logic 0 or logic 1) at the shift register's data entrance is shifted along the register by one bit whenever a clock pulse is received at the shift register's clock entrance (shift pulse entrance). The shift process happens at the leading edge of the clock pulse (figure 7-06). The clock pulses can be regular or irregular. But note! No shift takes place between clock pulses, and providing one uses HR or AR words all information in the register is stored if power to the PLC fails.

Each shift register "pocket" is given a bit number and the register's word number. For example, the register's fifth "pocket" from the right in figure 7-06 has bit number 04 and word number 05 in front. Together, this "pocket's" full label or operand is 0504. The zero of the word is not shown for simplicity, but the mnemonic list printout in figure 7-06 shows it with the zero in front (see address 0002). Thus, if the completely "empty" shift register of word 5 shown in figure 7-06 received a regular clock pulse every two seconds, and with every clock pulse a data value (1 or 0) of the following pattern were entered on the data input:

0 at clock pulse 9 (16 seconds)
1 at clock pulse 8 (14 seconds)
0 at clock pulse 7 (12 seconds)
1 at clock pulse 6 (10 seconds)
1 at clock pulse 5 (8 seconds)
1 at clock pulse 4 (6 seconds)
0 at clock pulse 3 (4 seconds)
0 at clock pulse 2 (2 seconds)
1 at clock pulse 1 (0 seconds)

the register would then have the following content:

0 0 0 0 0 0 0 1 0 0 1 1 1 0 1 0 ⇦ shift
515 500

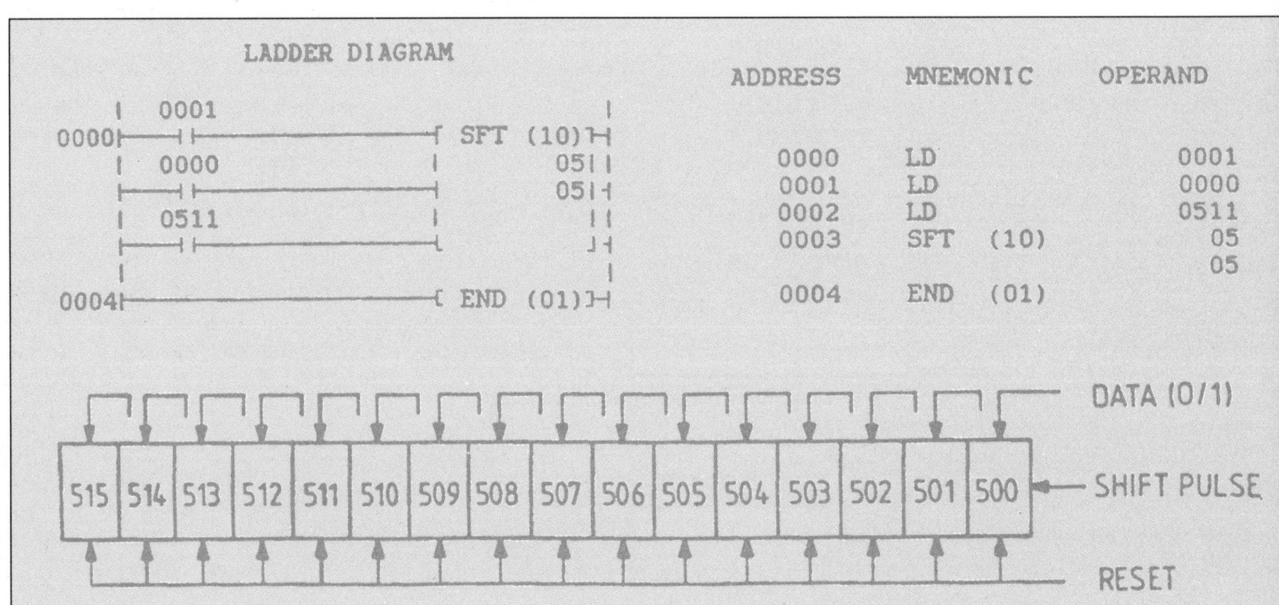

Figure 7-06. Basic shift register concept.

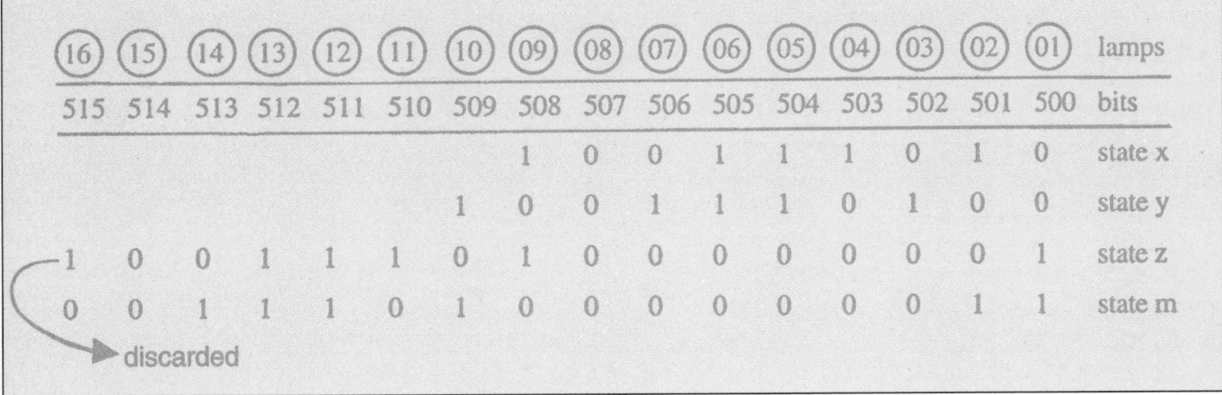

16	15	14	13	12	11	10	09	08	07	06	05	04	03	02	01	lamps
515	514	513	512	511	510	509	508	507	506	505	504	503	502	501	500	bits
							1	0	0	1	1	1	0	1	0	state x
						1	0	0	1	1	1	0	1	0	0	state y
1	0	0	1	1	1	0	1	0	0	0	0	0	0	0	1	state z
0	0	1	1	1	0	1	0	0	0	0	0	0	0	1	1	state m

Figure 7-07. State table for the loading process of a shift register.

This means that the shift register's nine pockets, counting from the right, have been filled with data (0 or 1). The other remaining seven pockets are not yet filled, which means that they are still empty or filled with zeros. For illustration purposes we call this register state x in figure 7-07.

If the 16 bits of this shift register were each connected to a lamp, then lamps 2, 4, 5, 6 and 9 would be illuminated. If with the next pulse input a logic 0 were presented at the data entrance, then lamps 3, 5, 6, 7 and 10 would be illuminated (register state y). If six more clock pulses were entered, but with every arriving clock pulse the data entrance read logic 0 except for the last one, then lamps 1, 9, 11, 12, 13 and 16 would be illuminated (register state z). By now, every "pocket" of the shift register is filled with a logic 1 or a logic 0, and the shift register is filled to capacity. If a further clock pulse arrives, all data are again shifted by one bit to the left, and the data bit that was in pocket 16 (515) is lost (discarded). With this latest clock pulse we assume a logic 1 was at the shift register's data entrance (register state m). The register's pockets are now filled as follows.

PROGRAMMING THE SHIFT REGISTER

Operation of a shift register is controlled by three conditions:

- The pulse input, which may be a fixed regular clock pulse from a pulsing bit or a pulse generator on a machine, or a totally irregular clock pulse given by an operator or a machine condition, whenever a shift must take place.
- The reset input signal, which causes all pockets to become logic 0, despite their content (1 or 0) before the arrival of the reset signal.
- The data input, which serves as an entrance to feed data of logic 1 or logic 0 into the shift register. Data pass the data entrance only if a clock pulse is present.

During programming of the shift register in ladder form (figure 7-06) on an Omron PLC, the first instruction loads the data bit. In the given example this is bit 001. The second instruction is the clock pulse bit. In the given example, this is input 000. The third instruction nominates the reset bit. In the given example, this is bit 511 of the shift register itself. This means that whenever pocket 511 of the shift register is logic 1, the register is being reset. The fourth programming instruction identifies the word (or words) being used for the shift register. If more than one word is used, the starting word and the end word must be nominated (figure 7-08). If the register consists of only one word, the end word is then the same as the starting word. This is the case with the shift register shown in figure 7-06, where the end word and the starting word are 05. For Omron PLCs, the shift register instruction is entered in the mnemonic column as function 10 (FUN 10). For further explanations of shift register operation, see the shift register application circuit in figures 7-11 and 7-13.

MULTIPLE WORD SHIFT REGISTER

Shift registers may be constructed from several words. The rightmost word is then called the starting word (figure 7-08). The leftmost word is called the end word, and any in-between words are called interim words. In the given example, word 05 is the starting word, word 06 is the interim word, and word 07 is the end word. Thus, the shift register has a total of 3 × 16 pockets or bits, which amounts to 48 bits. The multiple shift register in figure 7-08 is reset by its last bit (bit 0715). Its data entrance is bit 004 and the clock pulse stems from one of the PLC's three pulsing bits (bit 1902 with a frequency of 1 pulse per second). Pulsing can be enabled or disabled with input 000. Various shift register applications and comments pertaining to their operation are given in figures 4-08, 4-09, 7-11, 7-12, 7-13 and 7-16.

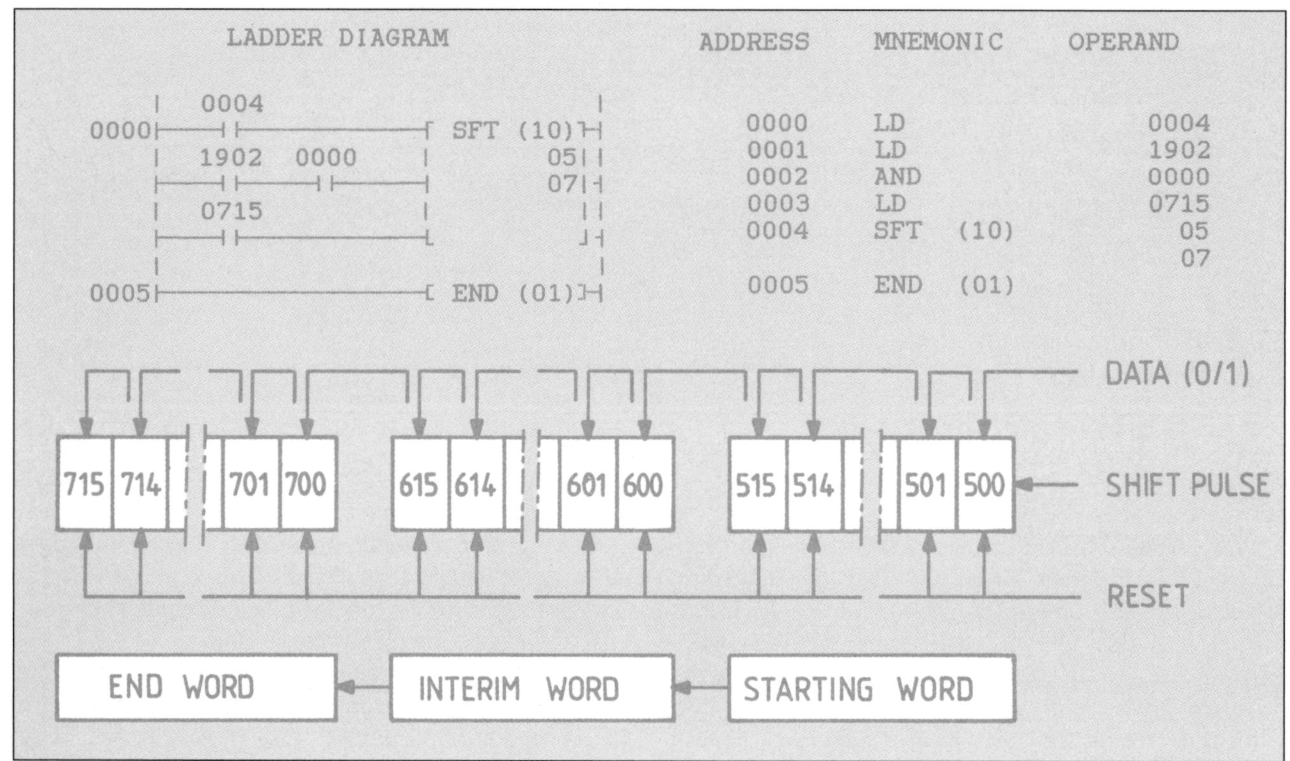

Figure 7-08. Multiple word shift register

SIMULTANEOUS EVENT SIGNALLING BLOCK

Where a machine function may be started only if two events have happened at the same time or, as is the case with the circuit in figure 7-09, within 0.3 second or a precisely limited time frame, then this circuit is an ideal building block for ready use. This building block has also proved its usefulness numerous times as a two-hand safety start control, which forces the machine operator to apply both hands and press the two start buttons within 0.3 second. If one button is illegally fixed down, the circuit will not work. For this reason, the building block shown here is sometimes also called the "no-tie-down starting block".

Caution! Industrial safety regulating bodies in some countries no longer permit the use of the simultaneous event-signalling block for machine start control as the only means of safety. Complete machine guarding may be mandatory! Please check your local or national machine design regulations and requirements before using such two-hand, safety start, control circuits. In Australia the supervising and regulating body is the Department of Labour and Industries in each State.

In the PLC ladder diagram of figure 7-09, the holding relay HR 300 is a reset-dominant type flip-flop, which means it has preferential reset characteristics if both the set and the reset conditions are true. Hence, with neither of the two inputs being "ON" (014, 015), their "OR"-connected, normally closed contacts are held closed and are therefore holding the flip-flop (HR 300) in its reset state. Once the two inputs (014, 015) are actuated, the flip-flop reset signal disappears and the two timer contacts, being of normally closed nature and being "AND" connected to inputs 014 and 015, have a chance to set the flip-flop. This is, of course, only the case if neither of the two timers has timed-out before the holding relay HR 300 (flip-flop) is set. In the event of one of the two inputs being "ON" permanently, or being actuated more than 0.3 second before the other input is turned "ON", its timer would time-out and would disrupt (inhibit) the logic line for the setting of the flip-flop. Output 510 signals start "FREE", whereas output 511 signals start "BLOCKED".

$$\text{Set HR 300} = \overline{\text{TIM 14}} \bullet \overline{\text{TIM 15}} \bullet 014 \bullet 015$$

$$\text{Reset HR 300} = \overline{014} + \overline{015}$$

The two inputs (014 and 015) in the flip-flop set line could theoretically be omitted and the circuit block would still work. But **FOR SAFETY REASONS, THEY MUST BE ADDED!** Without them, if the wires from both input switches to the PLC broke, the machine would immediately start, with perhaps disastrous effects to machine and operator. For this reason, PLC circuit designers designing programs for machine starting and stopping conditions must always consider such a "wire break event" and must take precautions to avoid unsafe circuits.

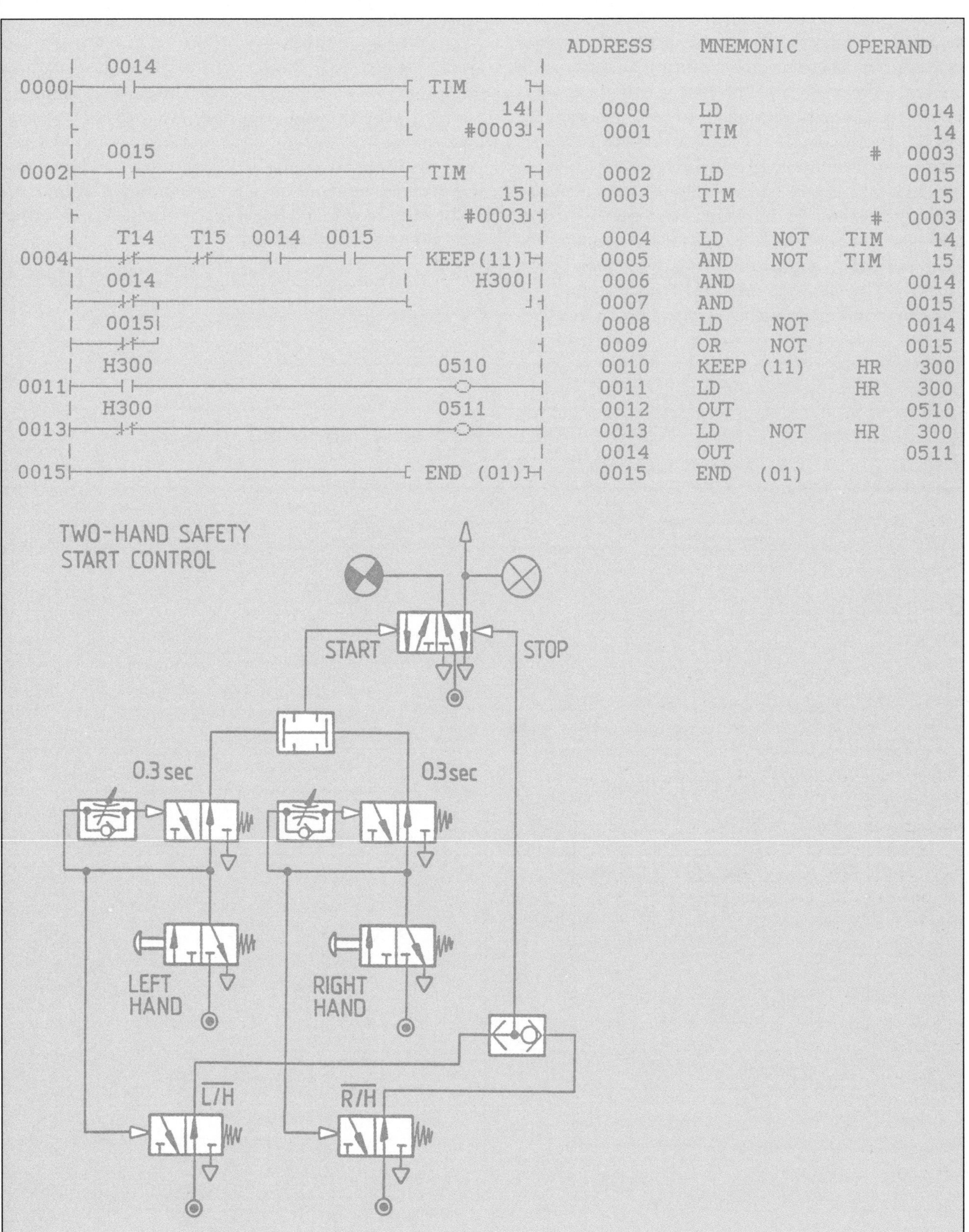

ADDRESS	MNEMONIC		OPERAND	
0000	LD			0014
0001	TIM			14
			#	0003
0002	LD			0015
0003	TIM			15
			#	0003
0004	LD	NOT	TIM	14
0005	AND	NOT	TIM	15
0006	AND			0014
0007	AND			0015
0008	LD	NOT		0014
0009	OR	NOT		0015
0010	KEEP	(11)	HR	300
0011	LD		HR	300
0012	OUT			0510
0013	LD	NOT	HR	300
0014	OUT			0511
0015	END	(01)		

Figure 7-09. Simultaneous event signalling block (two-hand safety start circuit).

AUTOMATIC ACTUATOR REVERSAL CIRCUIT BLOCK

For some machine control applications, when emergency stopping or cycling interruption is signalled, some actuators may need automatic motion reversal for whatever motion they are executing at the time of interference (figure 7-10). This means that if the actuator is extending, its motion must automatically be reversed to retraction. If the actuator is retracting, its motion must then automatically be reversed to extension. This turn-about has to occur instantaneously, and must happen the moment interference is signalled. Should the actuator be stationary in one of the two end positions, it must then remain there and must not make any movements until emergency stopping is cancelled. Although perhaps a rare design requirement, this circuit block proves to be a practical and automatically functioning subcircuit. It may be given only to some of the actuators in a control circuit, depending on machine sequence performance and job specification.

Holding relay HR 800 is selected to "ON" if the actuator is currently extending or to "OFF" (HR 800) if the actuator is retracting. While the actuator is in motion (not on 001 or 002), auxiliary relay 1000 is "ON". When the actuator is in one of the two end positions, auxiliary relay 1000 is "OFF" ($\overline{1000}$), which means that contact 1000 is opened. When stopping is signalled, the two inhibiting contacts controlled by the stopping relay HR 015 are opened, and the normal extending or retracting signal (A1, A0) is instantly cancelled (inhibited). Thus, any opposing commands to the solenoids are automatically eliminated and emergency reversal signalling may take place unhindered.

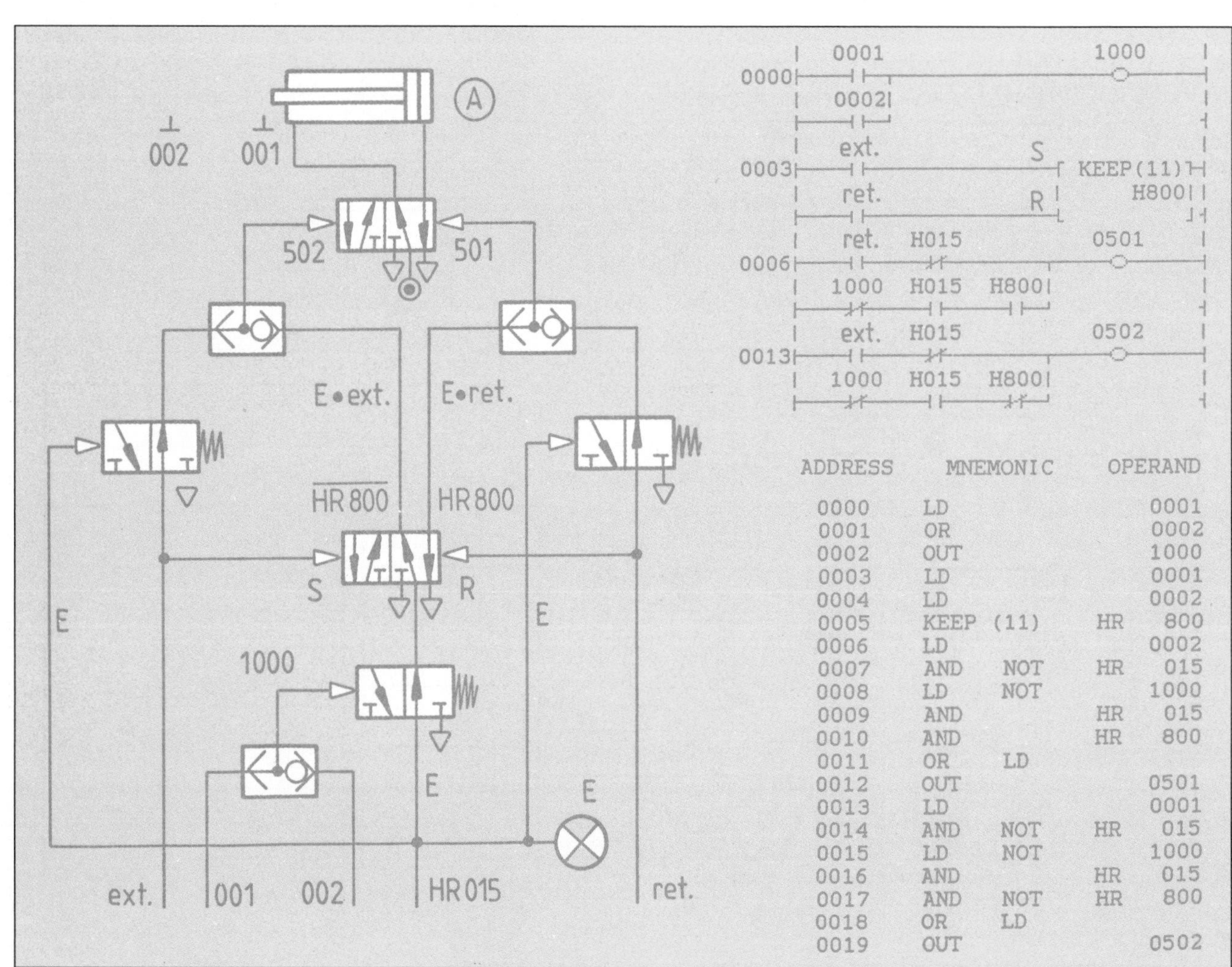

ADDRESS	MNEMONIC	OPERAND
0000	LD	0001
0001	OR	0002
0002	OUT	1000
0003	LD	0001
0004	LD	0002
0005	KEEP (11)	HR 800
0006	LD	0002
0007	AND NOT	HR 015
0008	LD NOT	1000
0009	AND	HR 015
0010	AND	HR 800
0011	OR LD	
0012	OUT	0501
0013	LD	0001
0014	AND NOT	HR 015
0015	LD NOT	1000
0016	AND	HR 015
0017	AND NOT	HR 800
0018	OR LD	
0019	OUT	0502

Figure 7-10. Circuit block for automatic actuator reversal.

COMPOUND BUILDING BLOCK APPLICATION CIRCUIT — CONTROL PROBLEM 1

A shift register is used to build a T flip-flop. Contact 1813 is one of the Omron PLC's special relays. Contact 1813 is an always "ON" bit, which means that as long as the PLC is powered up, bit 1813 is always logic 1 ("ON"). Hence, shift register word 12 is constantly given a logic 1 at its data input. Whenever a shift pulse is present, this logic 1 data is shifted one bit to the left in the shift register. Thus, after the first pulse of input 010, pocket 1 of shift register 12 is "ON" (bit 1200 is "ON"). With the second pulse of input 010, this logic 1 is shifted to pocket 2, and therefore bit 1201 is "ON" (logically bit 1200 is also "ON"). Since the reset input of this shift register is bit 1201 (pocket 2), the shift register resets itself whenever bit 1201 is logic 1. At that time, pocket 1 and pocket 2 are reset (logic 0). Pocket 1 (bit 1200) is used to turn output 500 "ON". The shift function of logic 1 data for a T flip-flop is shown in the truth table of figure 7-11.

COMPOUND BUILDING BLOCKS APPLICATION CIRCUIT — CONTROL PROBLEM 2

A bottle filling and capping machine is controlled by a PLC (figure 7-12). Conveyor motion is achieved with a to and fro movement of actuator Ⓐ, signalled by outputs 508 (extension) and 507 (retraction). This actuator's movements are fully automatic and cause the conveyor to move whenever the enabling relay HR 000 is "ON" and actuator Ⓐ extends (extension stroke). The conveyor is stationary when actuator Ⓐ retracts. Pulses for these to and fro motions are generated by a variable flicker signal block, as is shown and explained in figure 7-03. The pulses generated (output 500) trigger a T flip-flop (shift register 12). This T flip-flop controls and generates the signals for actuator Ⓐ to extend and retract (see figure 7-13, addresses 43 to 46 and 53 to 61).

Approaching bottles on the conveyor are checked for the presence of labels (figure 7-12). Signal 014 is rendered when no label on the bottle is detected. Bottles are filled in position 1 and capped in position 2. Label checking happens in position 0. Bottles without a label are not filled and not capped, and are rejected in position 3 by actuator Ⓑ's pushing them into a reject chute (signal 506). Actuator Ⓑ retracts automatically when fully extended (limit switch 006). A shift register is used to accompany the moving bottles without labels for them to be rejected in position 3 (bottle presence sensor 003 and shift register bit 1703, being logic 1 in addresses 0036 to 0042 in figure 7-13). Output 501 signals the filling process to begin and output 502 signals the capping process to begin.

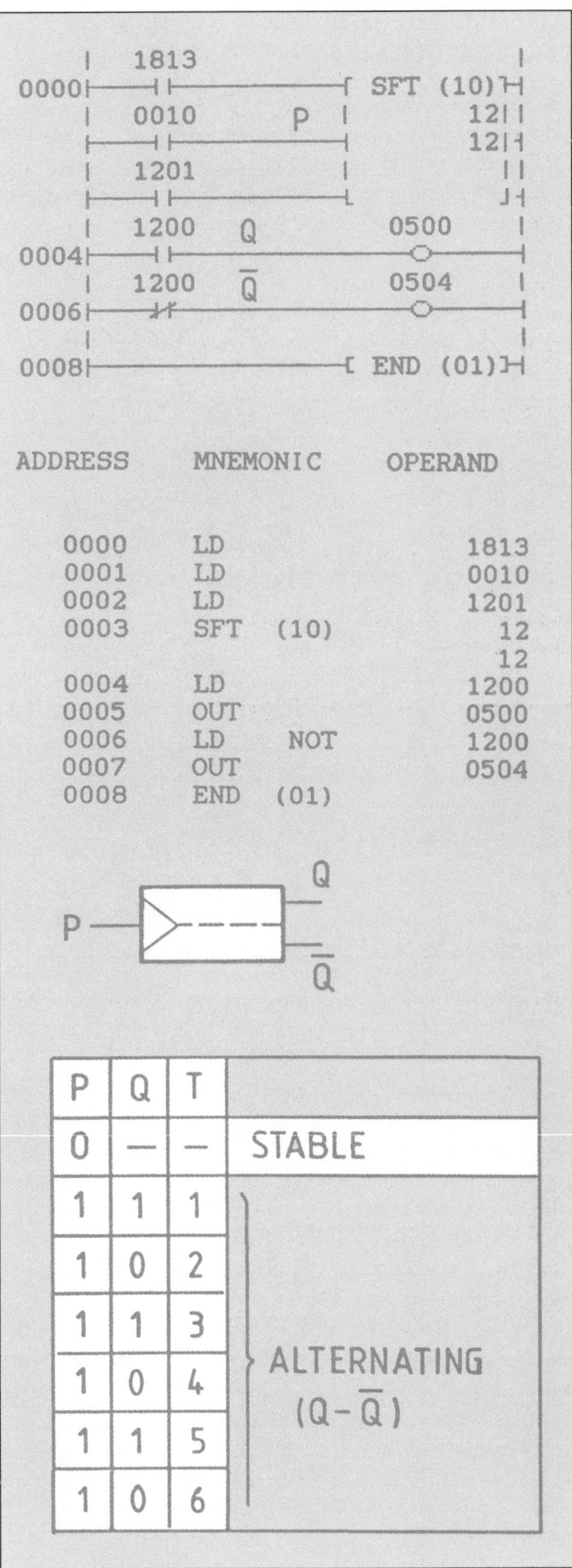

ADDRESS	MNEMONIC		OPERAND
0000	LD		1813
0001	LD		0010
0002	LD		1201
0003	SFT	(10)	12
			12
0004	LD		1200
0005	OUT		0500
0006	LD	NOT	1200
0007	OUT		0504
0008	END	(01)	

P	Q	T	
0	—	—	STABLE
1	1	1	ALTERNATING ($Q-\bar{Q}$)
1	0	2	
1	1	3	
1	0	4	
1	1	5	
1	0	6	

Figure 7-11. T flip-flop with shift register and truth table, showing data shift.

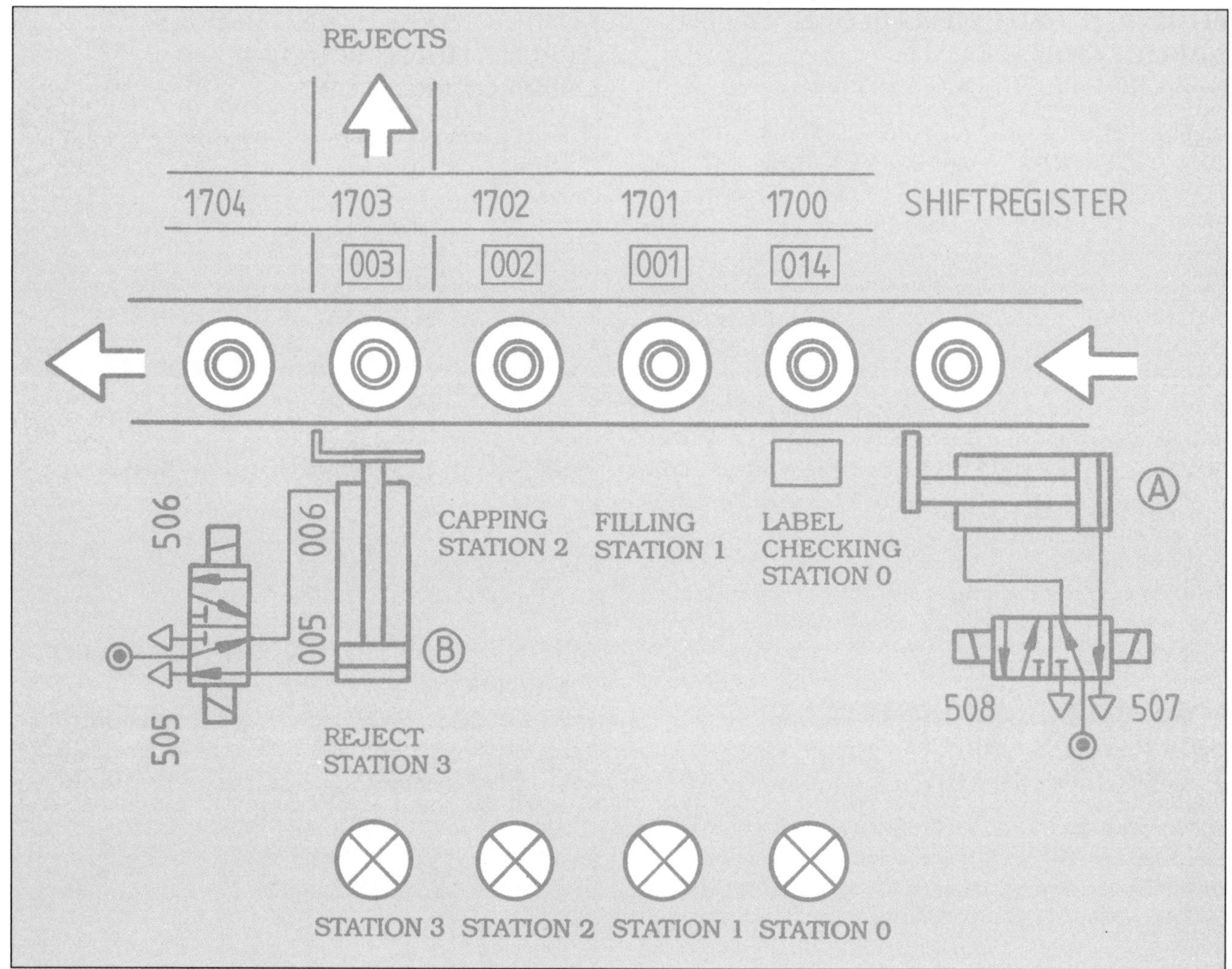

Figure 7-12. Bottle filling and capping machine (machine layout).

The pulsing of shift register 17 (word 17) is accomplished with bottle presence signals from position 0 (sensor 000), position 1 (sensor 001), position 2 (sensor 002) and position 3 (sensor 003). These sensors are active when a bottle is in front of them on the conveyor belt. An edge flag function (DIFU, FUN 13) is used to create short pulses of one scan length on the leading edge of these four signals (see DIFU and DIFD explanation in Chapter 5). This edge flag pulse 1000 pulses the shift register to move a logic 1 or a logic 0 through the 16-bit word of shift register 17. A logic 1 indicates a bottle without a label as detected by sensor 014.

Outputs 501 and 502 for the filling and capping processes are inhibited if a bottle without a label is in their respective process stations (addresses 0019 to 0026). After three rejected bottles, the machine is automatically shut down by counter 47, which counts rejected bottles (bottles without labels). For this see CNT 47 in addresses 0014 to 0016, and 0002, and HR 000 in addresses 0021, 0025, 0027 and 0053.

Input 013 enables (starts) the machine, and input 015 or counter contact 47 stops the machine. Input 015 is also used to reset shift register 17 and bottle reject counter 47. Outputs 509, 510 and 511 turn "ON" position lamps 1, 2 and 3, which indicate the presence of a faulty bottle on the conveyor in positions 0, 1 and 2. Output 506 has a dual function. It causes actuator Ⓑ to remove the faulty bottle and it also turns "ON" lamp 3, which indicates a faulty bottle in position 3 of the conveyor. Correct bottles travel over and beyond position 3 towards the packaging machine. This control problem shows a typical application for two shift registers, in which shift register 12 is used to create a T flip-flop and shift register 17 transfers the faulty bottle data (being a logic 1) to the reject position. This control problem also shows a practical application for a flicker signal block generating 0.3 second pulses every five seconds. Output 506 is a latched output relay and the machine is enabled or stopped with recorded signals in a preferential reset flip-flop HR 000 (reset-dominant flip-flop).

PLC INPUT/OUTPUT ASSIGNMENT LIST (OMRON C20)

PLC inputs:

000 = sensor in label checking position 0
001 = bottle sensor in filling position 1
002 = sensor in capping position 2
003 = bottle sensor in reject position 3
005 = actuator Ⓑ in retracted position
006 = actuator Ⓑ in extended position
013 = start machine (enable)
014 = sensor—"bottle without label"
015 = stop machine and reset machine

PLC outputs:

500 = flicker signal indication lamp
501 = start filling process
502 = start capping process
503 = machine enable lamp
504 = machine stopped lamp
505 = actuator Ⓑ retract solenoid
506 = actuator Ⓑ extend solenoid
506 = reject bottle in position 3 lamp
507 = actuator Ⓐ retract solenoid
508 = actuator Ⓐ extend solenoid
509 = reject bottle in position 0 lamp
510 = reject bottle in position 1 lamp
511 = reject bottle in position 2 lamp
HR000 = machine start/stop recorded
CNT47 = reject bottle counter #0003

COMPOUND BUILDING BLOCKS APPLICATION CIRCUIT — CONTROL PROBLEM 3

A reciprocating actuator's motion (stroke) needs to be instantly reversed to the opposite direction when an emergency reversal signal (013) is given (figures 7-15 and 7-10). The actuator moves between limit switches 001 (retracted) and 002 (extended). Reciprocation motions are automatically terminated by timer 01 (figure 7-15), but may prematurely be terminated by input 015 or extraordinarily be terminated by pressing the emergency stop/reversal button 013 (HR 015 records this signal). In case of the actuator being in one of the two end positions (001, 002), auxiliary relay 1000 is "ON" and actuator reversal is not necessary (see contacts in addresses 0019 and 0027 of figure 7-15). Holding relay HR 800 records the last actuator end position, and therefore its last motion direction. Outputs 501 and 502 signal actuator retraction or extension.

PLC INPUT/OUTPUT ASSIGNMENT LIST (OMRON C20)

PLC inputs:

001 = actuator Ⓐ in retracted position
002 = actuator Ⓐ in extended position
013 = emergency reversal signal
014 = start and emergency reset signal
015 = premature motion termination

PLC outputs:

501 = actuator Ⓐ retracted solenoid
502 = actuator Ⓐ extended solenoid
HR000 = start signal recording
HR015 = emergency stop recording

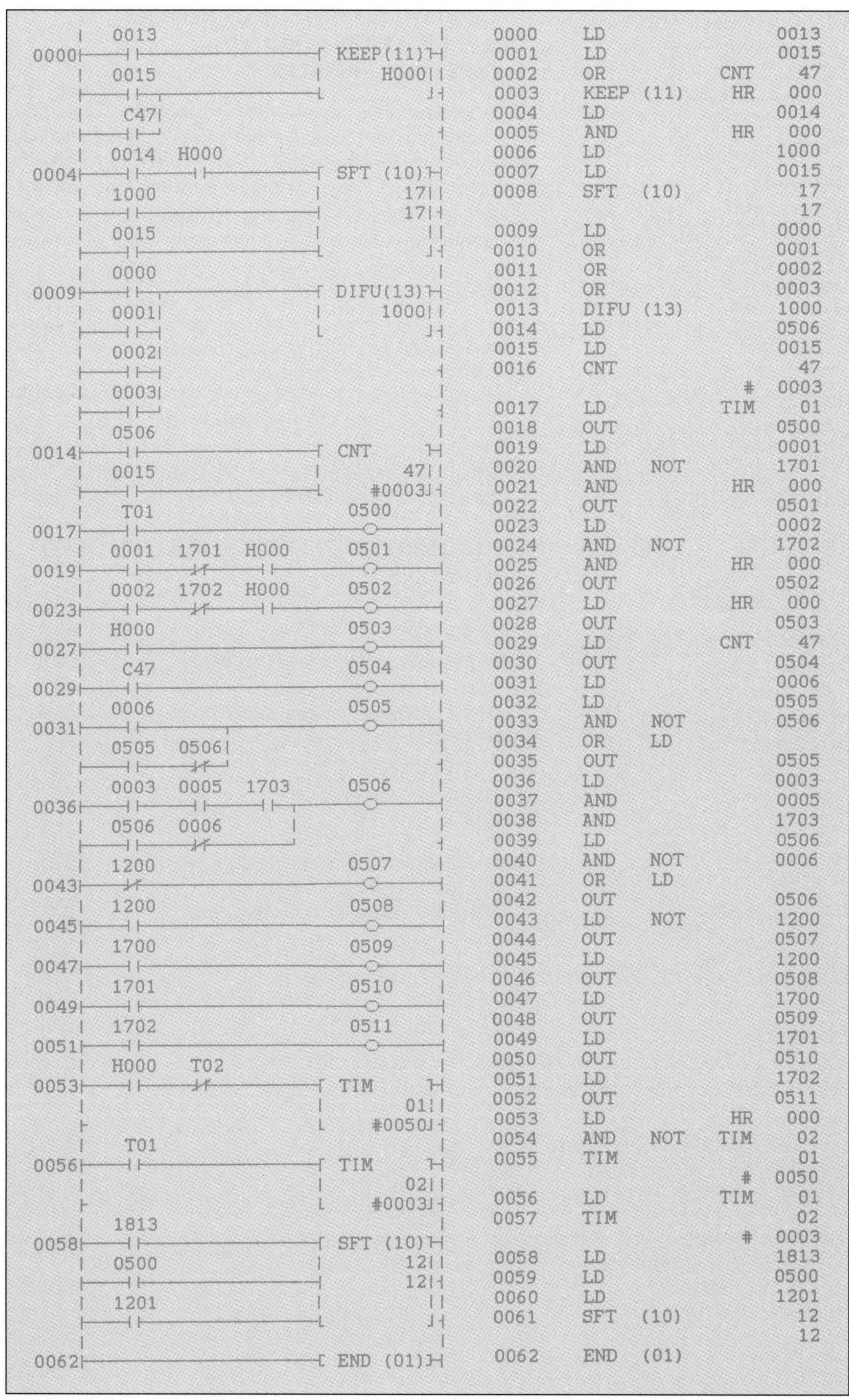

Address	Instruction		Operand	Data
0000	LD			0013
0001	LD			0015
0002	OR		CNT	47
0003	KEEP	(11)	HR	000
0004	LD			0014
0005	AND		HR	000
0006	LD			1000
0007	LD			0015
0008	SFT	(10)		17
				17
0009	LD			0000
0010	OR			0001
0011	OR			0002
0012	OR			0003
0013	DIFU	(13)		1000
0014	LD			0506
0015	LD			0015
0016	CNT			47
			#	0003
0017	LD		TIM	01
0018	OUT			0500
0019	LD			0001
0020	AND	NOT		1701
0021	AND		HR	000
0022	OUT			0501
0023	LD			0002
0024	AND	NOT		1702
0025	AND		HR	000
0026	OUT			0502
0027	LD		HR	000
0028	OUT			0503
0029	LD		CNT	47
0030	OUT			0504
0031	LD			0006
0032	LD			0505
0033	AND	NOT		0506
0034	OR	LD		
0035	OUT			0505
0036	LD			0003
0037	AND			0005
0038	AND			1703
0039	LD			0506
0040	AND	NOT		0006
0041	OR	LD		
0042	OUT			0506
0043	LD	NOT		1200
0044	OUT			0507
0045	LD			1200
0046	OUT			0508
0047	LD			1700
0048	OUT			0509
0049	LD			1701
0050	OUT			0510
0051	LD			1702
0052	OUT			0511
0053	LD		HR	000
0054	AND	NOT	TIM	02
0055	TIM			01
			#	0050
0056	LD		TIM	01
0057	TIM			02
			#	0003
0058	LD			1813
0059	LD			0500
0060	LD			1201
0061	SFT	(10)		12
				12
0062	END	(01)		

Figure 7-13. Ladder diagram and mnemonic list for bottle filling and capping machine (control problem 2).

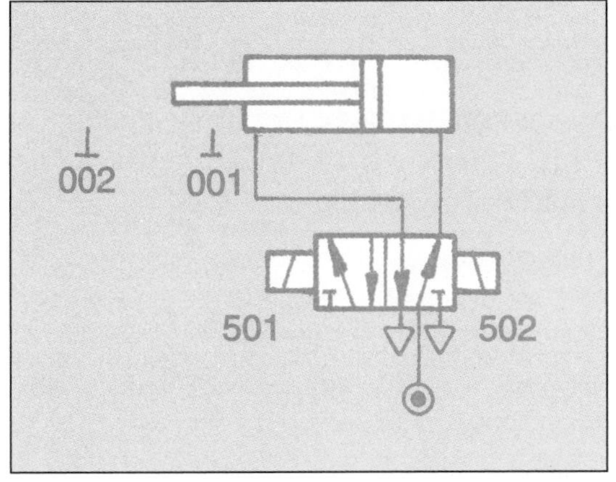

Figure 7-14. Fluid power circuit for reciprocating actuator Ⓐ in control problem 3.

COMPOUND BUILDING BLOCKS APPLICATION CIRCUIT — CONTROL PROBLEM 4

A safety-coding circuit must be designed. The circuit inputs are operated by key and push-buttons and allow the machine to be "Start-enabled" or "Start-disabled" depending on certain conditions. The circuit to be designed must provide the following operating conditions:

- For safety reasons, the machine can be started only if a specific, four-digit, secret code combination is entered with four of the ten numbered push-button keys on the key pad (figure 7-16).
- Machine operations can be disabled only if the same four-digit combination is entered again.

```
                LADDER DIAGRAM

     | 0014                                      |
0000 |--| |----------------------------[ KEEP(11) ]-|
     | 0015                            |    H000 |
     |--| |--+-------------------------|         |
     | T01   |                                   |
     |--| |--|                                   |
     | H015  |                                   |
     |--| |--'                                   |
     | 0013                                      |
0005 |--| |----------------------------[ KEEP(11) ]-|
     | 0014                            |    H015 |
     |--| |----------------------------|         |
     | H000                                      |
0008 |--| |----------------------------[ TIM      ]-|
     |                                 |      01 |
     |                                 |   #9999 |
     | 0001                           1000       |
0010 |--| |--+--------------------------( )------|
     | 0002  |                                   |
     |--| |--'                                   |
     | 0001                                      |
0013 |--| |----------------------------[ KEEP(11) ]-|
     | 0002                            |    H800 |
     |--| |----------------------------|         |
     | 0002   H000   H015             0501       |
0016 |--| |----| |----|/|--+-----------( )-------|
     | 1000   H015   H800  |                     |
     |--|/|----| |----| |--'                     |
     | 0001   H000   H015             0502       |
0024 |--| |----| |----|/|--+-----------( )-------|
     | 1000   H015   H800  |                     |
     |--|/|----| |----|/|--'                     |
     |                                           |
0032 |---------------------------------[ END (01) ]-|
```

ADDRESS	MNEMONIC		OPERAND	
0000	LD			0014
0001	LD			0015
0002	OR		TIM	01
0003	OR		HR	015
0004	KEEP	(11)	HR	000
0005	LD			0013
0006	LD			0014
0007	KEEP	(11)	HR	015
0008	LD		HR	000
0009	TIM			01
			#	9999
0010	LD			0001
0011	OR			0002
0012	OUT			1000
0013	LD			0001
0014	LD			0002
0015	KEEP	(11)	HR	800
0016	LD			0002
0017	AND		HR	000
0018	AND	NOT	HR	015
0019	LD	NOT		1000
0020	AND		HR	015
0021	AND		HR	800
0022	OR	LD		
0023	OUT			0501
0024	LD			0001
0025	AND		HR	000
0026	AND	NOT	HR	015
0027	LD	NOT		1000
0028	AND		HR	015
0029	AND	NOT	HR	800
0030	OR	LD		
0031	OUT			0502
0032	END	(01)		

Figure 7-15. Ladder diagram and mnemonic list for control problem 3.

- Tampering with the code selection keys to obtain unauthorised machine start must immediately prevent the machine from being started (disable start).

The secret key-in code must be entered with the push button keys being pressed in the following order:

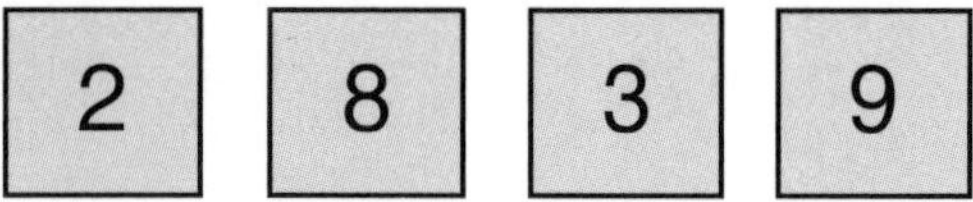

Any wrong keys being pressed cause internal auxiliary relay 1000 to turn "ON". This automatically resets the four step-counter flip-flops HR 901 to HR 904 (figure 7-16).

The first correct key selection, 2 (input 002), causes HR 901 flip-flop to set and record this selection. The second correct key selection, 8, causes HR 902 flip-flop to set, providing HR 901 is still set and is contributing (preparing) in "AND" connection to the set signal of HR 902. The third correct key selection, 3, causes HR 903 flip-flop to set, if HR 902 is still set, and provides HR 903 with the necessary preparation signal "AND" input HR 902. The fourth correct key selection, 9, causes HR 904 flip-flop to set, providing HR 903 is still set. By now, all four keys have been pressed (selected), and the secret combination 2-8-3-9 is entered in correct sequence. If a wrong key is pressed during the selection of the starting combination, all four flip-flops receive a reset command (1000). When reaching HR 904, an "always on" signal 1813 is entered into the data input of shift register 12, and timer 01 resets the four flip-flops automatically 0.5 second later.

HR 904 is the clock input, which moves this logic 1 data 1813 into pocket 1200 of shift register 12 (word 12, bit 00). (For further explanation of T flip-flop operation and shift register operation, see figures 7-6 and 7-11.) Shift register pocket 1200 is now logic 1. The bit contact of this pocket (1200) is used to set HR 915 (see address 0031). This in turn sets output relay 500 to "ON". Output relay 500 drives the machine-enable lamp. Contact 500 of this relay is included in the starting condition of the machine, but is not shown. The machine sequencing circuit is not shown either.

To disable the machine, the same four digit combination must be entered again. This then renders a second pulse of the HR 904 contact to the shift register, thus causing pocket 1201 of the shift register to become logic 1 (1201). As a result of this, the shift register resets, and pockets 1200 and 1201 (bits 1200 and 1201) are now logic 0. Bit 1201, as it was logic 1, did reset HR 915, thus causing output 500 to go "OFF" and output 511 to come "ON" (see addresses 0032 to 0037). Output 511 switches the disable lamp "ON". To operate correctly, HR 915 flip-flop needs to be a reset-dominant flip-flop (see "Reset-Dominant Flip-Flop" in Chapter 4 and figure 4-03).

COMPOUND BUILDING BLOCKS APPLICATION CIRCUIT — CONTROL PROBLEM 5

A combination lock for a laboratory entrance door must be designed to prevent unauthorised entry and render the following operating and entering conditions:

- Door may always be opened with a fitting door key, which actuates the door lock release solenoid 500 via normally closed switch 013 (figures 7-17 and 7-18).
- Door may be opened if push button keys 3-8-5 are selected within 30 seconds. The order of pressing these three keys must be as shown.
- Tampering with the push buttons, by pressing any of the seven incorrect keys, must be recorded.
- Twelve such recorded tampering attempts must set off an alarm, which causes alarm lamp 511 to blink with a frequency rate of 1 pulse per second.
- Incomplete or too slow progression of selecting the three correct push-button keys will automatically cancel the door opening process after 30 seconds (HR 000 and TIM 01).
- Four tampering attempts must prevent any further attempts, and the door may then only be opened with a key, or if the correct key-in combination is achieved within 5 seconds (1001 relay and counter 03 reset).
- The alarm system (lamp 511 blinking) is reset if a reset push button within the laboratory is pressed (input 012).
- The combination key-in circuit's last flip-flop (HR 003) is automatically reset if the door is opened (input 015, normally closed switch mounted in the door jamb), or if switch 015 is illegally removed, or if the wires between switch 015 and the PLC are broken.

The secret key-in code must be entered with the keys being pressed in the following order:

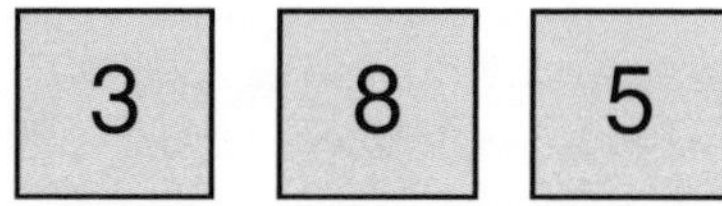

Any wrong keys being pressed cause relay 1000 to signal automatic reset of HR 001 to HR 003. For normal correct key-in function, see also the control problem 4 explanation in this chapter.

After entering the correct combination within 30 seconds, and thus HR 003 being set and counter 03 not

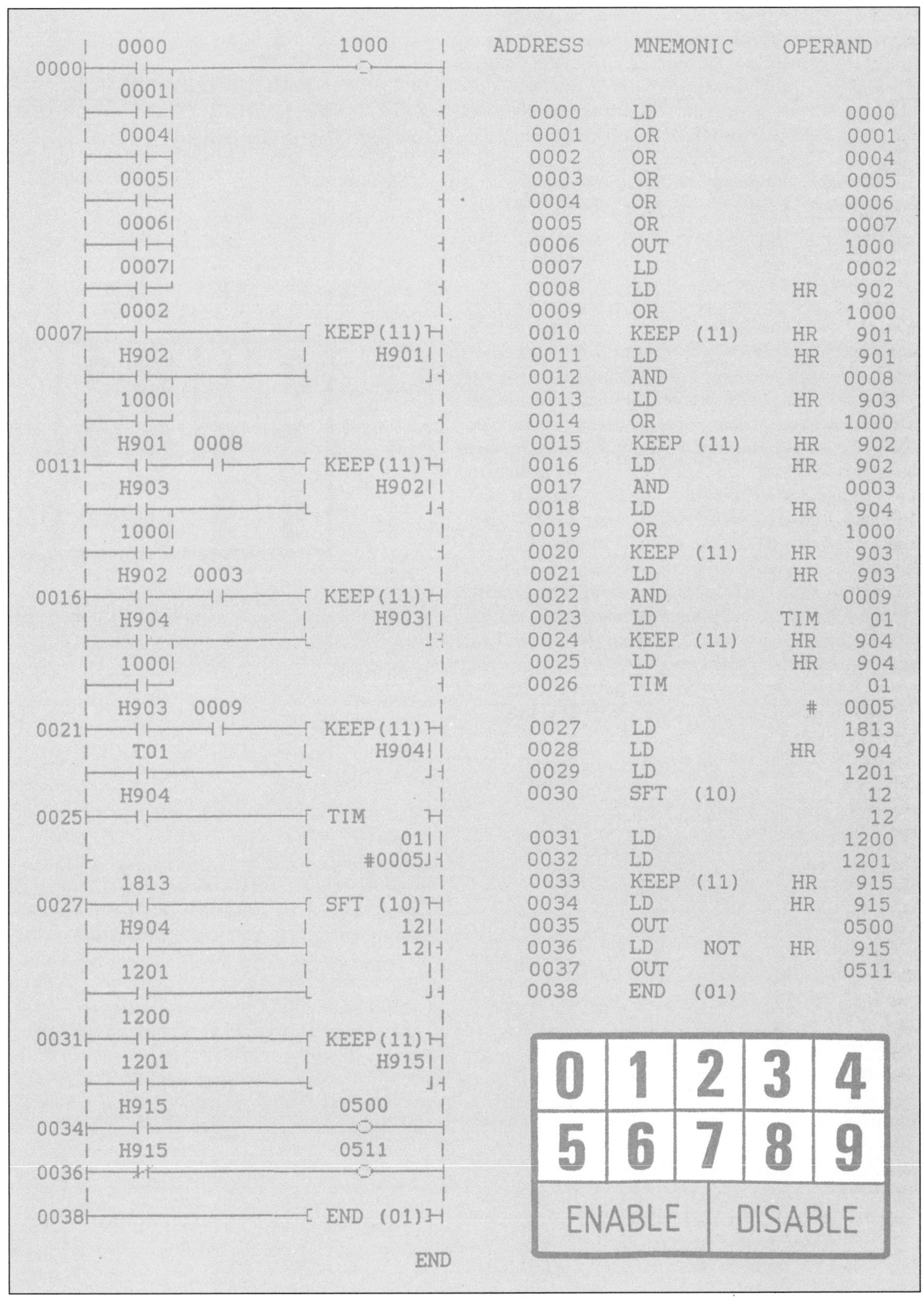

ADDRESS	MNEMONIC		OPERAND	
0000	LD			0000
0001	OR			0001
0002	OR			0004
0003	OR			0005
0004	OR			0006
0005	OR			0007
0006	OUT			1000
0007	LD			0002
0008	LD		HR	902
0009	OR			1000
0010	KEEP	(11)	HR	901
0011	LD		HR	901
0012	AND			0008
0013	LD		HR	903
0014	OR			1000
0015	KEEP	(11)	HR	902
0016	LD		HR	902
0017	AND			0003
0018	LD		HR	904
0019	OR			1000
0020	KEEP	(11)	HR	903
0021	LD		HR	903
0022	AND			0009
0023	LD		TIM	01
0024	KEEP	(11)	HR	904
0025	LD		HR	904
0026	TIM			01
			#	0005
0027	LD			1813
0028	LD		HR	904
0029	LD			1201
0030	SFT	(10)		12
				12
0031	LD			1200
0032	LD			1201
0033	KEEP	(11)	HR	915
0034	LD		HR	915
0035	OUT			0500
0036	LD	NOT	HR	915
0037	OUT			0511
0038	END	(01)		

Figure 7-16. Ladder diagram and mnemonic list for control problem 4.

having recorded at least four tampering attempts, the solenoid actuator in the door jamb retracts the locking pin. The door may now be opened. This causes contact 015 of the normally closed switch 015 in the door jamb to open and resets step-counter flip-flop HR 003. All three HRs are now reset (HR 002 and HR 003 automatically reset their previous holding relays).

Holding relay HR 000 records the pressing of any push-button keys and sets timer 01 in motion. Timer 01, if permitted to time out, causes the step-counter flip-flops HR 001 to HR 003 to reset immediately, thus limiting the opening procedures to 30 seconds.

Four tampering attempts cause counter flag CNT 03 to turn "ON". Opening of the door is now possible only with a laboratory key (switch 0013), or with a "fast opening" key-in approach, where the three correct keys must be pressed in less than 5 seconds. This is achieved if HR 003 is reached (set) before timer 02 opens contact T 02 in address 051 (see ladder diagram figure 7-18). By this, relay 1001 is caused to go "ON" and resets counter 03, thus causing contacts HR 003 and C 03 in logic line, starting at address 054, to become true, and therefore rendering output 500.

Twelve tampering attempts (1000) cause counter 04 to turn "ON" flag CNT 04. This counter flag's contact C 04 in address 058, "AND" connected to pulsing bit 1902, signals alarm lamp 511 to turn "ON". The alarm lamp is turned "OFF" with input 012.

PLC INPUT/OUTPUT ASSIGNMENT LIST (OMRON C20)

PLC inputs:

000–009 = key-in push-button switches
012 = alarm reset switch
013 = door key switch
014 = tampering lock disable switch
015 = coding circuit reset switch

PLC outputs:

500 = door unlock solenoid
511 = alarm lamp
1000 = wrong key selection relay
1001 = fast key-in counter reset relay
HR000 = opening attempt recording
HR001 = first key-in correct recording
HR002 = second key-in correct recording
HR003 = third key-in correct recording
TIM 01 = too slow key-in reset
TIM 02 = fast key-in limitation
CNT 03 = 4 tampering attempts disable
CNT 04 = 12 tampering attempts alarm

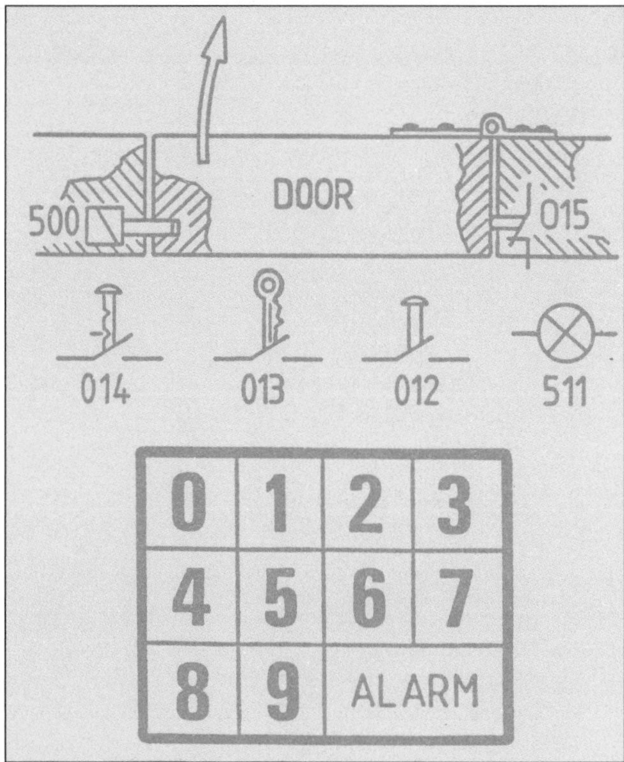

Figure 7-17. Input/output devices for control problem 5. Inputs 000 to 009 constitute the push-button switches to select secret door-opening code combination.

COMPOUND BUILDING BLOCKS SUMMARY

Compound building blocks, being proven subcircuits, are usually integrated into a large PLC program. Such building blocks are made up of logic functions, flip-flops, monoflops, timers, counters and shift registers. Having knowledge of such building blocks saves time when programming PLC user software, and enables better circuit solutions and design controls for complex and complicated control applications to be achieved. Compound building blocks may also expand the programming potential of PLCs with shortcomings (e.g. the variable flicker bit circuit is used where a PLC has no pulse bits). Some compound building blocks are of pure combinational nature, whereas some are mixed with sequential attributes.

Figure 7-18. Ladder diagram and mnemonic list for control problem 5.

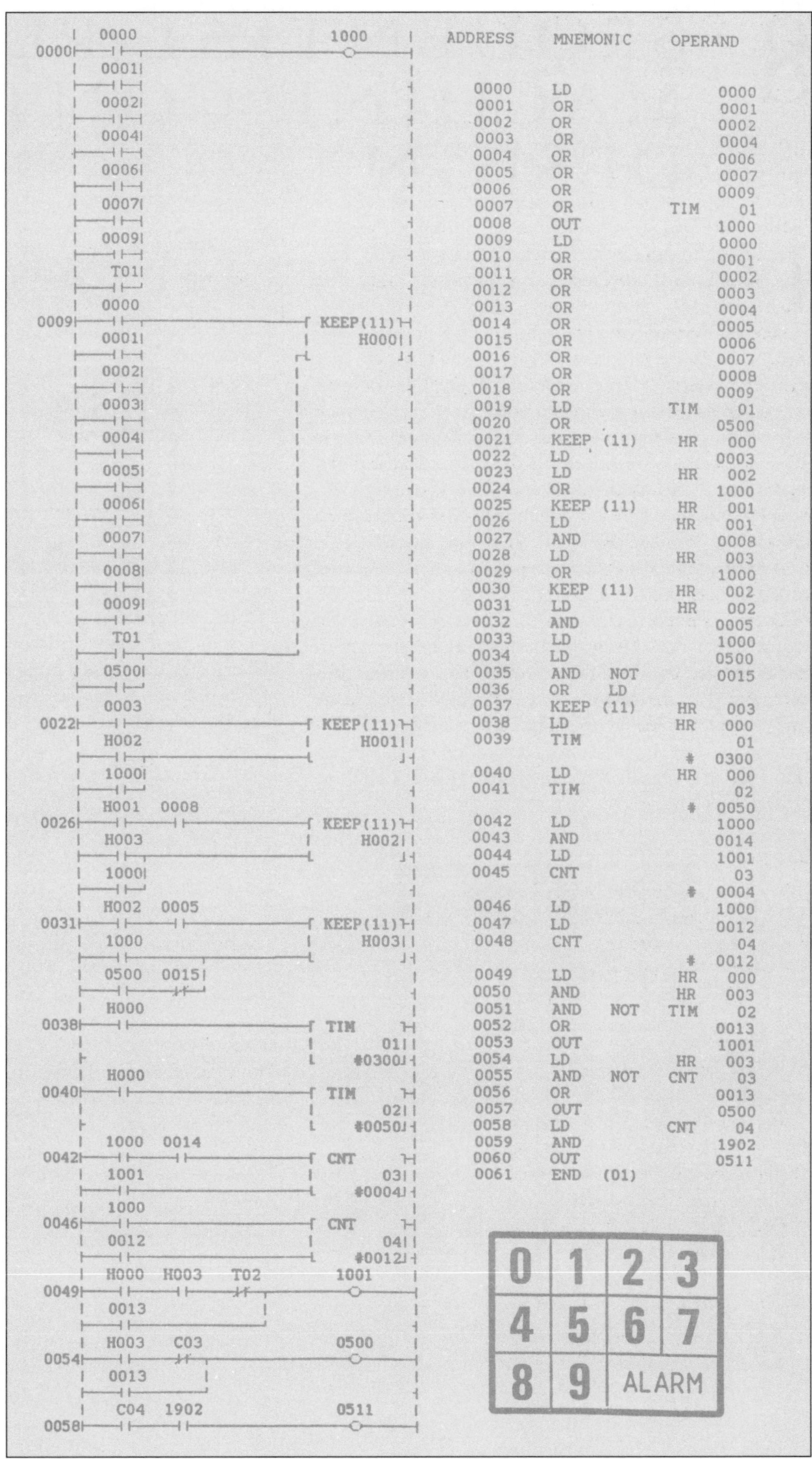

ADDRESS	MNEMONIC	OPERAND
0000	LD	0000
0001	OR	0001
0002	OR	0002
0003	OR	0004
0004	OR	0006
0005	OR	0007
0006	OR	0009
0007	OR	TIM 01
0008	OUT	1000
0009	LD	0000
0010	OR	0001
0011	OR	0002
0012	OR	0003
0013	OR	0004
0014	OR	0005
0015	OR	0006
0016	OR	0007
0017	OR	0008
0018	OR	0009
0019	LD	TIM 01
0020	OR	0500
0021	KEEP (11)	HR 000
0022	LD	0003
0023	LD	HR 002
0024	OR	1000
0025	KEEP (11)	HR 001
0026	LD	HR 001
0027	AND	0008
0028	LD	HR 003
0029	OR	1000
0030	KEEP (11)	HR 002
0031	LD	HR 002
0032	AND	0005
0033	LD	1000
0034	LD	0500
0035	AND NOT	0015
0036	OR LD	
0037	KEEP (11)	HR 003
0038	LD	HR 000
0039	TIM	01
		# 0300
0040	LD	HR 000
0041	TIM	02
		# 0050
0042	LD	1000
0043	AND	0014
0044	LD	1001
0045	CNT	03
		# 0004
0046	LD	1000
0047	LD	0012
0048	CNT	04
		# 0012
0049	LD	HR 000
0050	AND	HR 003
0051	AND NOT	TIM 02
0052	OR	0013
0053	OUT	1001
0054	LD	HR 003
0055	AND NOT	CNT 03
0056	OR	0013
0057	OUT	0500
0058	LD	CNT 04
0059	AND	1902
0060	OUT	0511
0061	END (01)	

8 SEQUENTIAL PLC MACHINE CONTROL DESIGN

Industrial control systems are commonly grouped into combinational systems and sequentially operating systems. In a combinational control system, the output signals depend entirely on the binary logic state of its input signals (see figures 7-01, 7-09 and 7-10). In a sequential system, however, some or all of its outputs depend on previous machine events or states as well as on current machine states. Thus, sequential systems require memory elements (flip-flops) that can record such previous events or entire machine states.

To view a sequential machine control system from a completely different angle, one may also say that in a sequentially operating machine, no action or sequence step will or should ever take place unless the immediately previous machine action or step is completed and confirmed. Action confirmation, therefore, requires feedback from the machine. This feedback in turn necessitates limit switches, flow rate switches, temperature switches, pressure switches and many other types of sensors, to confirm the successful completion of the desired and previously signalled action.

Sequential systems may also be subdivided into **synchronous** and **asynchronous** control systems. Synchronous control systems are usually stepped through their sequence by timing pulses, whereas asynchronous control systems are stepped through their sequence only when action feedback signals are received (essential confirmation signals). Since synchronous circuits in industrial automation are very rare, this book will concentrate only on asynchronous sequential controls.

For most industrial controls, there are also requirements to include "fringe conditions" into the sequential control. Such fringe conditions may include one or more modules (subcircuits) that take care of routines such as emergency stop initiation, automatic or manual mode cycling selection, alternative machine program selection provision, and stepping through the program. As these subroutines are usually in modular form, and are always attached at the fringe of the sequential control, they are named fringe conditions (see figures 8-01, 9-02, 9-04, 9-07 and 9-08). Chapter 9 explains design concepts, functions and integration of fringe condition modules.

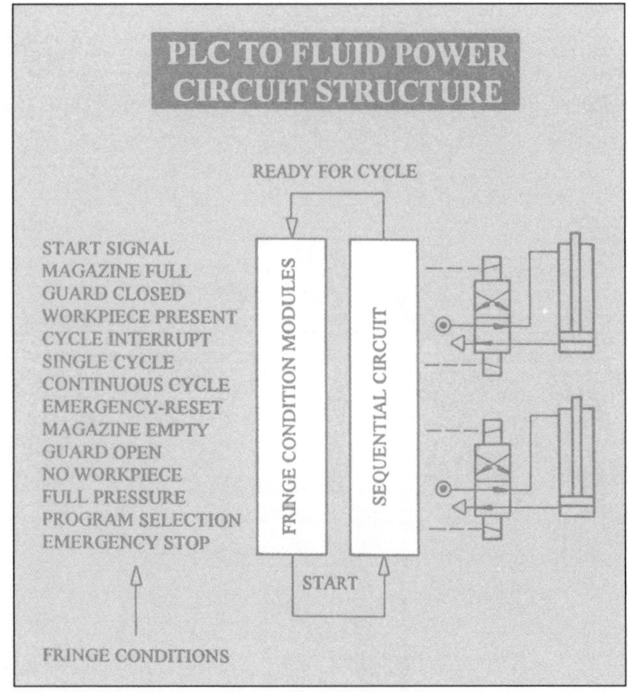

Figure 8-01. Sequential control circuit structure including fringe condition modules.

Designing a sequential control system for a PLC requires careful preparation and good knowledge of automation concepts. Firstly, the designer and programmer of a PLC user control program has to determine how many actuators are involved in the control. Where fluid power actuators are used, the designer must also find out their initial position (extended or retracted, for example). The designer also needs to find out the various positions of limit switches required to confirm any action completed by these linear or semi-rotary actuators, and how often an actuator repeats its motion during one cycle. All this is best expressed in a "step-action diagram" (figure 8-03). The step-action diagram is sometimes referred to as a traverse-time diagram or step-motion diagram. The step-action diagram is explained and illustrated in detail through the following pages.

ALLOCATION OF OUTPUT AND INPUT RELAY NUMBERS FOR SEQUENTIAL CONTROLS

For programming convenience and control uniformity, the labelling of PLC input and output relay points for connection to pneumatic and hydraulic valve solenoids and of signals from position limit switches for fluid power actuators follows a firmly set pattern in this book. This pattern, although not standardised yet, makes circuit design, circuit testing and fault diagnostics easy and methodical. Actuator extension commands on an Omron C20 or C200H PLC, for example, are labelled with even relay numbers: 502, 504, 506, 508 etc. Retraction commands are labelled with odd numbers: 501, 503, 505, 507 etc. (figure 8-02). On an Omron C28K PLC controller, which uses word 01 for the outputs, these outputs numbers would change to 102, 104, 106, 108 etc. for extension commands and 101, 103, 105, 107 etc. for retraction commands.

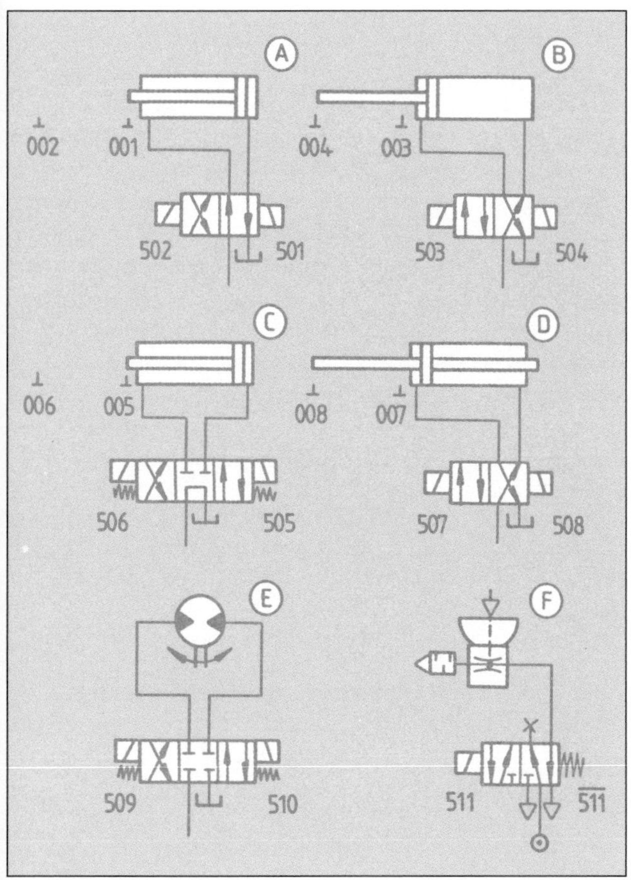

Figure 8-02. Allocation of PLC output and input relay numbers for actuators as used in fluid power sequential control systems.

Limit switches placed in the fully extended position on the hydraulic or pneumatic linear actuators are given an even number: 002, 004, 006, 008 etc. Limit switches placed in the fully retracted position on these actuators are given an odd number: 001, 003, 005, 007 etc. (figure 8-08). Hence, if actuator Ⓑ in figure 8-08 must extend, the PLC actuates output relay 504. Valve solenoid 504, which is connected to PLC output point 504, then causes actuator Ⓑ to extend. When actuator Ⓑ is fully extended, limit switch 004 signals to the PLC that the initiated action is completed. Thus, a 504 solenoid command, when executed, causes limit switch 004 to send a feedback signal to the PLC.

This system, unfortunately, does not work for woodpecker type sequential controls (see figure 8-14) or for any circuit where the actuator has interim stops as is shown in figures 8-14 and 8-12.

STEP-ACTION DIAGRAM

To explain the construction and features of a step-action diagram, it may be best to apply this diagram to a real industrial control situation.

A machine has two linear pneumatic actuators (cylinders). One actuator serves for clamping the workpiece in the machine vice; the other actuator moves the drill head up and down (figures 8-03 and 8-04). The machine cycle may be described as follows:

Step 1. Clamp the workpiece by extending actuator Ⓐ (moving from limit switch 0001 to limit switch 0002).

Step 2. Move the drill head down by extending actuator Ⓑ (moving from limit switch 0003 to limit switch 0004).

Step 3. Move the drill head up by retracting actuator Ⓑ (moving from limit switch 0004 to limit switch 0003).

Step 4. Unclamp the workpiece by retracting actuator Ⓐ (moving from limit switch 0002 to limit switch 0001).

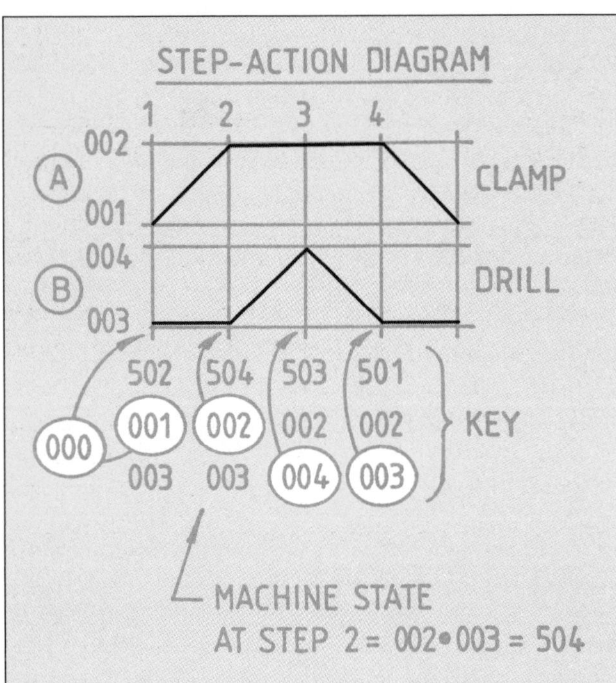

Figure 8-03. Step-action diagram for automated clamp and drill machine with four sequential steps and two actuators.

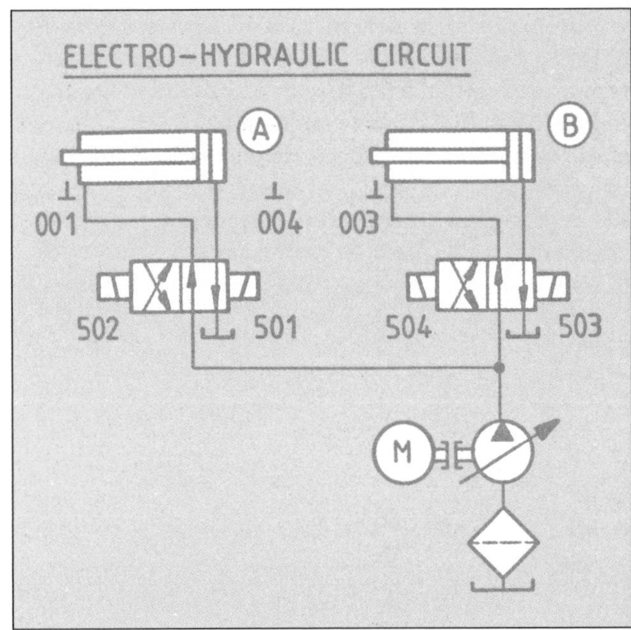

Figure 8-04. Electropneumatic circuit for the clamp and drill machine depicted in figure 8-03.

With confirmation of the fourth machine step being completed, the machine cycle automatically terminates. After the machine is manually unloaded and reloaded, the machine is ready for a new machining cycle. The step-action diagram in figure 8-03 shows the operations of this machine. The electropneumatic circuit in figure 8-04 illustrates how the PLC interacts with the pneumatic solenoid valves and pneumatic linear actuators.

This interaction demands that the PLC's output points be electrically connected to the four solenoids on the two pneumatic valves, and that the PLC's input points be electrically connected to the four limit switches placed on the stroke extremities of the two linear pneumatic actuators. The push-button start switch is connected to input point 0000 (figures 1-03, 8-02 and 8-03).

The vertical lines in the step-action diagram show the beginning and end of adjoining sequential steps. For example, the second vertical line from the left shows the completion of sequence step 1, which then becomes the beginning of sequence step 2. The numbers above these vertical lines indicate the start of their respective machine sequence steps. The rightmost line in the step-action diagram indicates the completion of step 4 (actuator Ⓐ having completed its retraction motion). Such vertical lines are also called time lines or action lines.

For each actuator in use, the step-action diagram shows two parallel horizontal lines. The upper of these two lines is the position where the actuator is extended; the lower one is where the actuator is retracted. Next, and to the left of each horizontal line, is the limit switch number listed. This number label indicates which limit switch is actuated when the actuator has reached that extended or retracted position. Hence, if actuator Ⓑ is fully extended, as is the case at the end of step 2 (action line 3), limit switch 0004 is actuated. When all motions and sequence steps are completed, the limit switches 0001 and 0003 are actuated. Each vertical pair of these limit switches indicates the "logic machine state" of the control circuit at that point. This machine state depicts all limit switches actuated and rendering a logic 1 signal at that point of sequence advancement position. Thus, the machine state at the end of sequence step 1 is:

0002 • 0003.

At the end of sequence step 2, the machine state is:

0002 • 0004.

At the end of sequence step 3, the machine state is:

0002 • 0003.

And finally, at the end of sequence step 4, the machine state is:

0001 • 0003.

The machine state under the rightmost action line is

not given, since it is the same machine state as is shown under action line 1. These machine states are of great importance when a machine accidentally stops and fault-finding is required. Such fault-finding procedures are explained in detail later in this chapter and in Chapter 11.

THE "ESSENTIAL CONFIRMATION SIGNAL"

To make a sequential circuit function properly and safely, one always needs to confirm the motion or action just completed. Such confirmation signals are indicated in the step-action diagram by circling the appropriate limit switch signal number (operand). Hence below time line 2, signal 0002 is circled, since limit switch 0002 confirms that sequence action 1 is now completed (figures 8-03 and 8-04). Here sequence action 1 is the extension of actuator Ⓐ. This circled feedback signal is the "essential confirmation signal". The essential confirmation signal under time line 4 is limit switch signal 0003. Such circled feedback signals are then fed into the sequential control program within the PLC (see figures 8-07 and 1-03).

It would be valid to use all feedback signals from each machine state, but essentially only the particular signal that confirms the action just completed is required. This logically assumes that no erratic motions took place and nobody tampered with the machine or moved actuators manually into irregular positions. For further step-action diagrams see figures 8-08, 8-12 and 8-14.

Where tampering with the machine could be expected or for absolute safety reasons, one could, of course, use all signals in the machine state to confirm completed actions and indicate present actuator positions.

The step-action diagram in figure 8-14 also shows a situation where an actuator moves to and stops in an interim position, as is the case for so-called "woodpecker" motion controls. Woodpecker motion controls are used mainly for machining operations where swarf removal is required. Such machine motions are explained in detail in the text discussing figures 8-11 and 8-12.

SUMMARY STEP-ACTION DIAGRAM

The step-action diagram is an absolutely essential tool for the design of any sequential machine control program. The step-action diagram, when properly drawn and labelled, reveals and illustrates the following machine control aspects:

- How many linear or rotary actuators are participating in the machine being controlled by the PLC (six actuators including the Venturi vacuum generator in figures 8-28 and 8-29)
- The initial position of these actuators, as in the case of linear pneumatic or hydraulic actuators, or swivel motors (figure 8-14 actuators Ⓐ and Ⓑ retracted at beginning of step 1)
- Position and number of limit switches per actuator and their operand labels (figures 8-02, 8-04, 8-08 and 8-14)
- Total number of sequence steps (action steps) for the entire machine cycle (figures 8-03 and 8-08)
- Any time-delay phases affecting the sequence and their respective time-delay set value (figures 8-08 and 8-12)
- Motion pattern for all actuators involved in the machine sequence as regards stroke repetition, interim stops (as is the case for woodpecker motion), and

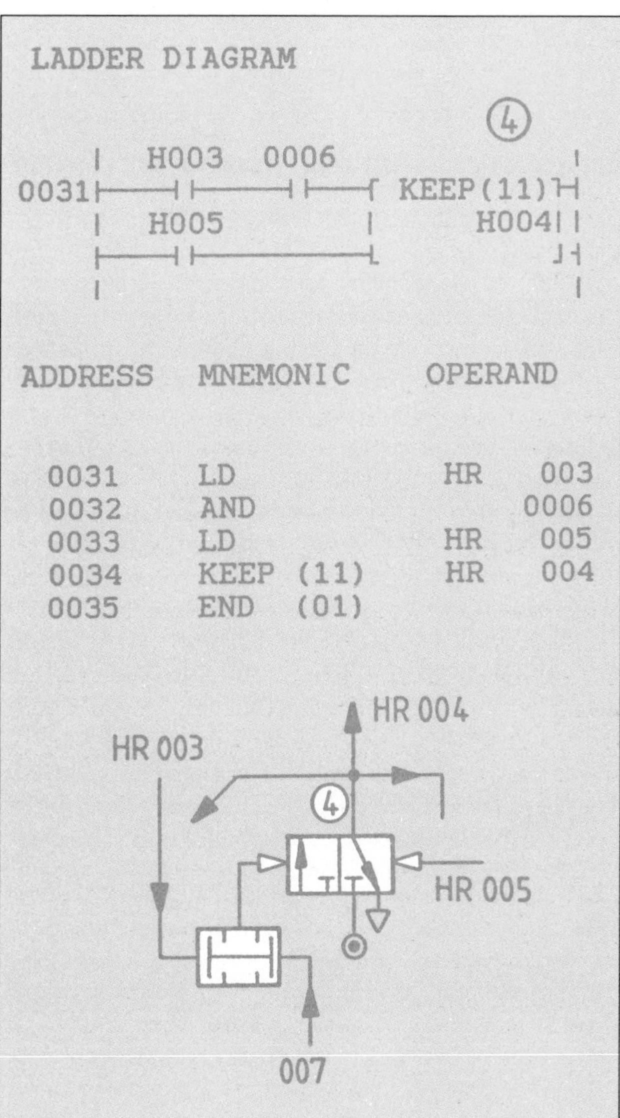

Figure 8-05. Step-counter module as programmed for PLC controllers and its pneumatic counterpart. This would be the fourth module within a step-counter chain (compare with figure 8-06).

left- or right-hand rotation of motors and swivel actuators (figures 8-08 and 8-12)

- Simultaneous motion of actuators (figures 8-08 and 8-14), or singular motion, where only one actuator travels per sequence step (figure 8-03)
- PLC output relay commands rendering the desired actuator motions (valve solenoid signals or operands, etc; see figure 8-14)
- Machine states signalled by the limit switches and sensors for each action step as well as on completion of a step (figures 8-03 and 8-14)
- Circled essential confirmation signals for each action step and necessary machine start conditions to initiate or permit start of cycling (figure 8-08)
- Label and action description for each participating actuator (Ⓐ = clamp, Ⓑ = drill, Ⓒ = air motor etc; see figures 8-03 and 8-14).

SEQUENTIAL CIRCUIT DESIGN METHODS

A circuit designer may choose from a variety of sequential circuit design concepts, but none of these design concepts comes even close to the simplicity and versatility of the **step-counter circuit design method**. This step-counter method or algorithm was presented in 1973 at a European engineering conference as an alternative to the then widely used cascade circuit design method. The step-counter as a concept, if properly applied, is totally safe for the sequencing of simple to even very complex machines because it automatically cancels any possible opposing output commands to the solenoids on the fluid power valves. It also prevents or interrupts the machine from sequencing if essential feedback signals are missing. It does this through not directly using the incoming confirmation signals to create outputs. Machine output commands leading to solenoid valves and motor starters, for example, are generated by modular and uniform, flip-flop-based building blocks (figure 8-05). Such building blocks (also called step-counter modules) are then connected into a chain of modules (figure 8-06). This chain of modules is easily expandable, easily changeable and completely flexible. For any control or machine retrofits, one simply adds or omits step-counter modules as applicable to suit the required number of machine sequencing steps.

Section A in figure 8-06 shows three interlinked step-counter modules as used in pneumatic control systems. Section B shows the same three modules with system flow chart symbols (called SFC). Section C represents these three modules in ladder diagram circuit presentation (called LAD). Section D depicts these same three modules in function plan symbolic (called FUP). For all four sections the "action" outputs represent electrical or pneumatic command signals leading to the fluid power directional control valves or similar action generators. The step-counter inputs are labelled x2, x3, x4 and so on and represent essential pneumatic or electrical confirmation signals coming from limit valves or limit switches on the machine.

The step-counter circuit design method as a concept has been in use for over 20 years in pneumatic sequential controls and has proven its safety, reliability, simplicity and versatility worldwide in millions of control applications in every facet of industrial automation. At one time there were over 15 manufacturers of pneumatic control equipment producing and selling pneumatic step-counter modules as valve blocks on the market. But with the upsurge of PLC controllers, pneumatic step-counter modules have been on a steady decline, and now there are only about five manufacturers still selling these pneumatic control blocks.

For the control program in figures 8-03 and 8-04, the machine requires four sequence steps. The PLC circuit therefore has four step-counter modules. The machine control shown in figure 8-08 needs seven sequence steps to complete its cycle, and its PLC circuit therefore has seven step-counter modules.

STEP-COUNTER ALGORITHM FOR SEQUENTIAL CIRCUIT DESIGN

The step-counter is a modular circuit design concept consisting of a chain of integrated and interdependent circuit blocks or modules in which each sequence step as depicted in the step-action diagram is allocated a module. The step-counter is sometimes also called step-sequencer or sequence controller. The simplicity of the step-counter stems from the unique construction of its modules and the integration of these modules into a sequence chain. Each module consists of a holding relay (RS flip-flop) and a pre-switched "AND" function in its logic "SET" line (figures 8-05 and 8-06). The two absolutely necessary signals required to make this "AND" function are:

- the essential feedback signal confirming the completion of the previous sequence step
- the preparation signal from the holding relay of the previous step-counter module.

The essential feedback signal confirms the completion of the previous sequence step. It is the signal circled in the key of the step-action diagram (see figures 8-03 and 8-12). If more than one actuator motion needs to be confirmed, as is shown in figures 8-08 and 8-23, one must then "AND" connect all essential confirmation signals that are circled below the action line of the step-action diagram, since they are all essential. Should one or several of these

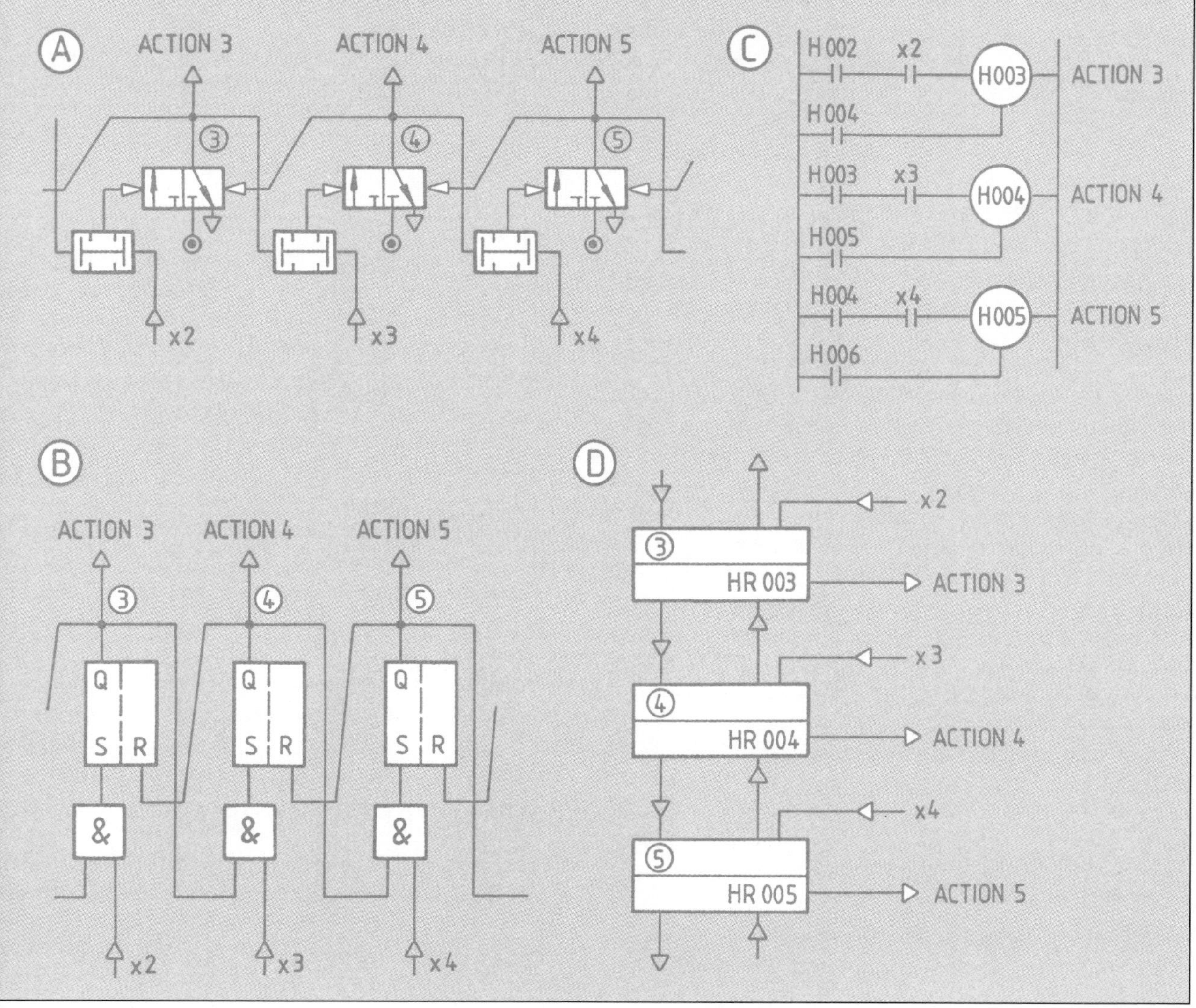

Figure 8-06. Various ways of symbolically depicting step-counter module circuits.

essential confirmation signals be missing, the sequence is then automatically interrupted until the required number of essential confirmation signals is present. This renders the circuit completely safe during machine malfunctions.

To illustrate: The two essential confirmation signals under action line 1 in figure 8-08 are limit switch signals 002 and 003. These two signals are essential since they confirm the completion of the actuator motions signalled by the previous step. This means actuator Ⓐ must be fully extended and actuator Ⓑ must be fully retracted before the new cycle can be started (see also logic line starting with address 000 in figure 8-09, where these two essential confirmation signals 002 and 003 are embedded in the set line of flip-flop HR 001).

The preparation signal, the second of the two signals listed above, originates from the holding relay of the previous step-counter module. By its "AND" connection to the essential confirmation signal it ensures that the machine sequence cannot get out of step if an essential confirmation signal should appear more than once within one sequence.

To illustrate: In figure 8-18, signal 003, which confirms actuator Ⓑ retraction, reappears in sequence step 7 under action line 7. Its second appearance, however, can no longer actuate step-counter module 4, as that module no longer has a preparation signal from step 3 module at its "AND" function in the set line of its flip-flop. Why? Because flip-flop 4 reset flip-flop 3, and thus cancelled the necessary preparation signal for flip-flop 4.

Besides the task of preparing the next holding relay (flip-flop), each holding relay (flip-flop) therefore also has the task of resetting the holding relay of the previous step-counter module in the chain into the "OFF" state (flip-flop reset command; see figures 8-06 and 8-07).

The preparation signal for flip-flop 4 (HR 004) in figure 8-07 is the flip-flop signal HR 003 shown at address 0013 of the ladder diagram and in address 0013 of the corresponding mnemonic list.

Opposing signals to the machine actuators thus cannot occur as any previously used essential feedback signals are automatically rendered inactive when the holding relay to which they are connected is being reset. It must also be noted that while the machine is cycling and step-counter switching is in progress, only one step-counter module is in the "ON" state, except for 0.1 millisecond during transition from one step to the next, where two step-counter modules (flip-flops) are briefly active. This is the transition time when module 3, for example, being "ON" at this time, resets module 2, which at this point is still "ON" (simultaneously "ON"). All sequential application circuits illustrated in this book are based on this step-counter algorithm.

To summarise: The flip-flop (holding relay) in a PLC step-counter program performs the same functions as the memory valve in a pneumatic step-counter (figures 8-06, 8-10 and 8-30). These three functions are:

- to switch the output relay in the PLC. The output relay then drives the solenoids on the fluid power valves (figure 8-07).
- to reset the holding relay (flip-flop) of the previous electronic step-counter module into the "OFF" state (figures 8-05 to 8-07).
- to provide the preparation signal at the "AND" function, which switches the holding relay (flip-flop) of the next electronic step-counter module into the "ON" state (figures 8-05 to 8-07).

STEP-COUNTER FOR CLAMP AND DRILL CIRCUIT

To further illustrate the operation and sequencing of a step-counter and the interaction and interlinking of its step-counter modules, we will again use the clamp and drill machine sequencing circuit, explained and illustrated in figures 8-03, 8-04 and 8-07.

From the step-action diagram in figure 8-03 and its explanations, it becomes obvious that the PLC program for this control requires a step-counter with four modules. A ladder diagram for these modules is given in figure 8-07. Where possible, this book adheres to the method of using holding relay operand numbers (flip-flop operand numbers) that correlate numerically with their sequence step number. In practice, this means that if a sequential control has 14 steps, we would use holding relay 001 to holding relay 014. This implies that holding relay 007 drives sequence step 7. Thus, for the clamp and drill circuit depicted in figure 8-07 we use holding relay 001 to holding relay 004. The first flip-flop, which is HR 001, has in its set line an "AND" function consisting of the preparation signal HR 004 coming from the previous step-counter module, and the essential confirmation signal 001, which is circled in the step-action diagram of figures 8-03 and 8-07. The start signal 000 is also "AND" connected to the logic conditions in that flip-flop's set line. In the reset line the HR 002 bit contact is used to reset HR 001. The second flip-flop, which is HR 002, has in its set line again an "AND" function, which consists of the preparation signal HR 001 from the previous step-counter module flip-flop HR 001, and the essential confirmation signal 002 circled below step 2 action line in the step-action diagram of figure 8-07.

Flip-flop 1 (HR 001) switches machine action 1
Flip-flop 2 (HR 002) switches machine action 2
Flip-flop 3 (HR 003) switches machine action 3
Flip-flop 4 (HR 004) switches machine action 4

Step 1 flip-flop (HR 001) is reset by flip-flop 2 (HR 002)
Step 2 flip-flop (HR 002) is reset by flip-flop 3 (HR 003)
Step 3 flip-flop (HR 003) is reset by flip-flop 4 (HR 004)
Step 4 flip-flop (HR 004) is reset by flip-flop 1 (HR 001)

Step 1 flip-flop (HR 001) is prepared by flip-flop 4 (HR 004)
Step 2 flip-flop (HR 002) is prepared by flip-flop 1 (HR 001)
Step 3 flip-flop (HR 003) is prepared by flip-flop 2 (HR 002)
Step 4 flip-flop (HR 004) is prepared by flip-flop 3 (HR 003)

The remaining two steps are similarly programmed and designed. Hence, one may again summarise:

Step 1 flip-flop (HR 001) is prepared by HR 004 (flip-flop 4) Step 2 flip-flop (HR 002) is prepared by HR 001 (flip-flop 1) Step 3 flip-flop (HR 003) is prepared by HR 002 (flip-flop 2) Step 4 flip-flop (HR 004) is prepared by HR 003 (flip-flop 3)	flip-flop preparation signals
Step 1 flip-flop (HR 001) is reset by HR 002 (flip-flop 2) Step 2 flip-flop (HR 002) is reset by HR 003 (flip-flop 3) Step 3 flip-flop (HR 003) is reset by HR 004 (flip-flop 4) Step 4 flip-flop (HR 004) is reset by HR 001 (flip-flop 1)	flip-flop reset signals
Step 1 essential confirmation signal is 0001 Step 2 essential confirmation signal is 0002 Step 3 essential confirmation signal is 0004 Step 4 essential confirmation signal is 0003	confirmation signals
Solenoid output signal 501 is activated by HR 004 Solenoid output signal 502 is activated by HR 001 Solenoid output signal 503 is activated by HR 003 Solenoid output signal 504 is activated by HR 002	solenoid output activation flip-flop bits

If this machine were controlled by a pneumatic step-counter, one could directly use the pneumatic flip-flops to issue signals to the four pilot ports on the two pneumatic power valves (see figures 8-06, 8-10 and 8-30; and Chapter 8 in the book *Pneumatic Control for Industrial Automation* by Peter Rohner and Gordon Smith). With a PLC step-counter, however, these flip-flops are an integral part of the PLC's microprocessor, and are by nature minute internal logic circuits, which issue equally minute electronic signals. These minute signals would, therefore, never suffice to activate a pneumatic or hydraulic solenoid directly! The PLC therefore sends these bit signals to the PLC's output relays, which in fact are miniature contactor relays within the PLC (figures 1-01 to 1-03). The current used on the valve solenoid is then switched through these miniature output relays. For the clamp and drill circuit, these relays in an Omron C20 PLC are 501, 502, 503 and 504 (see also figure 1-14 for the memory location of such output relays). The relays and their actuating flip-flop bits are shown at the bottom end of the PLC ladder diagram (figure 8-07).

As explained in Chapter 1, a PLC constantly scans the image pattern of its input and output signals and makes logic decisions if this image pattern changes. It is therefore of no importance where the output relays are placed in the ladder diagram, nor is the order of flip-flops of any importance. For design and circuit fault-finding, however, it is of benefit if a ladder diagram is orderly. Therefore the author suggests that circuits be arranged as shown in figures 8-07, 8-09, 8-26 and 9-12, which serve as examples.

TESTING THE PLC PROGRAM WITH THE "DRY-RUN TEST"

Once the program is entered into the user memory of the PLC it is mandatory to check the program not only for syntax errors but also for program design errors and program entering errors. This is accomplished with the so-called "dry-run test". This test eliminates machine function problems during the machine commissioning phase, and may even prevent damage to the machine or, worse still, harm to an operator. A step-by-step procedure for a correctly executed dry-run test is given below.

Preliminary procedures

Set the last holding relay (flip-flop) in the step-counter chain with the "force instruction" into the "ON" state. Modern PLCs allow the programmer to force outputs, timers, counters and flip-flops into the "ON" or "OFF" state. Forcing procedures are also explained and illustrated in Chapter 11 (see figure 11-09).

> **Note:** PLC inputs cannot be forced. To turn an input "ON", one has to activate the sensor or switch connected to the PLC's input rail.

For an Omron C20 PLC, forcing of outputs (output contacts) is accomplished in the "monitor" mode. To force the last holding relay of the PLC step-counter (HR 004), the following keys on the programming console need to be pressed in the order:

CLR, SHIFT, CONT, HR, 4, MONTR, SET.

Figures 8-07 and 8-10 illustrate why the last holding relay needs to be set (forced). For the given clamp and drill circuit, HR 004 needs to be set in the same way as one would set the last memory valve in a pneumatic step-counter (figures 8-10 and 8-30).

To execute the dry-run test correctly and speedily, one needs the step-action diagram and the ladder diagram as support material. The step-action diagram shows all the relevant essential confirmation signals that need to be manually actuated during the dry-run test. It also shows the relevant output commands for each step of the machine being turned "ON" by the PLC, and which holding relays are responsible for the simulated machine actions. The ladder diagram supports the step-action diagram if one had to make a correction to the entered program.

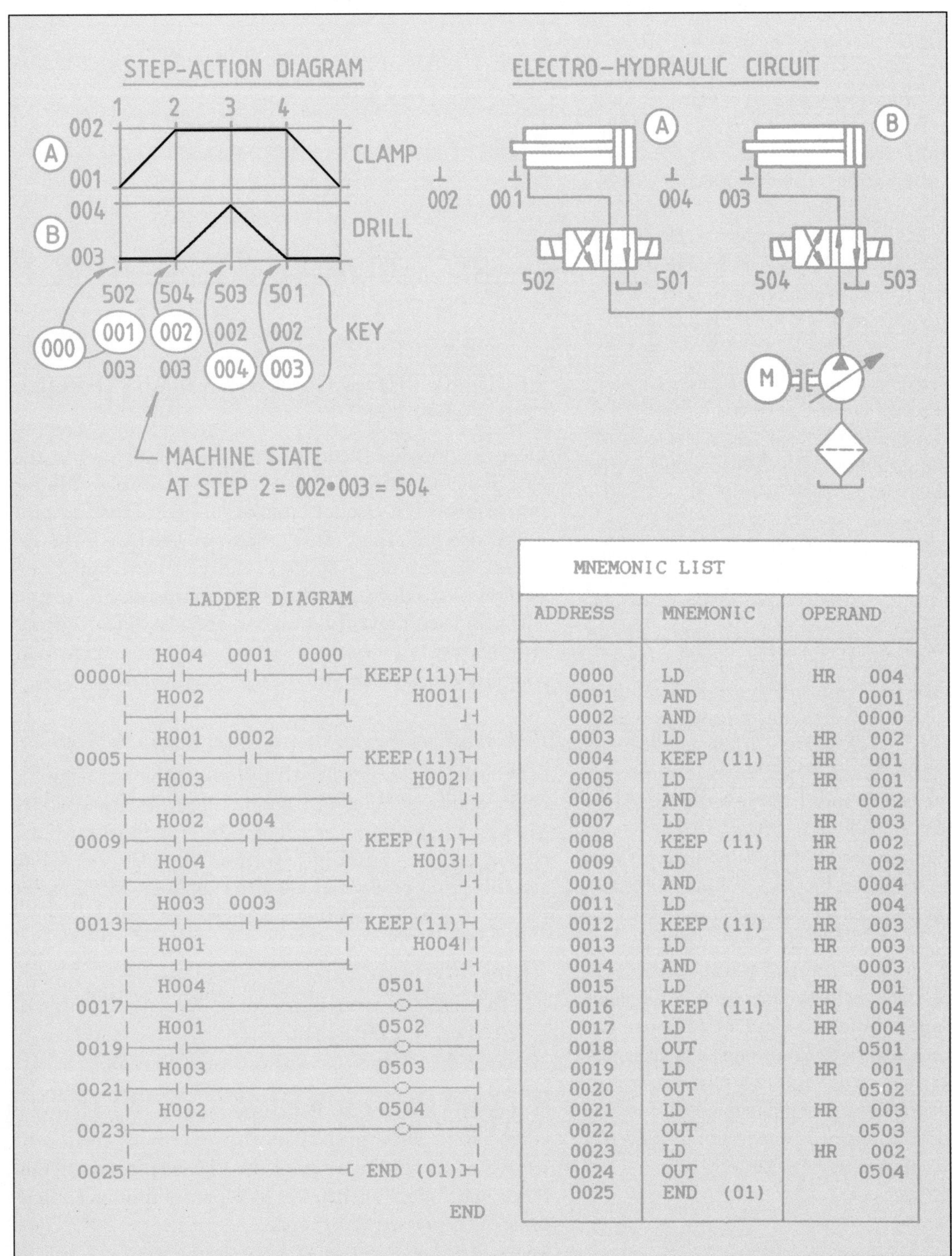

MNEMONIC LIST

ADDRESS	MNEMONIC	OPERAND
0000	LD	HR 004
0001	AND	0001
0002	AND	0000
0003	LD	HR 002
0004	KEEP (11)	HR 001
0005	LD	HR 001
0006	AND	0002
0007	LD	HR 003
0008	KEEP (11)	HR 002
0009	LD	HR 002
0010	AND	0004
0011	LD	HR 004
0012	KEEP (11)	HR 003
0013	LD	HR 003
0014	AND	0003
0015	LD	HR 001
0016	KEEP (11)	HR 004
0017	LD	HR 004
0018	OUT	0501
0019	LD	HR 001
0020	OUT	0502
0021	LD	HR 003
0022	OUT	0503
0023	LD	HR 002
0024	OUT	0504
0025	END (01)	

Figure 8-07. PLC ladder diagram for the clamp and drill circuit explained in the text.

Figure 8-08. Step-action diagram and electropneumatic circuit for a machine cycle with seven sequential steps and repeated action of actuator Ⓑ, and simultaneous actuator motions at sequence step 7.

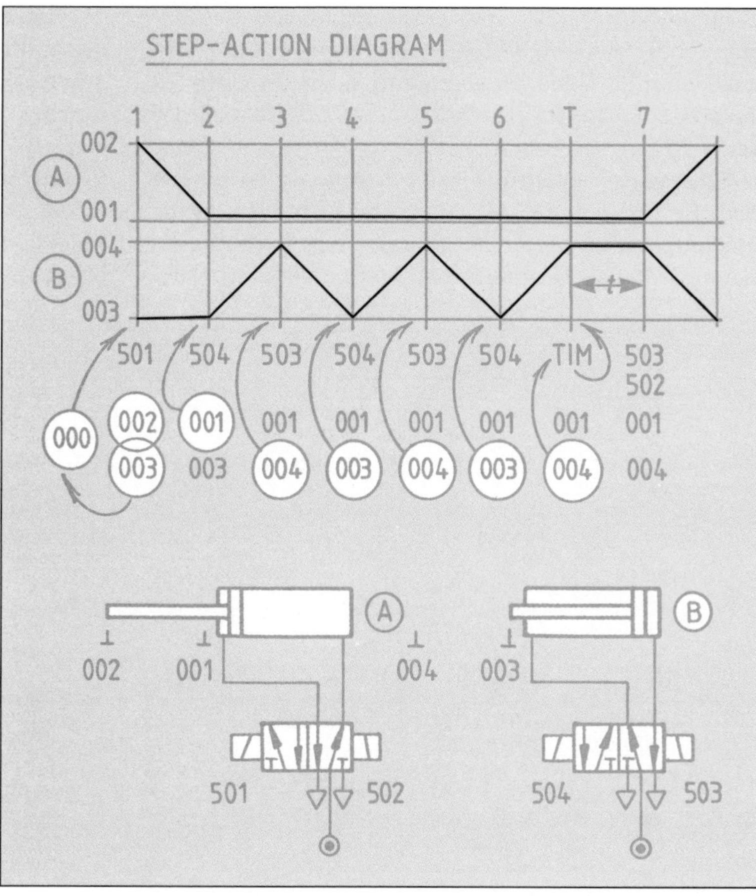

Step 1

Forcing the HR 004 flip-flop as the last flip-flop in this step-counter chain causes output 501 to turn "ON". This is displayed with the output 501 LED (light-emitting diode) on the diagnostics panel of the PLC. Now briefly actuate in "AND" connection the two switches circled in the step-action diagram of figure 8-07 (inputs 000 and 001, under time line 1). This causes output 501 to disappear and output 502 LED to light up. Output 502 causes the clamping action on the machine (actuator Ⓐ extension).

Step 2

Now briefly actuate input 002 (essential) confirmation signal circled below time line 2 in the step-action diagram. This will cause output 504 to light up and output 502 to cease. Output 504 is the solenoid signal for sequence step 2, causing drill extension (actuator Ⓑ extension).

Step 3

Now briefly actuate input 004, as circled below step 3 time line in the step-action diagram. This causes output 503 to light up. When the PLC is connected to the machine, output 503 will cause drill retraction (actuator Ⓑ retraction).

Step 4

Now briefly actuate input 003, as circled below time line 4 of the step-action diagram. This causes output 501 to light up (unclamping action on the machine, actuator Ⓐ retraction).

The dry-run test is now completed. HR 004 should remain "ON" and the other holding relays from HR 001 to HR 003 should be "OFF". This may be checked with the monitoring operations explained in Chapter 11 (figure 11-11). Should the program in the PLC have any hidden programming errors, either the dry-run test would not perform as intended or the step-counter could not proceed beyond certain steps. In that case the control circuit would need corrections, namely the alteration, deletion or insertion of a programming instruction. These three procedures are explained in figures 11-03, 11-05 and 11-06.

REPEATED ACTION OF ONE OR SEVERAL ACTUATORS

Where an actuator needs to retract or extend more than once during a single machine cycle, the output command signal from the step-counter chain needs to be "OR" connected in the PLC program before leaving the PLC on its output points for the solenoid valves (figure 8-09). For the machine control cycle presented in the step-action diagram of figure 8-08, actuator Ⓑ needs to extend in sequence step 2 and retract immediately thereafter in sequence step 3. It then extends a second time in sequence step 4 and retracts immediately upon completion of this stroke in sequence step 5. Then it extends a third time in sequence step 6 and retracts also for a third time in the same machine cycle in step 7, but not until 3.5 seconds has expired. This final retraction motion of actuator Ⓑ, as is shown in figure 8-08, happens in unison with the extension motion of actuator Ⓐ. To accomplish this triple extension and triple retraction signalling, the output commands of HR 002, HR 004 and HR 006 step-counter flip-flops need to be "OR" connected in the logic line leading to output relay 504 (figure 8-09). For the same reason, the output commands of HR 003, HR 005 and HR 007 step-counter flip-flops are also "OR" connected before their actuation of output 503.

LADDER DIAGRAM

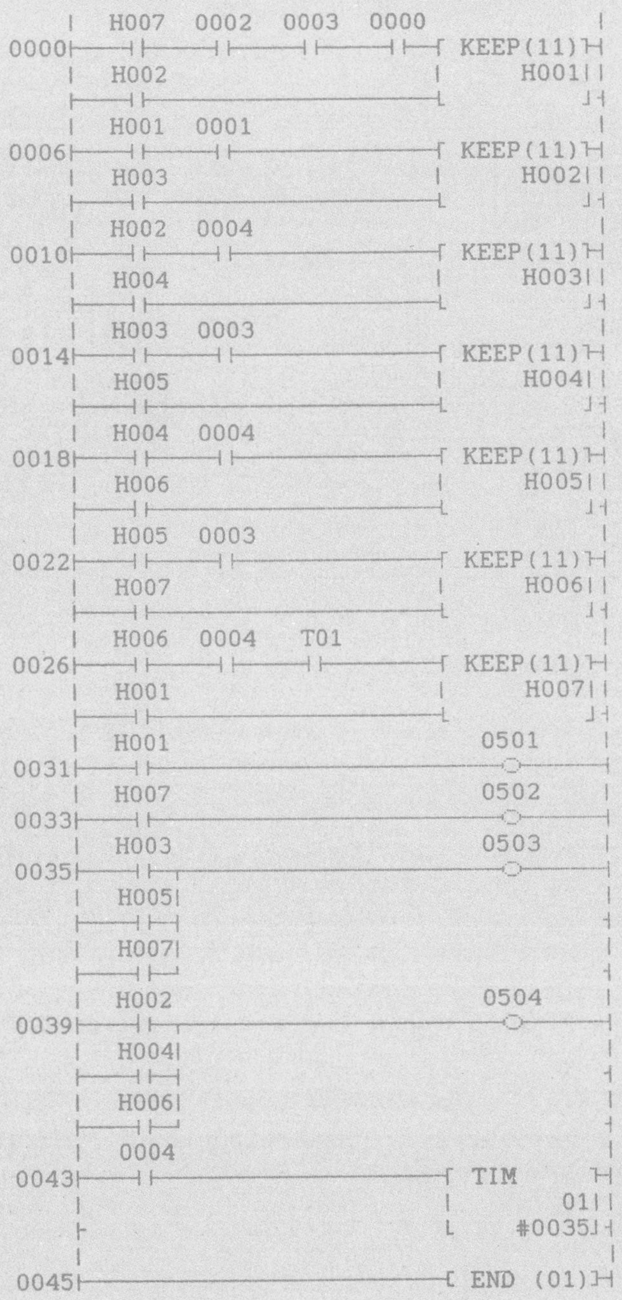

```
    | H007  0002  0003  0000          |
0000|--| |---| |---| |---| |--[ KEEP(11)]
    | H002                        |   H001|
    |--| |------------------------+       |
    | H001  0001                          |
0006|--| |---| |---------------[ KEEP(11)]
    | H003                        |   H002|
    |--| |------------------------+       |
    | H002  0004                          |
0010|--| |---| |---------------[ KEEP(11)]
    | H004                        |   H003|
    |--| |------------------------+       |
    | H003  0003                          |
0014|--| |---| |---------------[ KEEP(11)]
    | H005                        |   H004|
    |--| |------------------------+       |
    | H004  0004                          |
0018|--| |---| |---------------[ KEEP(11)]
    | H006                        |   H005|
    |--| |------------------------+       |
    | H005  0003                          |
0022|--| |---| |---------------[ KEEP(11)]
    | H007                        |   H006|
    |--| |------------------------+       |
    | H006  0004   T01                    |
0026|--| |---| |---| |---------[ KEEP(11)]
    | H001                        |   H007|
    |--| |------------------------+       |
    | H001                         0501   |
0031|--| |--------------------------( )---|
    | H007                         0502   |
0033|--| |--------------------------( )---|
    | H003                         0503   |
0035|--| |-+------------------------( )---|
    | H005|                               |
    |--| |-+                              |
    | H007|                               |
    |--| |-+                              |
    | H002                         0504   |
0039|--| |-+------------------------( )---|
    | H004|                               |
    |--| |-+                              |
    | H006|                               |
    |--| |-+                              |
    | 0004                                |
0043|--| |---------------------[ TIM     ]
    |                             |     01|
    |                             | #0035 |
    |                                     |
0045|--------------------------[ END (01)]
```

MNEMONIC LIST

ADDRESS	MNEMONIC	OPERAND
0000	LD	HR 007
0001	AND	0002
0002	AND	0003
0003	AND	0000
0004	LD	HR 002
0005	KEEP (11)	HR 001
0006	LD	HR 001
0007	AND	0001
0008	LD	HR 003
0009	KEEP (11)	HR 002
0010	LD	HR 002
0011	AND	0004
0012	LD	HR 004
0013	KEEP (11)	HR 003
0014	LD	HR 003
0015	AND	0003
0016	LD	HR 005
0017	KEEP (11)	HR 004
0018	LD	HR 004
0019	AND	0004
0020	LD	HR 006
0021	KEEP (11)	HR 005
0022	LD	HR 005
0023	AND	0003
0024	LD	HR 007
0025	KEEP (11)	HR 006
0026	LD	HR 006
0027	AND	0004
0028	AND	TIM 01
0029	LD	HR 001
0030	KEEP (11)	HR 007
0031	LD	HR 001
0032	OUT	0501
0033	LD	HR 007
0034	OUT	0502
0035	LD	HR 003
0036	OR	HR 005
0037	OR	HR 007
0038	OUT	0503
0039	LD	HR 002
0040	OR	HR 004
0041	OR	HR 006
0042	OUT	0504
0043	LD	0004
0044	TIM	01
		# 0035
0045	END (01)	

Figure 8-09. Ladder diagram and mnemonic list for the machine control depicted in figure 8-08.

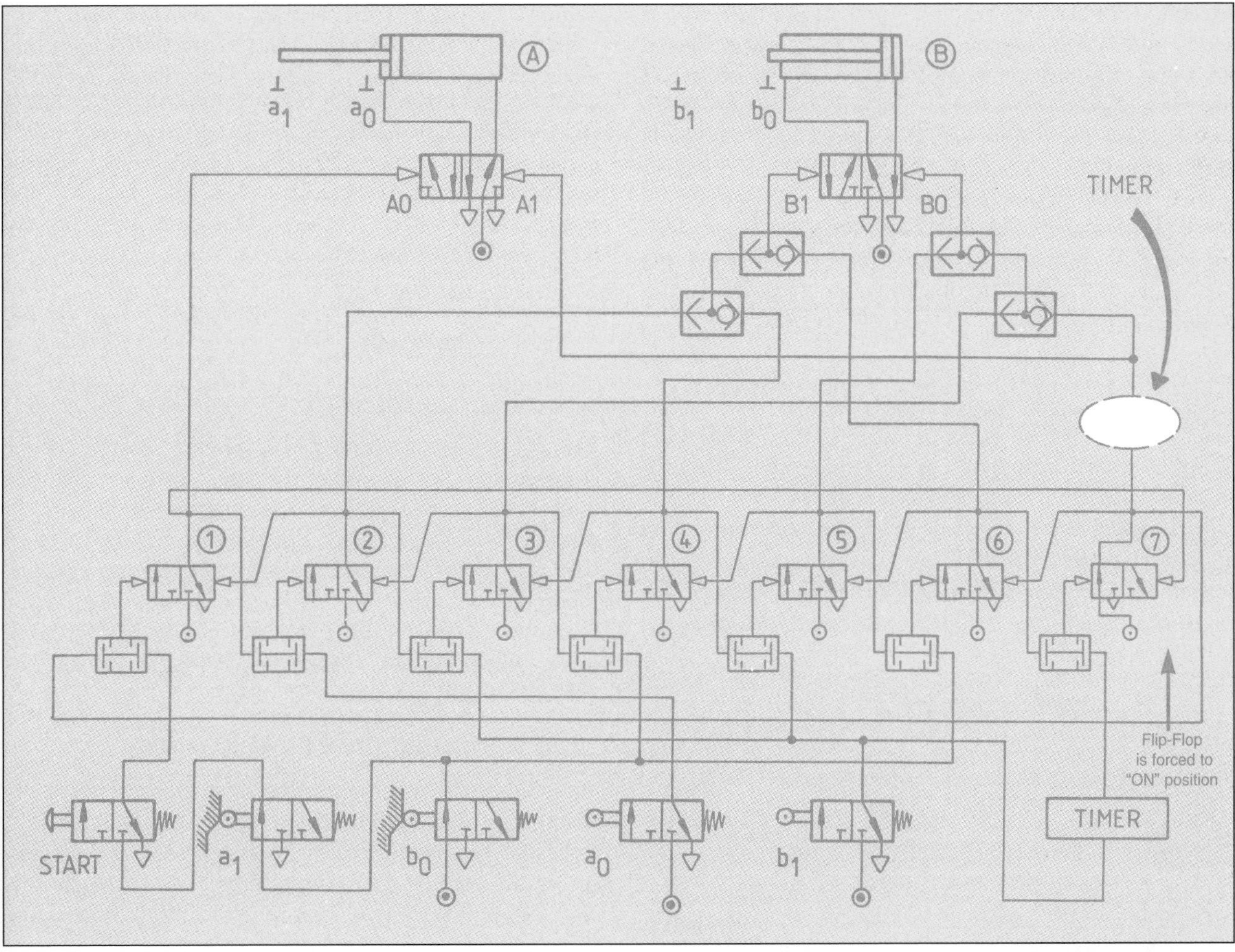

Figure 8-10. Pneumatic step-counter circuit equivalent to the machine control depicted in figure 8-08. Optional timer placement is shown by the arrow if the timer were placed into the step-counter output instead of the step-counter input, as shown with the boxed timer label.

For this control program it has been decided that the timer is best placed into the input of the step-counter (see also equivalent pneumatic circuit in figure 8-10). Hence the location of the timer is to be found in the confirmation signal line leading to the set point of flip-flop 7 at address 0028 (compare figures 8-09, 8-10 and 8-11). For this reason the timer starts to "count down" as soon as limit switch 004 becomes activated (address 0043 in figure 8-09). The contact bit activated by this timer (T 01) is therefore placed into the set rung leading to flip-flop 7 (HR 007).

Further illustrations and explanations regarding the importance and selection criteria for correct timer placement and its consequences are given in "Timer Placement in PLC Sequential Control Programming", below.

Disregarding the consequences of timer placement on machine performance, as an option the timer could by choice be connected into the output of the step-counter, as is shown by the arrow in the pneumatic circuit of figure 8-10. However, the timer contact would then have to be placed into the "OR" functions leading to outputs 502 and 503, and the timer enabling signal would no longer be limit switch contact 004, but flip-flop 7 (HR 007). A partial ladder diagram circuit for this option is given in figure 8-11.

SIMULTANEOUS ACTUATOR MOTIONS

Where a single step-counter module signals more than one action (this is the case where two or more actuators are caused to move simultaneously, as is shown in figures 8-08 and 8-14), the internal output of the step-counter flip-flop may then be used as often as required to signal these actions. In the ladder diagram of figure 8-09, HR 007 internal output contact in address 0033 drives output 502. Simultaneously it also drives output 503 through the "OR" function in address 0037. The next step-counter module, which follows the step that caused simultaneous actuator motions (in this example this is step 1 in the step-counter chain), must not switch until all these motions are completed and their respective essential confirmation signals are present at the PLC's input points. For this reason limit switch signals 002 and 003 are regarded as essential in step 1 and are therefore circled in the step-action diagram of sequence step 1 (under time line 1) in figure 8-08. For the same reason, input signals 002 and 003 are "AND" connected in the "set" line leading to HR 001 (flip-flop 1) in the ladder diagram of figure 8-09.

TIMER PLACEMENT IN PLC SEQUENTIAL CONTROL PROGRAMMING

The PLC control program example presented in figures 8-08 to 8-10 uses a timer to delay the two actions signalled by sequence step 7 of the PLC step-counter. As explained before, the PLC programmer often has the option to place the timer into either the step-counter input or the step-counter output. If placed into the step-counter input, the timer contact is usually found in the logic set line (among other "AND"-connected set conditions) that causes the flip-flop to switch into the "ON" state (figure 8-09). This timer contact therefore delays the set action of that flip-flop and, consequently, it also automatically delays all three basic tasks performed by that flip-flop. Briefly again, these three tasks are:

- reset previous flip-flop
- prepare next flip-flop
- switch action on machine.

Should the timer be placed into the output signal line of the step-counter for reasons of machine performance, the flip-flop in question will then immediately switch and perform the first two of the three aforelisted tasks. It will, however, still delay the switching action on the machine (see figure 8-11). PLC control circuit designers therefore need to evaluate the merits of both timer placement options. To illustrate the importance of such an evaluation, a control problem is given in figures 8-12 and 8-13, where, if the timer were wrongly placed, the machine would malfunction.

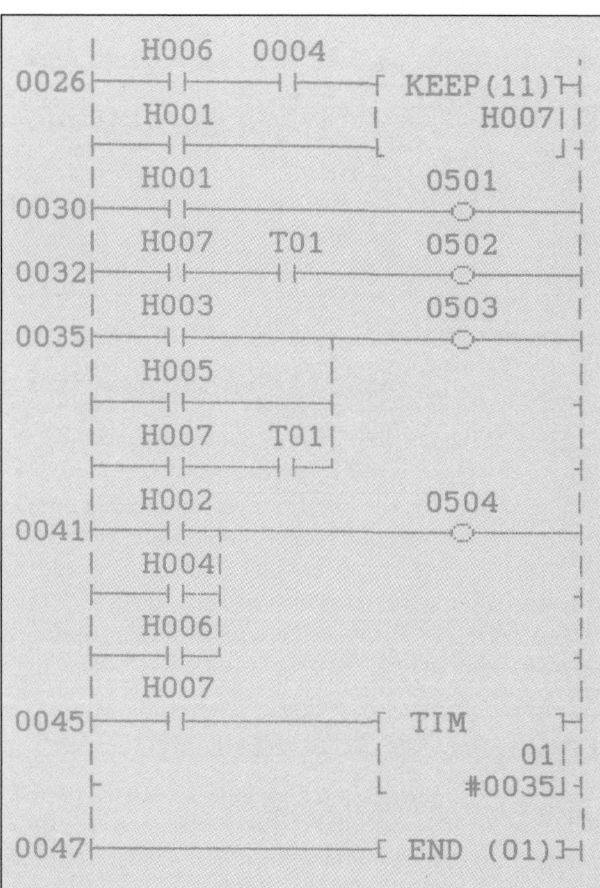

Figure 8-11. Partial ladder diagram for the control program illustrated and explained in figures 8-08 and 8-09, but with the optional timer placement in the step-counter output leading to output relays 503 and 502. Here the enabling signal for the timer is now HR 007.

TIMER PLACEMENT EVALUATION

With the timer contact being hypothetically and wrongly placed into the input of the step-counter, as is indicated by the arrows in figures 8-12 and 8-13, the incoming confirmation signal from limit switch 004 could then not proceed beyond the timer contact. The flip-flop of step 3 would therefore not switch into the "ON" state until 3 seconds on the timer has elapsed. Thus flip-flop 3 could not reset flip-flop 2, and signal 504 in figure 8-13, given by flip-flop 2, would not be cancelled when actuator Ⓑ has reached limit switch 004 (limit valve 004 in stroke middle position). This would logically permit the piston to proceed on its extension stroke without stopping in mid position, as is requested by the step-action diagram in figure 8-12!

As a general rule, program designers must be aware that timer placement for circuits with non-memory type fluid power valves need special consideration (all three-position-type hydraulic or pneumatic valves with centring springs fall into this category).

In the circuit presented in figures 8-12 and 8-13, such a self-assumed and spring-dictated position is, however, essential for the proper sequential functioning of this control.

> Valves with springs do not have memory characteristics and will therefore assume the switching position dictated by the spring when the PLC signal is cancelled.

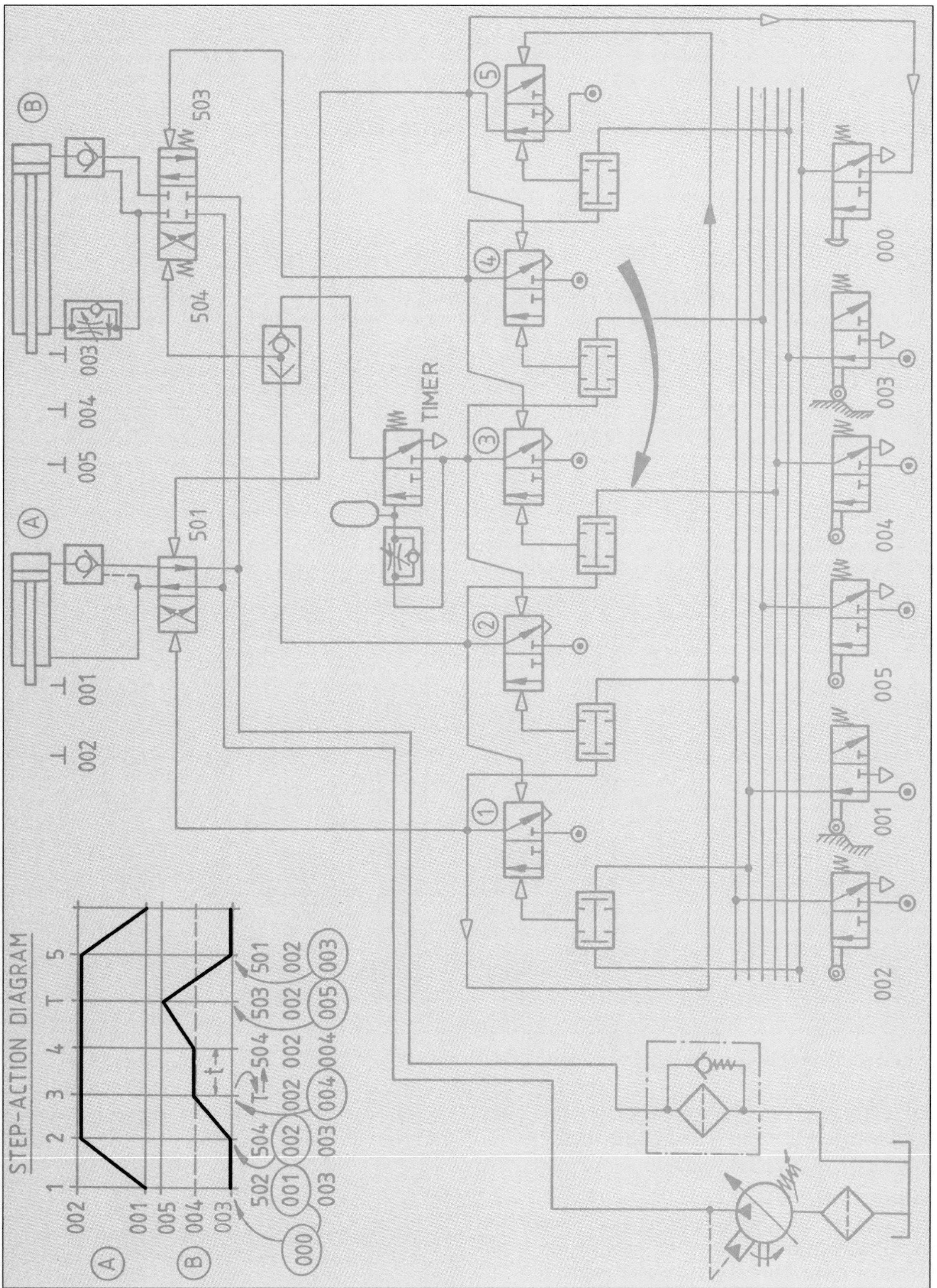

Figure 8-12. Timer placement circuit for hydraulic control with three-position directional control valve.

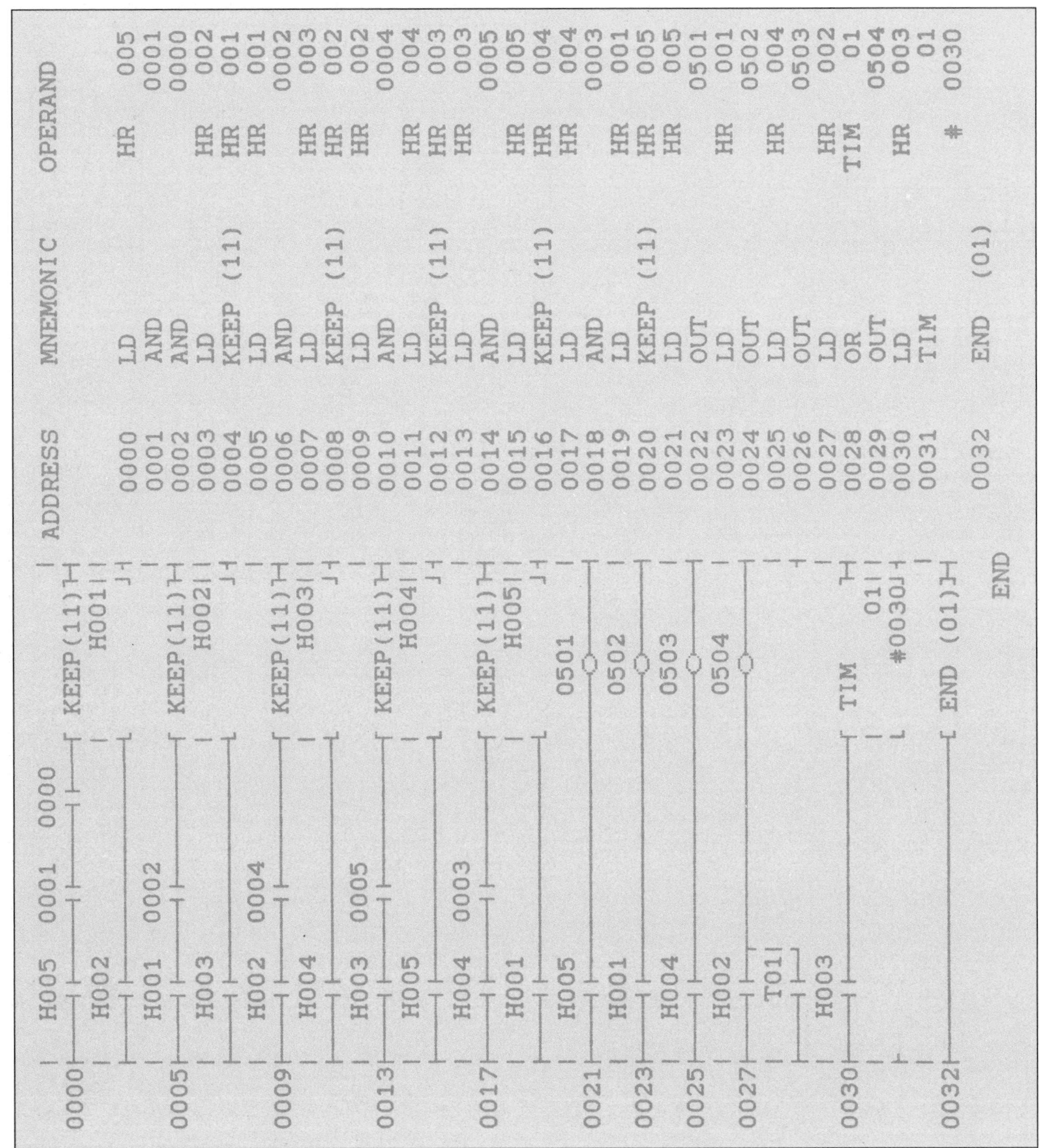

ADDRESS	MNEMONIC	OPERAND
0000	LD	HR 005
0001	AND	0001
0002	AND	0000
0003	LD	HR 002
0004	KEEP (11)	HR 001
0005	LD	HR 001
0006	AND	0002
0007	LD	HR 003
0008	KEEP (11)	HR 002
0009	LD	HR 002
0010	AND	0004
0011	LD	HR 004
0012	KEEP (11)	HR 003
0013	LD	HR 003
0014	AND	0005
0015	LD	HR 005
0016	KEEP (11)	HR 004
0017	LD	HR 004
0018	AND	0003
0019	LD	HR 001
0020	KEEP (11)	HR 005
0021	LD	HR 005
0022	OUT	0501
0023	LD	HR 001
0024	OUT	0502
0025	LD	HR 004
0026	OUT	0503
0027	LD	HR 002
0028	OR	TIM 01
0029	OUT	0504
0030	LD	HR 003
0031	TIM	01
		# 0030
0032	END (01)	

Figure 8-13. Ladder diagram and mnemonic list for control presented in figure 8-12.

PLC STEP-COUNTER PROGRAMMING FOR NON-MEMORY TYPE DIRECTIONAL CONTROL VALVES

As emphasised in the previous section on sequential control programming, non-memory type fluid-power control valves can create problems for inexperienced programmers or programmers who are not familiar with the peculiarities of fluid power control componentry. For this reason a further control application is presented to clarify matters and enhance understanding (see also control problems 3 and 4 in Chapter 9).

The machine to be controlled by a PLC (figures 8-14 and 8-15) taps screw holes into a casting used for lawnmower carburettors. Actuator Ⓐ clamps the casting, actuator Ⓑ moves the tapping head up and down and actuator Ⓒ, being a rotary type air motor, drives the tapping head. The pneumatic valve used for the air motor activates and powers the rotations of the tapping head.

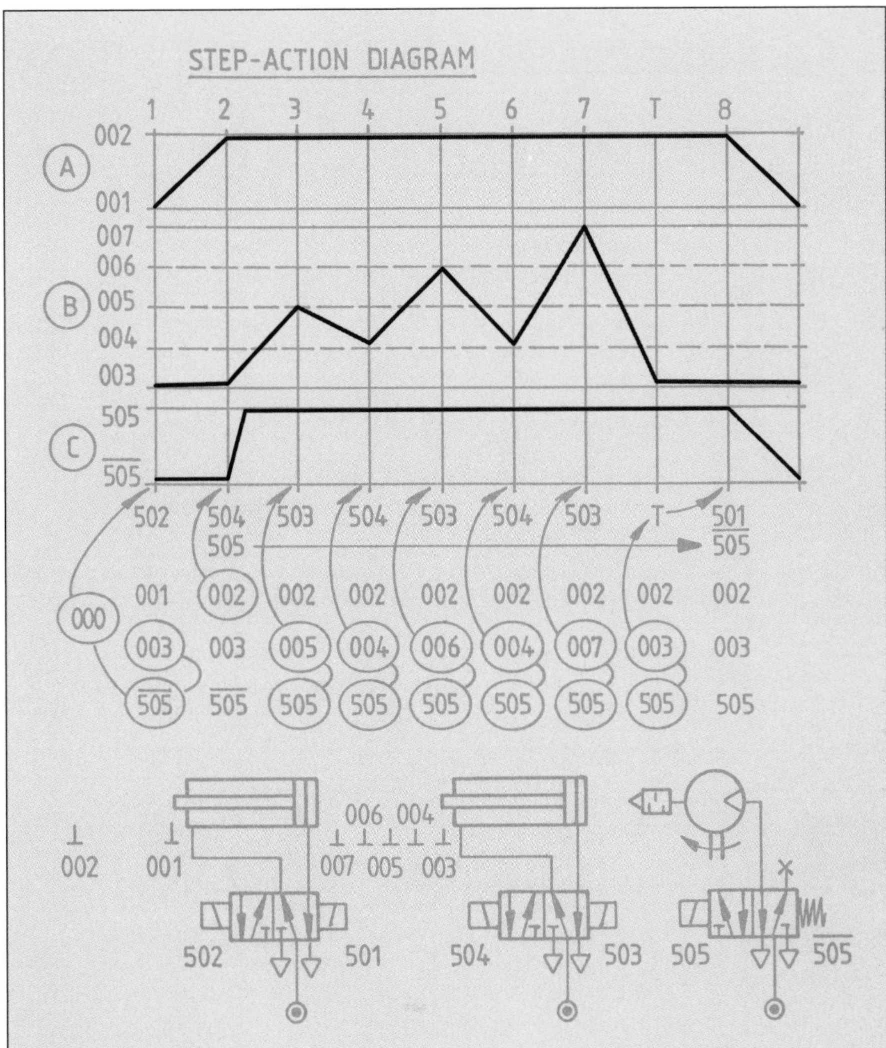

Figure 8-14.
Step-action diagram for woodpecker circuit.

Note: This valve is a single solenoid-operated valve that is reset by a spring. Such valves are often chosen to save costs. As illustrated in the step-action diagram for this control (figure 8-14), the air motor needs to start rotating at sequence step 2. Rotating action is cancelled by sequence step 8. To accomplish this task, the step-counter needs to maintain output signal 505 to the air motor valve solenoid for the duration of sequencing through steps 2, 3, 4, 5, 6 and 7, despite the fact that each step-counter flip-flop cancels the previous step-counter module flip-flop. If signal 505 were not maintained during these steps, the valve's reset spring would automatically reset the valve when step 3 flip-flop resets step 2 flip-flop at time line 3, hence the motor would run only for the duration of sequence step 2. The output signals of all these six steps are therefore "OR" connected so that they maintain the signal to solenoid 505 until step 8 automatically cancels step 7 and thus signal 505 disappears. There are, of course, other methods of maintaining such signals. Two such optional methods are shown in figure 8-17 and are described below.

Method A in figure 8-17 converts an output relay into an RS flip-flop. It does this by using function 11 before nominating output 505 during programming of the PLC. This method is not completely reliable as power restoration following an interruption will not return the output relay to the set state it had before.

Method B eliminates the disadvantages explained for method A as holding relays programmed in conjunction with function 11 are memory-retentive during power failure (see figures 1-14 and 4-03).

WOODPECKER ACTUATOR MOTION PROGRAMMING

The tapping machine depicted in figures 8-14 and 8-15 and described in the previous paragraph shows another control feature, called woodpecker control. This type of control is predominantly used on machine tool automation controls where the drill, tap or reamer makes a penetrating movement but frequently needs to retract to remove swarf. The name has been borrowed from the woodpecker bird, which hammers a hole in a tree trunk

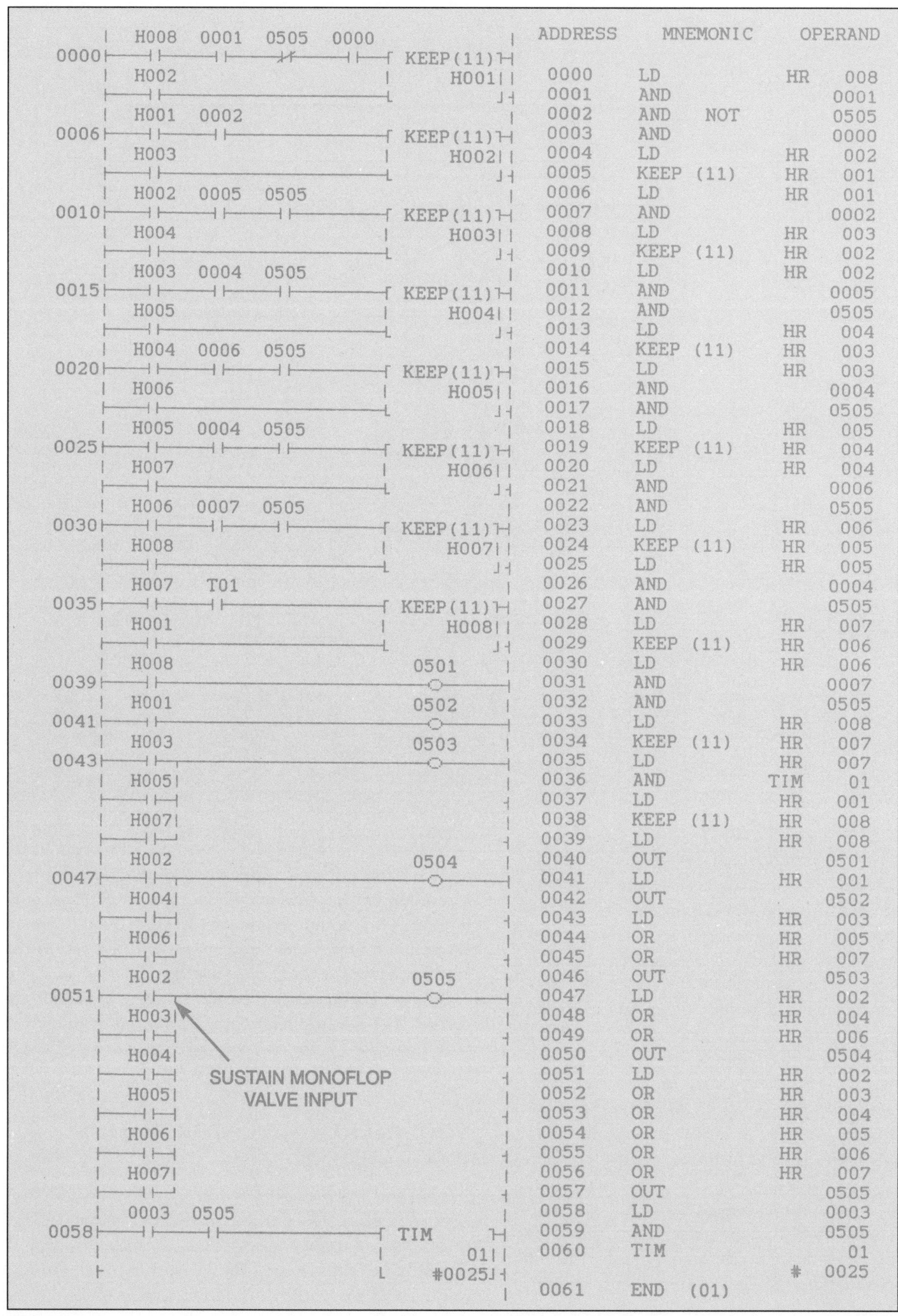

ADDRESS	MNEMONIC		OPERAND	
0000	LD		HR	008
0001	AND			0001
0002	AND	NOT		0505
0003	AND			0000
0004	LD		HR	002
0005	KEEP	(11)	HR	001
0006	LD		HR	001
0007	AND			0002
0008	LD		HR	003
0009	KEEP	(11)	HR	002
0010	LD		HR	002
0011	AND			0005
0012	AND			0505
0013	LD		HR	004
0014	KEEP	(11)	HR	003
0015	LD		HR	003
0016	AND			0004
0017	AND			0505
0018	LD		HR	005
0019	KEEP	(11)	HR	004
0020	LD		HR	004
0021	AND			0006
0022	AND			0505
0023	LD		HR	006
0024	KEEP	(11)	HR	005
0025	LD		HR	005
0026	AND			0004
0027	AND			0505
0028	LD		HR	007
0029	KEEP	(11)	HR	006
0030	LD		HR	006
0031	AND			0007
0032	AND			0505
0033	LD		HR	008
0034	KEEP	(11)	HR	007
0035	LD		HR	007
0036	AND		TIM	01
0037	LD		HR	001
0038	KEEP	(11)	HR	008
0039	LD		HR	008
0040	OUT			0501
0041	LD		HR	001
0042	OUT			0502
0043	LD		HR	003
0044	OR		HR	005
0045	OR		HR	007
0046	OUT			0503
0047	LD		HR	002
0048	OR		HR	004
0049	OR		HR	006
0050	OUT			0504
0051	LD		HR	002
0052	OR		HR	003
0053	OR		HR	004
0054	OR		HR	005
0055	OR		HR	006
0056	OR		HR	007
0057	OUT			0505
0058	LD			0003
0059	AND			0505
0060	TIM			01
			#	0025
0061	END	(01)		

Figure 8-15. Ladder diagram and mnemonic list for woodpecker circuit shown in figure 8-14. The timer in this ladder diagram is placed into the input of the step-counter.

```
     | H008  0001  0505  0000                  |
0000 |--| |---| |---|/|---| |--+--[ KEEP(11) ]-|
     | H002                    |     H001      |
     |--| |--------------------+               |
     | H001  0002                              |
0006 |--| |---| |--------------+--[ KEEP(11) ]-|
     | H003                    |     H002      |
     |--| |--------------------+               |
     | H002  0005  0505                        |
0010 |--| |---| |---| |--------+--[ KEEP(11) ]-|
     | H004                    |     H003      |
     |--| |--------------------+               |
     | H003  0004  0505                        |
0015 |--| |---| |---| |--------+--[ KEEP(11) ]-|
     | H005                    |     H004      |
     |--| |--------------------+               |
     | H004  0006  0505                        |
0020 |--| |---| |---| |--------+--[ KEEP(11) ]-|
     | H006                    |     H005      |
     |--| |--------------------+               |
     | H005  0004  0505                        |
0025 |--| |---| |---| |--------+--[ KEEP(11) ]-|
     | H007                    |     H006      |
     |--| |--------------------+               |
     | H006  0007  0505                        |
0030 |--| |---| |---| |--------+--[ KEEP(11) ]-|
     | H008                    |     H007      |
     |--| |--------------------+               |
     | H007  0003  0505                        |
0035 |--| |---| |---| |--------+--[ KEEP(11) ]-|
     | H001                    |     H008      |
     |--| |--------------------+               |
     |  T01                        0501        |
0040 |--| |------------------------( )---------|
     | H001                        0502        |
0042 |--| |------------------------( )---------|
     | H003                        0503        |
0044 |--| |-+----------------------( )---------|
     | H005|                                   |
     |--| |-+                                  |
     | H007|                                   |
     |--| |-+                                  |
     | H002                        0504        |
0048 |--| |-+----------------------( )---------|
     | H004|                                   |
     |--| |-+                                  |
     | H006|                                   |
     |--| |-+                                  |
     | H002   T01                  0505        |
0052 |--| |-+--|/|-----------------( )---------|
     | H003|                                   |
     |--| |-+                                  |
     | H004|                                   |
     |--| |-+                                  |
     | H005|                                   |
     |--| |-+                                  |
     | H006|                                   |
     |--| |-+                                  |
     | H007|                                   |
     |--| |-+                                  |
     | H008|                                   |
     |--| |-+                                  |
     | H008                                    |
0061 |--| |--------------------+--[ TIM      ]-|
     |                         |     01        |
     |                         +     #0025     |
```

ADDRESS	MNEMONIC		OPERAND	
0000	LD		HR	008
0001	AND			0001
0002	AND	NOT		0505
0003	AND			0000
0004	LD		HR	002
0005	KEEP	(11)	HR	001
0006	LD		HR	001
0007	AND			0002
0008	LD		HR	003
0009	KEEP	(11)	HR	002
0010	LD		HR	002
0011	AND			0005
0012	AND			0505
0013	LD		HR	004
0014	KEEP	(11)	HR	003
0015	LD		HR	003
0016	AND			0004
0017	AND			0505
0018	LD		HR	005
0019	KEEP	(11)	HR	004
0020	LD		HR	004
0021	AND			0006
0022	AND			0505
0023	LD		HR	006
0024	KEEP	(11)	HR	005
0025	LD		HR	005
0026	AND			0004
0027	AND			0505
0028	LD		HR	007
0029	KEEP	(11)	HR	006
0030	LD		HR	006
0031	AND			0007
0032	AND			0505
0033	LD		HR	008
0034	KEEP	(11)	HR	007
0035	LD		HR	007
0036	AND			0003
0037	AND			0505
0038	LD		HR	001
0039	KEEP	(11)	HR	008
0040	LD		TIM	01
0041	OUT			0501
0042	LD		HR	001
0043	OUT			0502
0044	LD		HR	003
0045	OR		HR	005
0046	OR		HR	007
0047	OUT			0503
0048	LD		HR	002
0049	OR		HR	004
0050	OR		HR	006
0051	OUT			0504
0052	LD		HR	002
0053	OR		HR	003
0054	OR		HR	004
0055	OR		HR	005
0056	OR		HR	006
0057	OR		HR	007
0058	OR		HR	008
0059	AND	NOT	TIM	01
0060	OUT			0505
0061	LD		HR	008
0062	TIM			01
			#	0025

Figure 8-16. Ladder diagram for woodpecker circuit with timer being placed into the output signal from the step-counter (compare with figure 8-15).

searching for food or to build its nest. The circuit in figures 8-14 and 8-15 is an apt representation of the woodpecker control concept found frequently in industrial automation. As the step-action diagram shows, actuator Ⓑ reciprocates from limit switch 003 to 005, back to 004, forward to 006, back again to 004 and then a final extension to limit switch 007, followed by a final complete retraction to limit switch 003. During these to and fro motions, air motor Ⓒ is always operating to drive the tapping head. Figure 8-16 is an alternative solution to the control depicted by the step-action diagram in figure 8-14. Here the timer is placed into the step-counter output signal for solenoid 505. Compare this circuit with the ladder diagram given in figure 5-15.

```
      |  H002                        |
 0051 |---| |----[ KEEP(11) ]        |   (A)
      |  H008   |        0505        |
      |---| |---+                    |
      |                              |

      |  H800        0505            |
 0051 |---| |---------( )------------|
      |  H002                        |
 0053 |---| |----[ KEEP(11) ]        |   (B)
      |  H008   |        H800        |
      |---| |---+                    |
      |                              |
```

Figure 8-17. Optional methods to sustain output signals for non-memory type directional control valves.

ALTERNATE SEQUENCING PROGRAM WITH PLC STEP-COUNTER

The principle of step-counter operation has amply been explained in this chapter (figures 8-06 and 8-07). But little has been said about the speed by which a step-counter advances through its stepping chain. As a rule, the tempo by which the step-counter advances depends entirely on the velocity of the actuators' moving from limit switch to limit switch, and on the speed of solenoid valve response to the signals given by the PLC. With these concepts in mind, we now explore a programming structure concept that has been successfully used many times to select and control selective multiple programs from a single step-counter chain. To explain this rather complex concept, it may be advantageous to make use of a simple control application. The task at hand is to develop a PLC con-

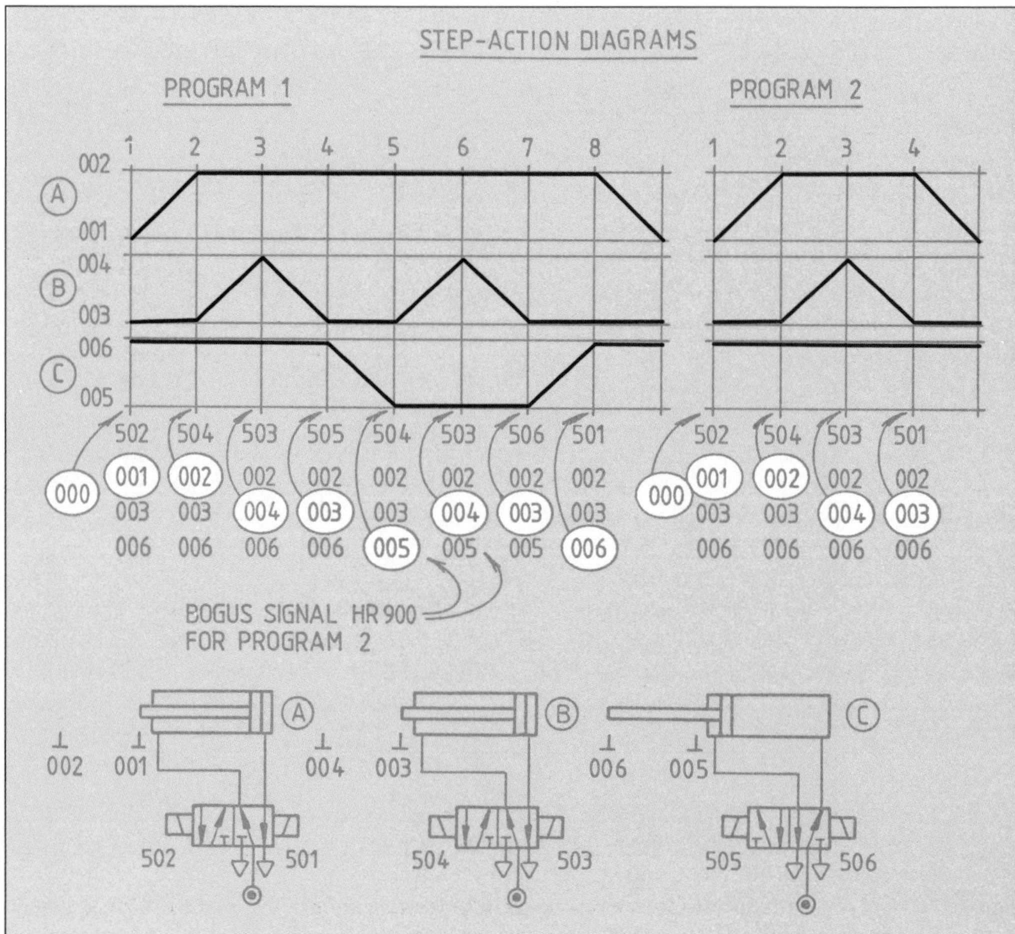

Figure 8-18. Step-action diagrams for alternate sequencing programs. Program 1 contains all the steps, whereas program 2 needs only steps 1, 2, 3 and 8 of program 1.

trol program using only one step-counter chain to cause a machine to cycle as per program 1 (see figure 8-18) if program 1 is selected or alternatively as per program 2, if program 2 has been selected.

A close investigation of the two step-action diagrams in figure 8-18 reveals that sequence steps 1, 2 and 3 are identical for both programs. Similarly, sequence step 8 of program 1 is identical to sequence step 4 of program 2. Furthermore, since actuator Ⓒ remains stationary throughout program 2, both programs have identical start positions and finishing positions for all their actuators. For this reason it seems logical that if the step-counter could be forced to skip sequence steps 4 to 7, had program 1 been selected, one could then accomplish this control task with only one step-counter of eight stepping modules. Step-counters can be forced to skip steps, but an easier solution is found by furnishing the step-counter with so-called "bogus" inputs. These bogus inputs are "OR" connected to the normal essential confirmation signals (see figures 8-18 to 8-20). They are active (logic 1) only when action of certain step-counter modules needs to be suppressed, as is the case in program 2 in this control example. Here, the actions signalled by sequence steps 4, 5 and 6 need to be suppressed or permanently cancelled.

BOGUS SIGNAL STEP RACING CONCEPT

Assuming program 2 has been selected, then right from the start of program 2, the bogus signal (HR 900) will be present on stepping modules 5 and 6. When the step-counter has advanced to step 4, and step 4 flip-flop prepares step 5 flip-flop through the "AND" function on its set input, flip-flop 5 will then switch at almost the same time as step 4, since the bogus signal is already present. The enabled set equation for flip-flop 5 under these conditions is:

HR 004 • HR 900 = SET FLIP-FLOP 5.

Thus step 5 flip-flop need not wait until action 4 is naturally completed. Hence, step 5 switches rapidly and at the same time resets step 4. This suppresses the electrical output signal 505 of step 4 in less than 0.000099 second! A PLC output signal 505 of such short duration has no significant effect on solenoid 505 of the valve that causes actuator Ⓒ to retract. Actuator Ⓒ therefore does not respond and does not retract at all at sequence step 4.

The same procedures and principles also affect the signalled extension of actuator Ⓑ at step 5. Rapid switching of step 6 flip-flop by the bogus signal causes step 5 also to reset prematurely. This suppresses output signal 504. Again, the output signal given at step 5 is so short that actuator Ⓑ does not respond and remains in the retracted position. At this point step 6 flip-flop prepares step 7 flip-flop at the "AND" function on its set line. With actuator Ⓑ not having moved at all at step 5, limit switch 003 has been actuated for the last three steps (time lines 4, 5 and 6). Thus the essential confirmation signal 003 contributing to the "AND" function in the set line of flip-flop 7 is also present. Hence flip-flop 7 switches. But again, no action takes place because actuator Ⓒ is already in the extended position (it never moved in program 2). By now the step-counter switching process has arrived at step 8, where it finds essential confirmation signal 006 already present (actuator Ⓒ remained on limit switch 006 throughout program 2). Flip-flop 8 is therefore caused to switch and retract actuator Ⓐ. The cycle for program 2 is now completed.

In summary, one may say that the bogus signals at flip-flop 5 and 6 and the resulting inaction of actuators Ⓑ and Ⓒ at sequence steps 4 and 5 has caused the step-counter to "race" through sequence steps 4, 5, 6 and 7. This racing causes the machine to perform as per program 2 and gives the observer the impression that the step-counter skips steps 4, 5, 6 and 7. But in reality the step-counter did not skip these four steps at all, rather it raced through them. Flip-flop HR 900 is used to store the selection of program 1 or 2 and its output HR 900 signal is used to produce the bogus inputs for flip-flops HR 005 and HR 006. Hence input 0014 selects program 2 (the short program), whereas input 0015 selects program 1 (figures 8-18 to 8-20).

> Note: A bogus signal given to a particular step-counter stepping module will always suppress the action signalled by the previous step-counter stepping module.

For the purpose of program selection indication, one could use contact HR 900 to cause output 507 to illuminate a program 2 lamp and $\overline{\text{HR 900}}$ contact to cause output 508 to illuminate a program 1 indication lamp (see also figure 8-29 for similar applications).

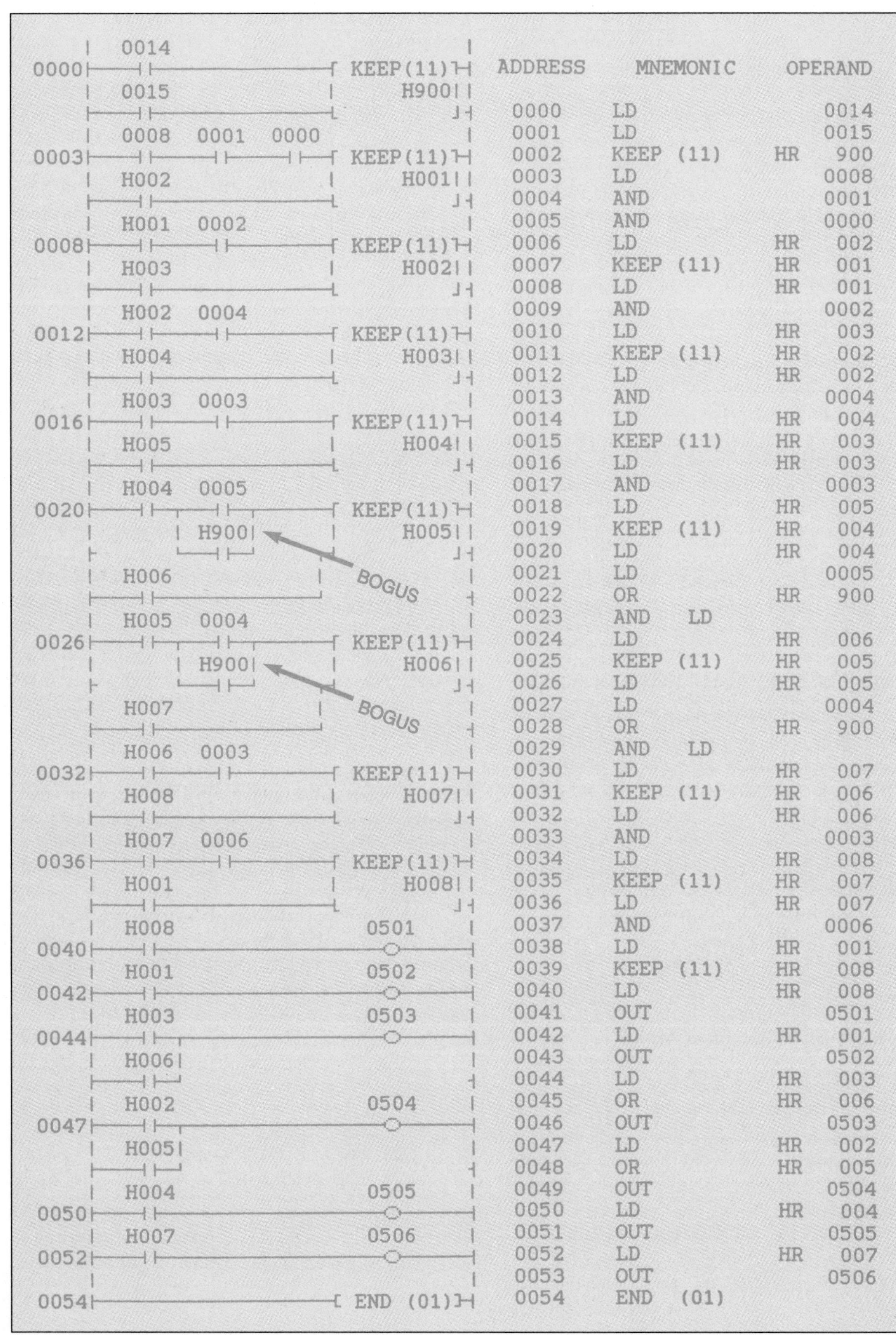

ADDRESS	MNEMONIC	OPERAND
0000	LD	0014
0001	LD	0015
0002	KEEP (11)	HR 900
0003	LD	0008
0004	AND	0001
0005	AND	0000
0006	LD	HR 002
0007	KEEP (11)	HR 001
0008	LD	HR 001
0009	AND	0002
0010	LD	HR 003
0011	KEEP (11)	HR 002
0012	LD	HR 002
0013	AND	0004
0014	LD	HR 004
0015	KEEP (11)	HR 003
0016	LD	HR 003
0017	AND	0003
0018	LD	HR 005
0019	KEEP (11)	HR 004
0020	LD	HR 004
0021	LD	0005
0022	OR	HR 900
0023	AND LD	
0024	LD	HR 006
0025	KEEP (11)	HR 005
0026	LD	HR 005
0027	LD	0004
0028	OR	HR 900
0029	AND LD	
0030	LD	HR 007
0031	KEEP (11)	HR 006
0032	LD	HR 006
0033	AND	0003
0034	LD	HR 008
0035	KEEP (11)	HR 007
0036	LD	HR 007
0037	AND	0006
0038	LD	HR 001
0039	KEEP (11)	HR 008
0040	LD	HR 008
0041	OUT	0501
0042	LD	HR 001
0043	OUT	0502
0044	LD	HR 003
0045	OR	HR 006
0046	OUT	0503
0047	LD	HR 002
0048	OR	HR 005
0049	OUT	0504
0050	LD	HR 004
0051	OUT	0505
0052	LD	HR 007
0053	OUT	0506
0054	END (01)	

Figure 8-19. Ladder diagram and mnemonic list for alternate sequencing program depicted in figure 8-18.

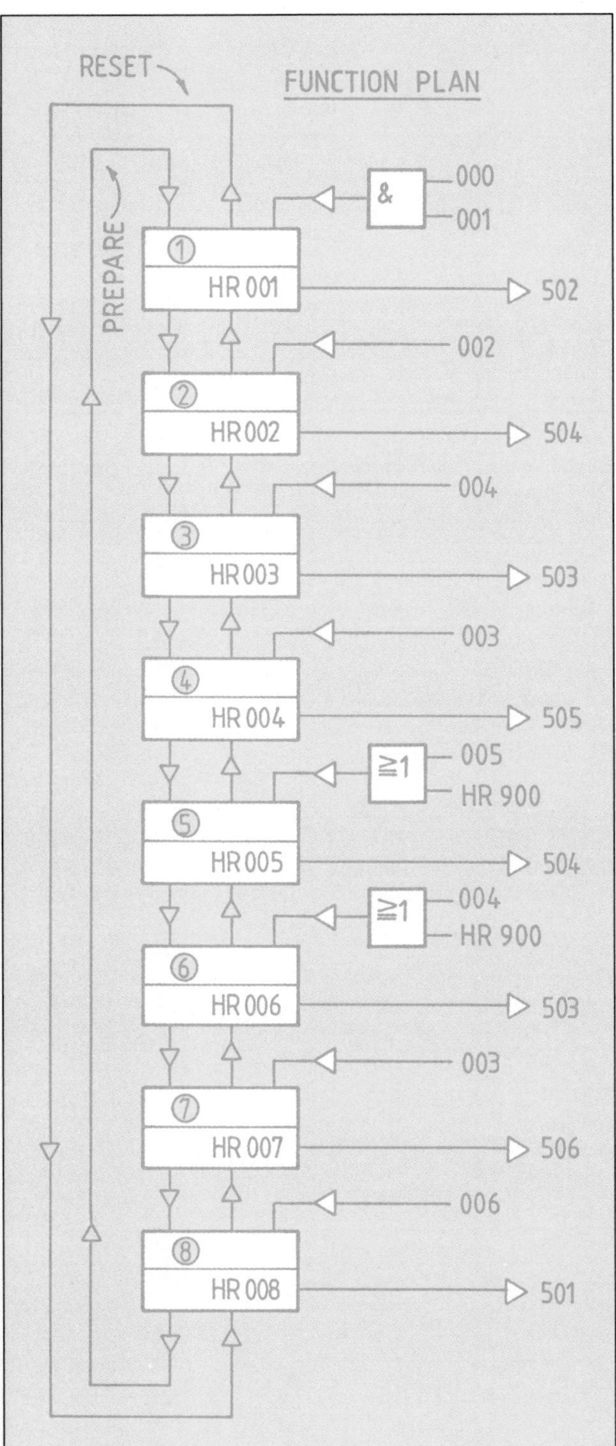

Figure 8-20. Function plan control illustration for the alternate sequencing program depicted in figures 8-18 and 8-19.

STEP-COUNTER WITH DEFINED SKIPPING

For alternative sequencing program design, real skipping of certain sequence steps can provide a simple solution if two machine programs such as those shown in figure 8-18 have steps in common. To illustrate this, program 1 and program 2 in figure 8-18 as explained before share the first three steps and the last step. For this reason, if skipping is used instead of the "bogus signal racing" concept explained before, and steps 4, 5, 6 and 7 are bypassed, one can cause the step-counter to signal machine actions as per program 2. Thus, one step-counter may be used to control both programs. It must be noted that actuator © never moves in program 2, but needs to extend and retract in program 1. Figures 8-21 and 8-22 show how such a step-skipping action is accomplished.

For the long program (program 1), program selector flip-flop HR 900 is reset. Hence the essential confirmation signal 003 at the "inhibition" gate on stepping module 4 is permitted to pass the gate and set flip-flop HR 004. This causes the step-counter to progressively switch through all of its eight stepping modules. (The inhibition function is explained in figures 3-05 and 3-09.)

If the short program is selected, HR 900 flip-flop is set. This causes the essential confirmation signal 003 at stepping module 4 to be barred from passing through the inhibition gate (contact $\overline{\text{HR 900}}$ is opened). Hence stepping module 4 cannot be set. At the same time the "AND" function consisting of inputs HR 003 and HR 900 and limit switch signals 003 and 006 is made at stepping module 8 (flip-flop 8). For the better understanding of this switching process see also the function plan given in figure 8-22. The result is that step 8 flip-flop switches and causes actuator A to retract (output signal 501). This skipping action would not, however, have been complete if flip-flop HR 003 had remained set. For this reason, flip-flop HR 008 has the task of resetting flip-flop 3 via the "OR" gate at the reset rung of flip-flop 3 in address 0015 (see ladder diagram in figure 8-21 and function plan circuit in figure 8-22). The Boolean logic equation for the step 8 flip-flop set signal (HR 008) reads:

$$[\text{HR } 007 + (\text{HR } 003 \bullet 003 \bullet \text{HR } 900)] \bullet 006 = \text{HR } 008$$

A comparison of the two ladder diagrams in figures 8-19 and 8-21 reveals that both output signal circuits are identical but their step-counters differ.

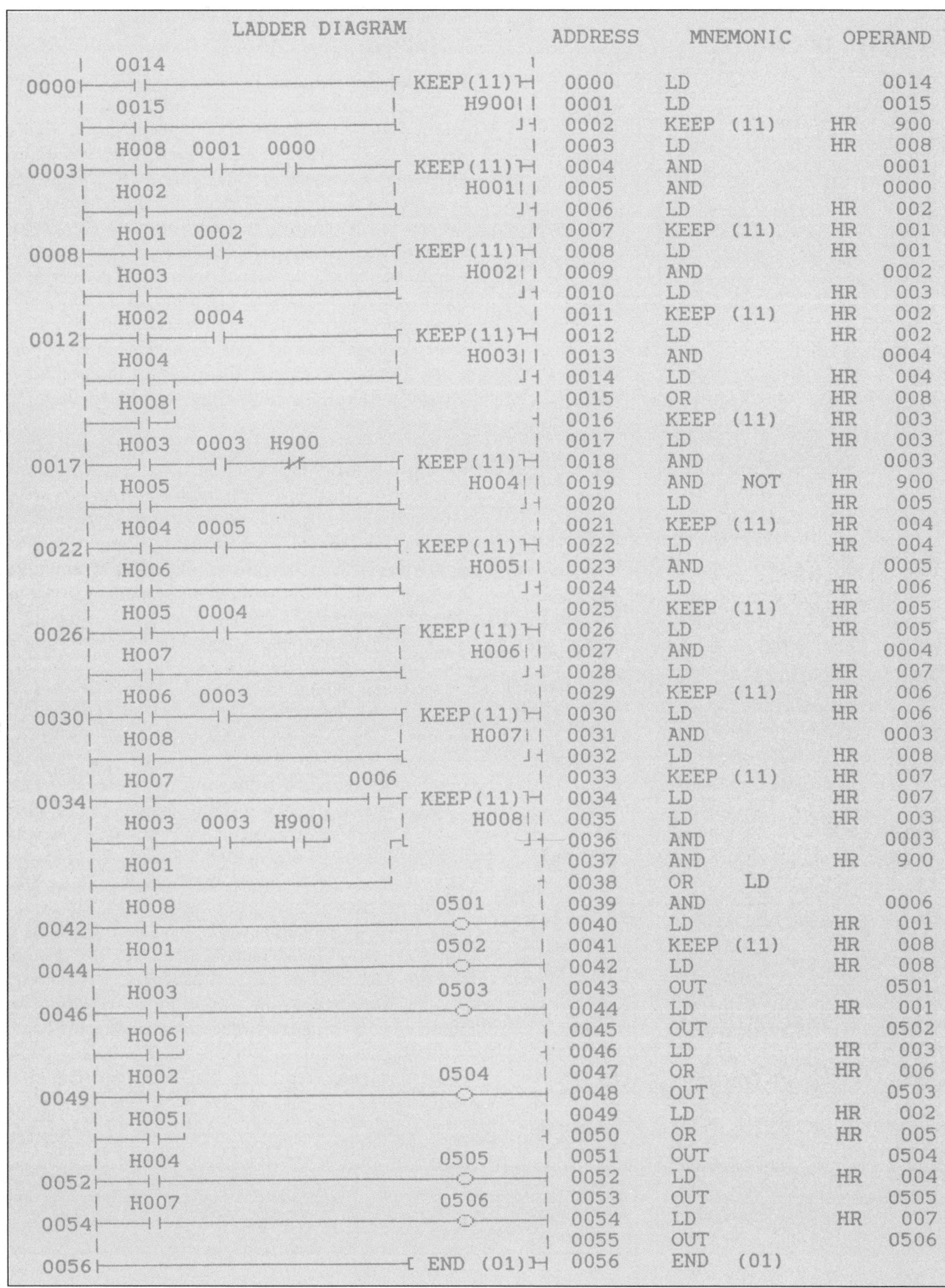

ADDRESS	MNEMONIC	OPERAND
0000	LD	0014
0001	LD	0015
0002	KEEP (11)	HR 900
0003	LD	HR 008
0004	AND	0001
0005	AND	0000
0006	LD	HR 002
0007	KEEP (11)	HR 001
0008	LD	HR 001
0009	AND	0002
0010	LD	HR 003
0011	KEEP (11)	HR 002
0012	LD	HR 002
0013	AND	0004
0014	LD	HR 004
0015	OR	HR 008
0016	KEEP (11)	HR 003
0017	LD	HR 003
0018	AND	0003
0019	AND NOT	HR 900
0020	LD	HR 005
0021	KEEP (11)	HR 004
0022	LD	HR 004
0023	AND	0005
0024	LD	HR 006
0025	KEEP (11)	HR 005
0026	LD	HR 005
0027	AND	0004
0028	LD	HR 007
0029	KEEP (11)	HR 006
0030	LD	HR 006
0031	AND	0003
0032	LD	HR 008
0033	KEEP (11)	HR 007
0034	LD	HR 007
0035	LD	HR 003
0036	AND	0003
0037	AND	HR 900
0038	OR LD	
0039	AND	0006
0040	LD	HR 001
0041	KEEP (11)	HR 008
0042	LD	HR 008
0043	OUT	0501
0044	LD	HR 001
0045	OUT	0502
0046	LD	HR 003
0047	OR	HR 006
0048	OUT	0503
0049	LD	HR 002
0050	OR	HR 005
0051	OUT	0504
0052	LD	HR 004
0053	OUT	0505
0054	LD	HR 007
0055	OUT	0506
0056	END (01)	

Figure 8-21. Ladder diagram and mnemonic list for alternate sequencing programs illustrated in the step-action diagrams of figure 8-18. In this control the step skipping concept is used.

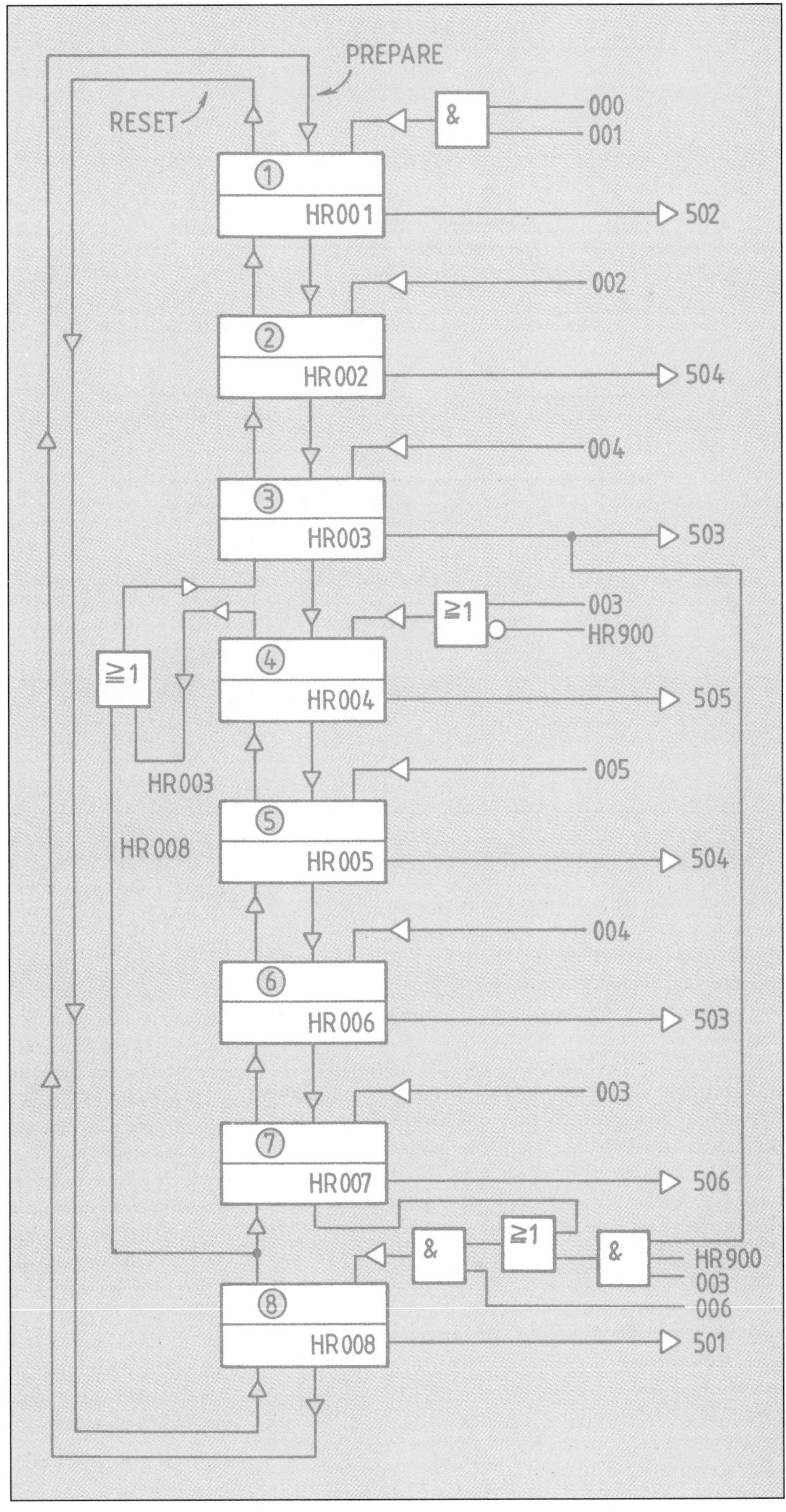

Figure 8-22. Function plan control illustration for the alternate sequencing program depicted in figures 8-18 and 8-21.

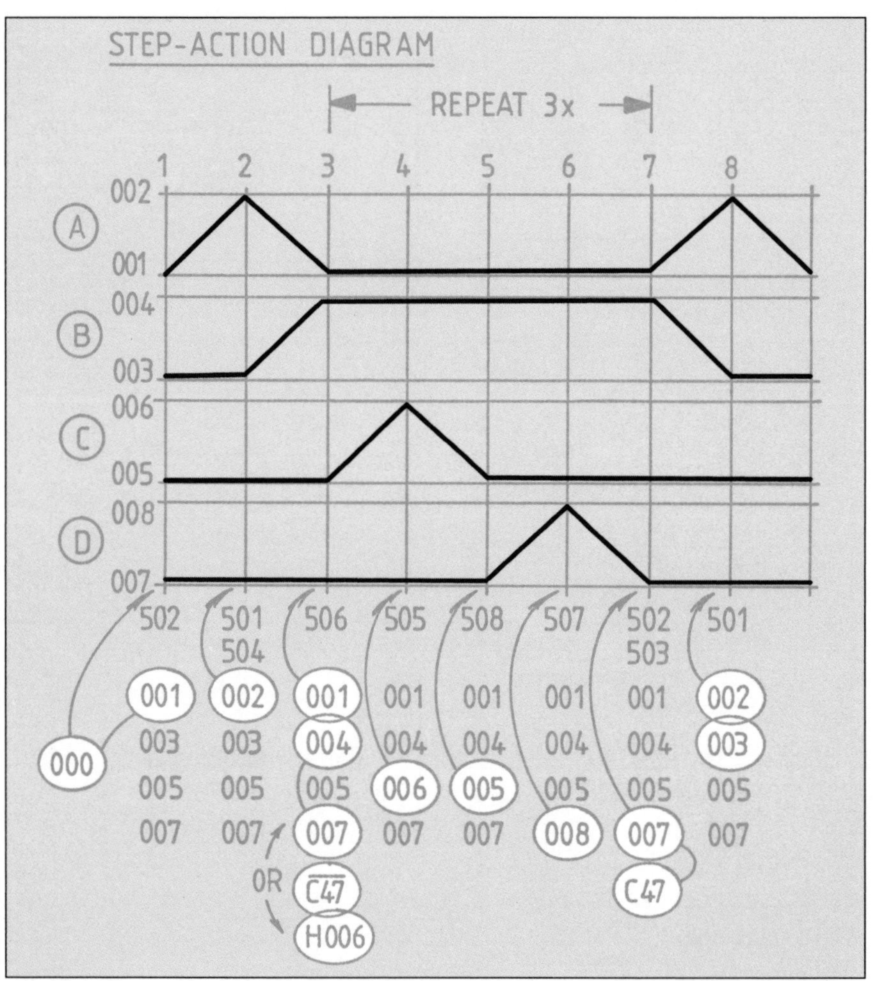

Figure 8-23. Step action diagram for step-counter with repeated steps. Steps 3, 4, 5 and 6 need to be repeated.

STEP-COUNTER FOR REPEATED STEPS

Step repetition for a cluster of steps in a sequential program is a versatile and frequently used method. Its application can simplify and shorten an otherwise long program considerably and thus save step-counter steps. It therefore also shortens the PLC program, which means fewer mnemonic instructions and a faster scan cycle. To illustrate this repeat concept, a sequence control program is given for a disk to be machined. The step-action diagram for this machining program is illustrated in figure 8-23 and the ladder diagram for this PLC program is given in figure 8-24. The disk is loaded automatically onto a revolving indexing table that at the same time acts as a machining platform for the disks. The disk is loaded by actuator Ⓐ and is clamped by actuator Ⓑ. The first machining function is performed with actuator Ⓒ. Indexing is accomplished with actuator Ⓓ. Repeated machining and indexing with actuators Ⓒ and Ⓓ must occur four times, thus machining a slot every 90°. After completion of the last indexing and machining motions, the workpiece is unclamped with actuator Ⓑ and automatically removed with actuator Ⓐ. The sequence is now completed.

STEP-COUNTER FOR ALTERNATE SEQUENCING PROGRAMS BUT DIFFERENT ACTUATOR START POSITIONS

PLC control programs of this nature are almost identical to the controls illustrated in figures 8-18 to 8-22. The starting position of actuator Ⓒ does, however, differ, and actuator Ⓒ does not move at all in program 2 (compare figure 8-18 with figure 8-25). To move actuator Ⓒ into its new starting position (retracted), the program selector circuit must be modified as is shown in figure 8-26. When program change is signalled, actuator Ⓒ will receive a correction command causing it to move into the opposite starting position. This correction command is then confirmed via the "OR"-connected "AND" functions at the set side of stepping module 1 (flip-flop 1) in the step-counter of figure 8-26. These "AND" functions consist of signals HR 900 and limit switch 005 for program 2 selection, and $\overline{\text{HR 900}}$ and limit switch 006 for program 1 selection. The Boolean switching equation for the set command of HR 001 (step-counter flip-flop 1) therefore reads:

$$\text{H008} \bullet 001 \bullet [(\text{H 900} \bullet 005) + (\overline{\text{H 900}} \bullet 006)] \bullet 000 = \text{H 001}$$

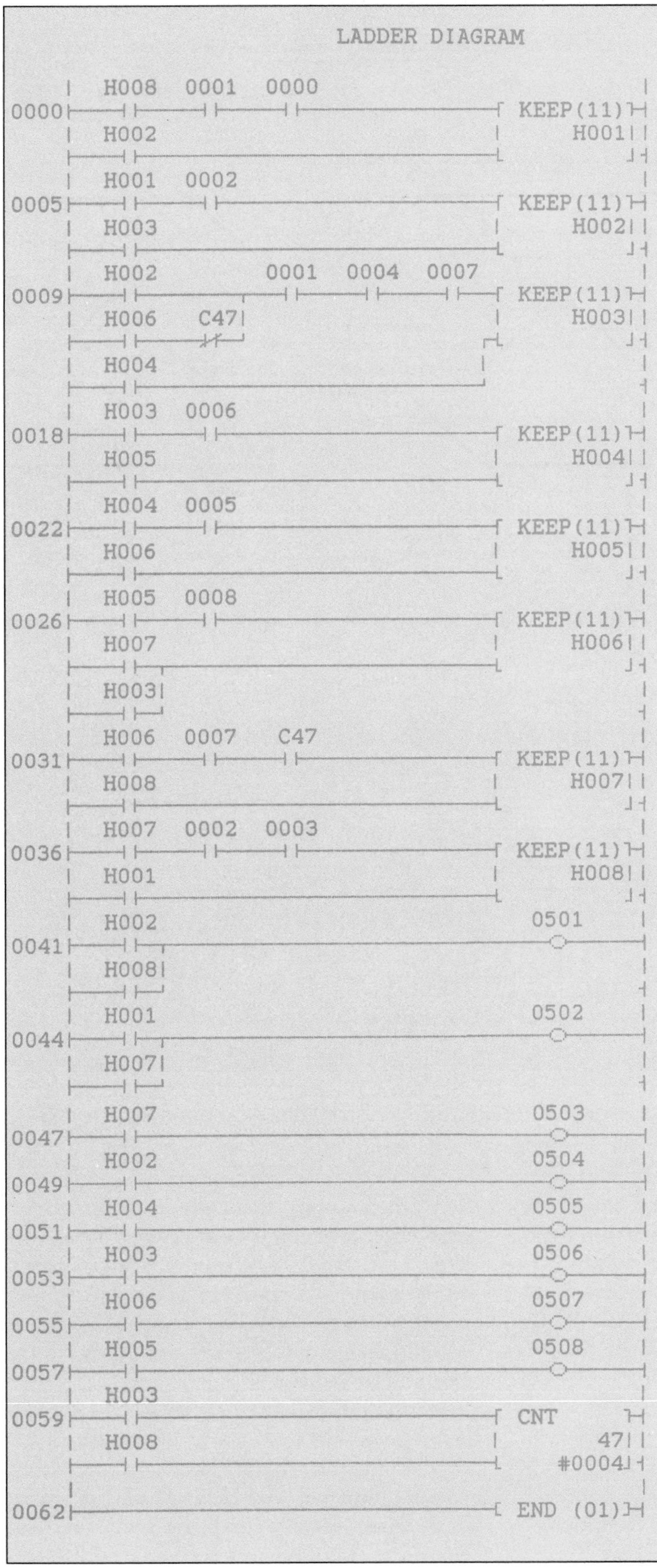

```
                         LADDER DIAGRAM

     | H008   0001   0000                         |
0000 |--| |---| |----| |-----------------[ KEEP(11) ]
     | H002                                   | H001 |
     |--| |-----------------------------------|      |
     | H001   0002                                   |
0005 |--| |---| |------------------------[ KEEP(11) ]
     | H003                                   | H002 |
     |--| |-----------------------------------|      |
     | H002          0001   0004   0007              |
0009 |--| |-------+--| |---| |---| |-----[ KEEP(11) ]
     | H006  C47 |                            | H003 |
     |--| |--|/|-|                          |-|      |
     | H004                                 |        |
     |--| |---------------------------------|        |
     | H003   0006                                   |
0018 |--| |---| |------------------------[ KEEP(11) ]
     | H005                                   | H004 |
     |--| |-----------------------------------|      |
     | H004   0005                                   |
0022 |--| |---| |------------------------[ KEEP(11) ]
     | H006                                   | H005 |
     |--| |-----------------------------------|      |
     | H005   0008                                   |
0026 |--| |---| |------------------------[ KEEP(11) ]
     | H007                                   | H006 |
     |--| |-+---------------------------------|      |
     | H003 |                                        |
     |--| |-|                                        |
     | H006   0007   C47                             |
0031 |--| |---| |----| |-----------------[ KEEP(11) ]
     | H008                                   | H007 |
     |--| |-----------------------------------|      |
     | H007   0002   0003                            |
0036 |--| |---| |----| |-----------------[ KEEP(11) ]
     | H001                                   | H008 |
     |--| |-----------------------------------|      |
     | H002                              0501        |
0041 |--| |-+---------------------------( )---------|
     | H008 |                                        |
     |--| |-|                                        |
     | H001                              0502        |
0044 |--| |-+---------------------------( )---------|
     | H007 |                                        |
     |--| |-|                                        |
     | H007                              0503        |
0047 |--| |-----------------------------( )---------|
     | H002                              0504        |
0049 |--| |-----------------------------( )---------|
     | H004                              0505        |
0051 |--| |-----------------------------( )---------|
     | H003                              0506        |
0053 |--| |-----------------------------( )---------|
     | H006                              0507        |
0055 |--| |-----------------------------( )---------|
     | H005                              0508        |
0057 |--| |-----------------------------( )---------|
     | H003                                          |
0059 |--| |-------------------------------[ CNT     ]
     | H008                                   |  47  |
     |--| |-----------------------------------| #0004|
     |                                               |
0062 |-----------------------------------[ END (01) ]
```

ADDRESS	MNEMONIC		OPERAND	
0000	LD		HR	008
0001	AND			0001
0002	AND			0000
0003	LD		HR	002
0004	KEEP	(11)	HR	001
0005	LD		HR	001
0006	AND			0002
0007	LD		HR	003
0008	KEEP	(11)	HR	002
0009	LD		HR	002
0010	LD		HR	006
0011	AND	NOT	CNT	47
0012	OR	LD		
0013	AND			0001
0014	AND			0004
0015	AND			0007
0016	LD		HR	004
0017	KEEP	(11)	HR	003
0018	LD		HR	003
0019	AND			0006
0020	LD		HR	005
0021	KEEP	(11)	HR	004
0022	LD		HR	004
0023	AND			0005
0024	LD		HR	006
0025	KEEP	(11)	HR	005
0026	LD		HR	005
0027	AND			0008
0028	LD		HR	007
0029	OR		HR	003
0030	KEEP	(11)	HR	006
0031	LD		HR	006
0032	AND			0007
0033	AND		CNT	47
0034	LD		HR	008
0035	KEEP	(11)	HR	007
0036	LD		HR	007
0037	AND			0002
0038	AND			0003
0039	LD		HR	001
0040	KEEP	(11)	HR	008
0041	LD		HR	002
0042	OR		HR	008
0043	OUT			0501
0044	LD		HR	001
0045	OR		HR	007
0046	OUT			0502
0047	LD		HR	007
0048	OUT			0503
0049	LD		HR	002
0050	OUT			0504
0051	LD		HR	004
0052	OUT			0505
0053	LD		HR	003
0054	OUT			0506
0055	LD		HR	006
0056	OUT			0507
0057	LD		HR	005
0058	OUT			0508
0059	LD		HR	003
0060	LD		HR	008
0061	CNT			47
			#	0004
0062	END	(01)		

Figure 8-24. Ladder diagram and mnemonic list for the program depicted in figure 8-23.

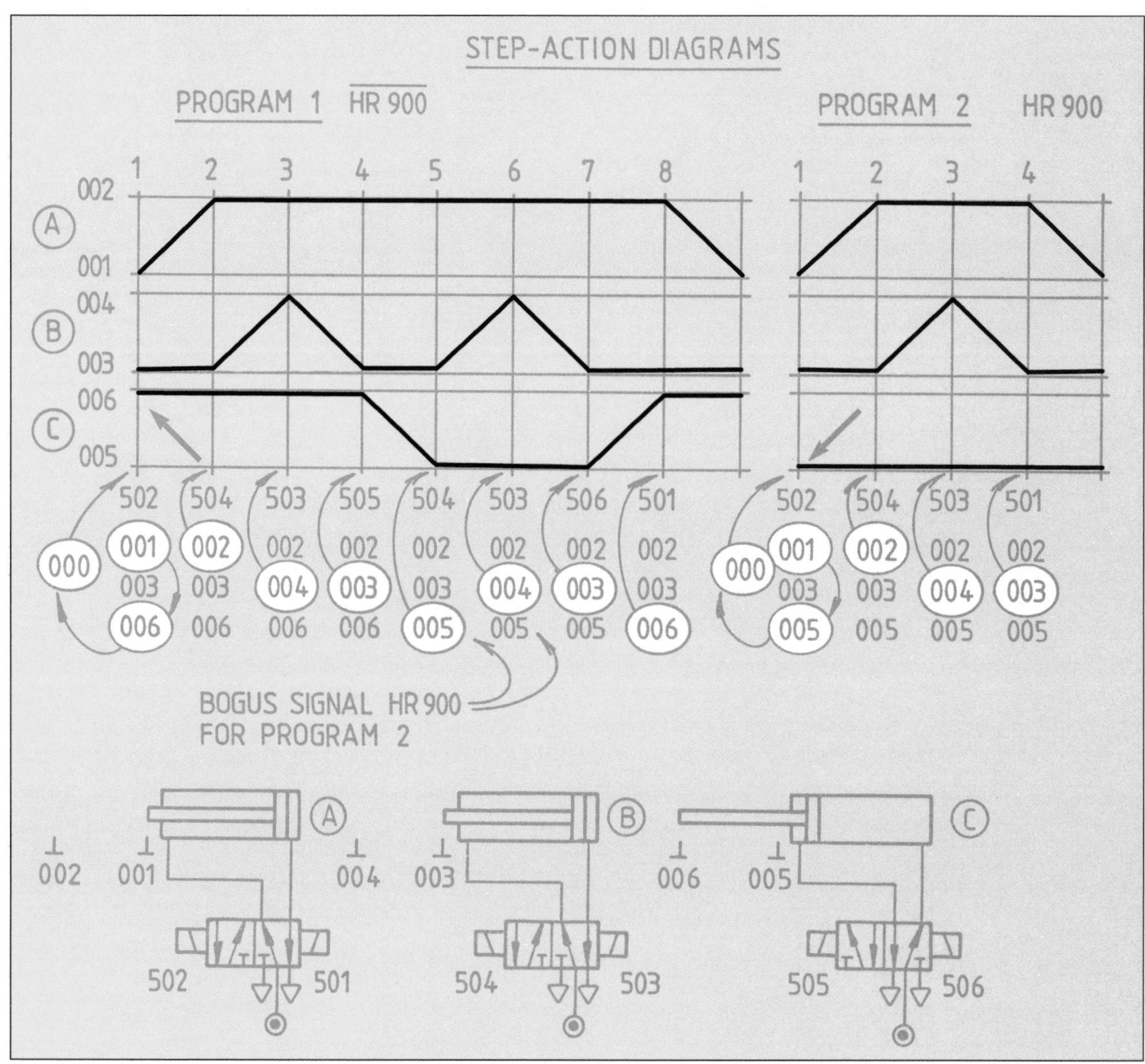

Figure 8-25. Step-action diagram for alternate programs with different starting positions for actuator Ⓒ.

To implement a program 2 correction movement for actuator Ⓒ before cycle start, solenoid 505 is given a correction signal from push-button switch 014, which is "OR" connected to the normal output command signal HR 004 (see addresses 0058 to 0060 in figure 8-20). Similarly, solenoid 506 receives a correction signal for program 1 operation from push-button switch 0015, which is "OR" connected to the normal output command signal HR 007 in address 0064. Sequence step 8 in the step-counter also needs a programming change as it is looking for essential confirmation signal 006 if program 1 is selected, or 005 if program 2 is selected. Hence the set command for flip-flop 8 in Boolean expression reads:

$$\text{HR 007} \bullet [(\text{HR 900} \bullet 005) + (\overline{\text{HR 900}} \bullet 006)] = \text{HR 008}$$

A programming change is also required for solenoid output 506, which must not cause actuator Ⓒ to extend at sequence step 7 if program 1 is selected. For this reason the equation section HR 007 • HR 900 • 005 pertaining to the set command of step 8 flip-flop resets step 7 flip-flop prematurely and thus suppresses the extension command of actuator Ⓒ.

HR 008 (flip-flop 8) signal is "AND" connected to the program selector flip-flop set and reset logic lines. A program change can therefore be signalled only if the cycle is completed, and the machine can be started only if the correction is accomplished and the last sequence step in the two programs is completed. For more elaborate program selector circuits, see Chapter 9.

LADDER DIAGRAM

```
     | 0014   H008                                 |
0000 |--| |----| |-----------------------------[ KEEP(11) ]-|
     | 0015   H008                                 |  H900  |
     |--| |----| |-----------------------------L          ]-|
     | H008   0001    H900    0005    0000         |
0005 |--| |----| |--+--| |----| |--+--| |------[ KEEP(11) ]-|
     |              | H900    0006|              |  H001  |
     |              +--|/|----| |-+             -+          ]-|
     | H002                                      |
     |--| |--------------------------------------+          -|
     | H001   0002                                 |
0016 |--| |----| |-----------------------------[ KEEP(11) ]-|
     | H003                                        |  H002  |
     |--| |----------------------------------------L          ]-|
     | H002   0004                                 |
0020 |--| |----| |-----------------------------[ KEEP(11) ]-|
     | H004                                        |  H003  |
     |--| |----------------------------------------L          ]-|
     | H003   0003                                 |
0024 |--| |----| |-----------------------------[ KEEP(11) ]-|
     | H005                                        |  H004  |
     |--| |----------------------------------------L          ]-|
     | H004   0005                                 |
0028 |--| |----| |-----------------------------[ KEEP(11) ]-|
     | H006                                        |  H005  |
     |--| |----------------------------------------L          ]-|
     | H005   0004                                 |
0032 |--| |--+--| |--+---------------------------[ KEEP(11) ]-|
     |        | H900|                              |  H006  |
     |        +--| |-+                            -+          ]-|
     | H007                                      |
     |--| |--------------------------------------+          -|
     | H006   0003                                 |
0038 |--| |----| |-----------------------------[ KEEP(11) ]-|
     | H008                                        |  H007  |
     |--| |----------------------------------------L          ]-|
     | H007   H900    0005                         |
0042 |--| |--+--| |----| |--+-------------------[ KEEP(11) ]-|
     |        | H900    0006|                    |  H008  |
     |        +--|/|----| |-+                   -+          ]-|
     | H001                                      |
     |--| |--------------------------------------+          -|
     | H008                                  0501           |
0051 |--| |-------------------------------------( )----------|
     | H001                                  0502           |
0053 |--| |-------------------------------------( )----------|
     | H003                                  0503           |
0055 |--| |--+----------------------------------( )----------|
     | H006|                                                |
     |--| |-+                                               -|
     | H002                                  0504           |
0058 |--| |--+----------------------------------( )----------|
     | H005|                                                |
     |--| |-+                                               -|
     | H004                                  0505           |
0061 |--| |--+----------------------------------( )----------|
     | 0014|                                                |
     |--| |-+                                               -|
     | H007                                  0506           |
0064 |--| |--+----------------------------------( )----------|
     | 0015|                                                |
     |--| |-+                                               -|
     |                                                      |
0067 |-----------------------------------------[ END (01) ]-|
                                                        END
```

ADDRESS	MNEMONIC			OPERAND
0000	LD			0014
0001	AND		HR	008
0002	LD			0015
0003	AND		HR	008
0004	KEEP	(11)	HR	900
0005	LD		HR	008
0006	AND			0001
0007	LD		HR	900
0008	AND			0005
0009	LD	NOT	HR	900
0010	AND			0006
0011	OR	LD		
0012	AND	LD		
0013	AND			0000
0014	LD		HR	002
0015	KEEP	(11)	HR	001
0016	LD		HR	001
0017	AND			0002
0018	LD		HR	003
0019	KEEP	(11)	HR	002
0020	LD		HR	002
0021	AND			0004
0022	LD		HR	004
0023	KEEP	(11)	HR	003
0024	LD		HR	003
0025	AND			0003
0026	LD		HR	005
0027	KEEP	(11)	HR	004
0028	LD		HR	004
0029	AND			0005
0030	LD		HR	006
0031	KEEP	(11)	HR	005
0032	LD		HR	005
0033	LD			0004
0034	OR		HR	900
0035	AND	LD		
0036	LD		HR	007
0037	KEEP	(11)	HR	006
0038	LD		HR	006
0039	AND			0003
0040	LD		HR	008
0041	KEEP	(11)	HR	007
0042	LD		HR	007
0043	LD		HR	900
0044	AND			0005
0045	LD	NOT	HR	900
0046	AND			0006
0047	OR	LD		
0048	AND	LD		
0049	LD		HR	001
0050	KEEP	(11)	HR	008
0051	LD		HR	008
0052	OUT			0501
0053	LD		HR	001
0054	OUT			0502
0055	LD		HR	003
0056	OR		HR	006
0057	OUT			0503
0058	LD		HR	002
0059	OR		HR	005
0060	OUT			0504
0061	LD		HR	004
0062	OR			0014
0063	OUT			0505
0064	LD		HR	007
0065	OR			0015
0066	OUT			0506

Figure 8-26. Partial ladder diagram and mnemonic list for control depicted in figure 8-25.

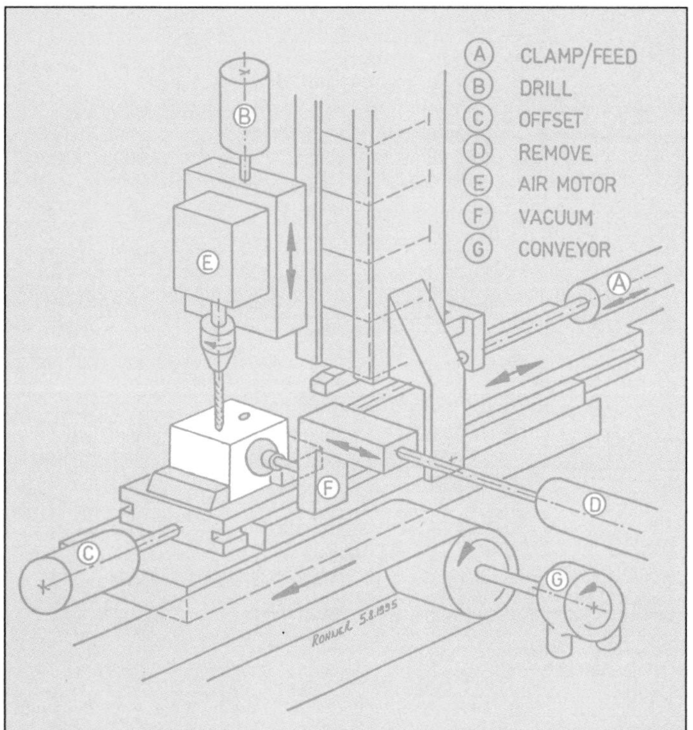

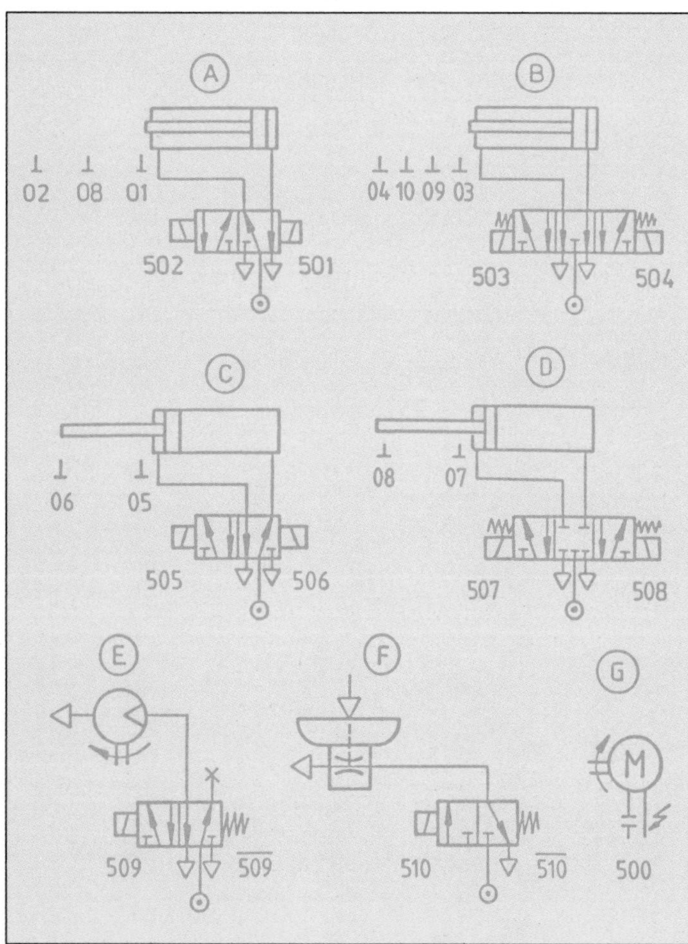

Figure 8-27. Isometric conceptual sketch for a machine for drilling dowel holes as used in control problem 1.

CONTROL PROBLEM 1

A machine for drilling dowel holes, as illustrated by the isometric conceptual sketch in figure 8-27, needs to be controlled by a PLC. Functions of the various fluid power actuators are explained in the legend in this illustration. The fluid power actuators and their initial starting position are shown in the electropneumatic circuit of the same illustration. The machine sequencing actions are depicted in the step-action diagram of figure 8-28. This PLC program has to be designed to drill, with the same machine, components that need one or two dowel holes. To accomplish this, the bogus signal step racing concept as explained with figures 8-18, 8-19 and 8-20 is used once again. For program 2, selected with input 014, flip-flop HR 900 is logic 1, and thus produces the bogus signal to cause step racing for sequence steps 6, 7, 8 and 9. Step 10 needs no racing action, as actuator Ⓑ has not moved since the step racing started and it is therefore still in retracted position.

If program 1 is selected, the machine drills two dowel holes and the step-counter must therefore not omit any sequencing actions.

The ladder diagram for the control problem solution is given in figure 8-29. Drilled components are removed by actuator Ⓓ with the attached vacuum suction cup. Once the component is above the conveyor, the vacuum is stopped and the component falls onto the moving conveyor. The conveyor driven by motor Ⓖ then removes the machined parts.

The motor starts to drive the conveyor Ⓖ at the start of sequence step 14 (HR 014). Since step-counter flip-flop 15 resets step-counter flip-flop 14, both of these contacts are "OR" connected to keep the motor running (see address 091 in figure 8-29). The motor is automatically stopped by timer 01 after a four-second running time. The timer also needs an "OR"-connected enabling signal for the same reasons as the motor output 500 needs one.

The explanations given in the paragraph "PLC Step-Counter Programming for Non-Memory Type Directional Control Valves" are also applicable to the solenoid output signals 509 and 510 in the circuit given for this control problem (see figure 8-29).

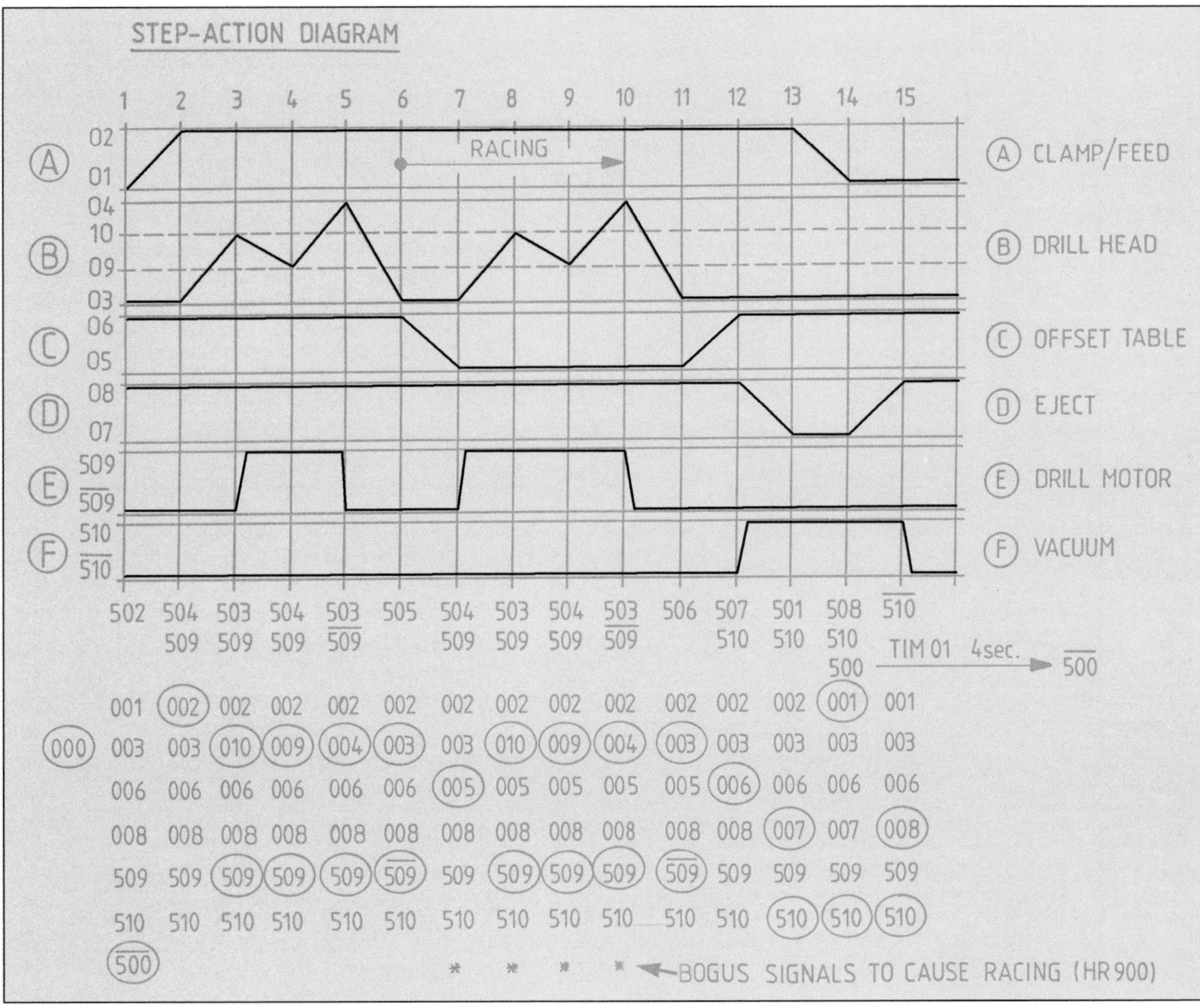

Figure 8-28. Step-action diagram for a machine for drilling dowel holes.

INPUT/OUTPUT ASSIGNMENT LIST

PLC inputs:

000 = start signal
001 = actuator Ⓐ retracted
002 = actuator Ⓐ extended
003 = actuator Ⓑ retracted
004 = actuator Ⓑ extended
005 = actuator Ⓒ retracted
006 = actuator Ⓒ extended
007 = actuator Ⓓ retracted
008 = actuator Ⓓ extended
009 = actuator Ⓑ 1/3 extended
010 = actuator Ⓑ 2/3 extended
014 = select short program
015 = select long program

PLC outputs:

500 = conveyor motor run signal
501 = actuator Ⓐ retract solenoid
502 = actuator Ⓐ extend solenoid
503 = actuator Ⓑ retract solenoid
504 = actuator Ⓑ extend solenoid
505 = actuator Ⓒ retract solenoid
506 = actuator Ⓒ extend solenoid
507 = actuator Ⓓ retract solenoid
508 = actuator Ⓓ extend solenoid
509 = drill motor run solenoid Ⓔ
510 = vacuum venturi air supply solenoid Ⓕ
511 = short program selected lamp
600 = long program selected lamp

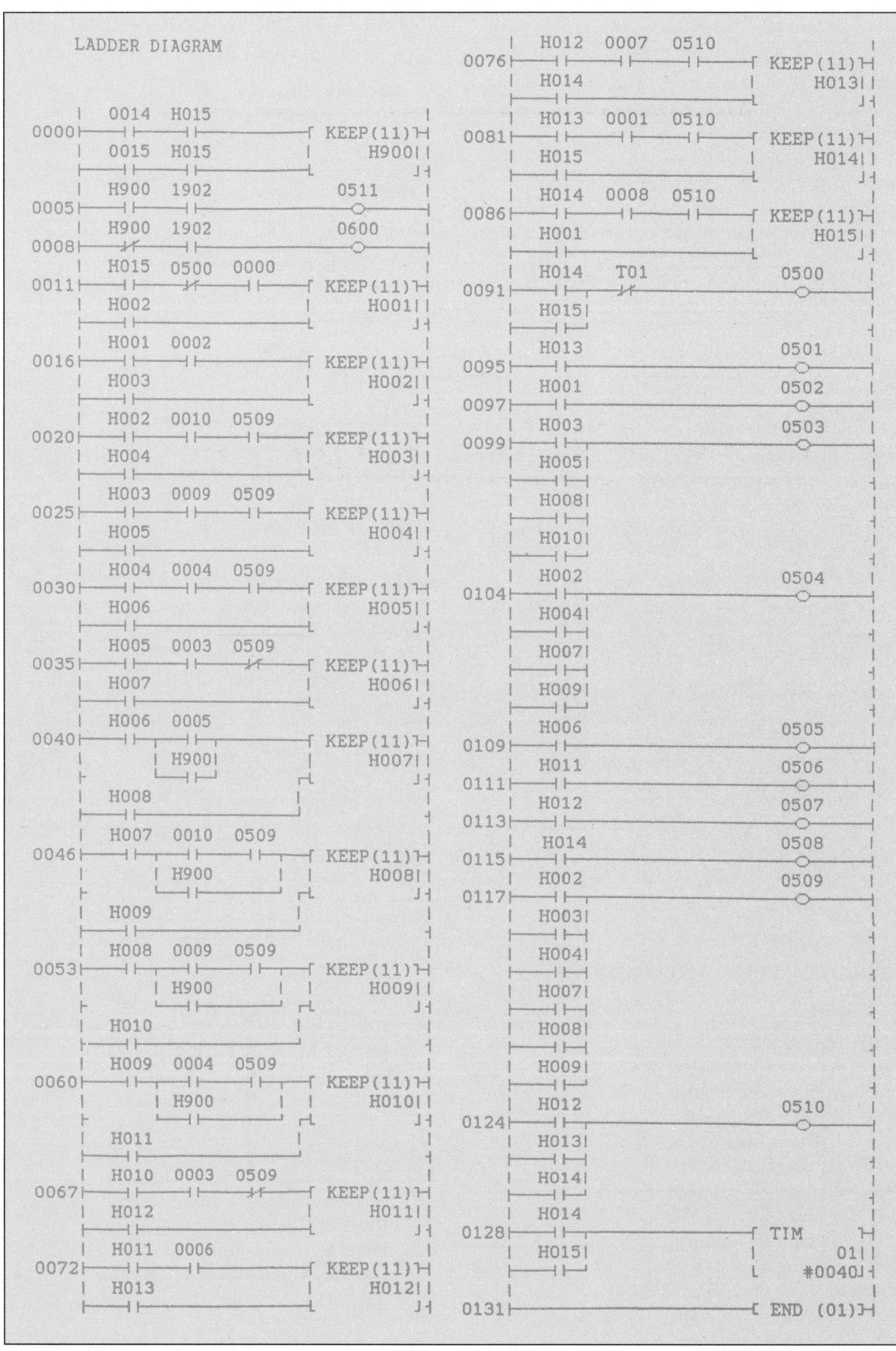

Figure 8-29. Ladder diagram for control problem 1.

CONTROL PROBLEM 2

A fully pneumatically powered and controlled machine needs to be refurbished. Its program has to be changed so that the second time-delay span for sequence step 7 may be selected to achieve a 4-second or a 9-second delay. By actuating push-button input 014, the machine operator can select the 4-second delay, or by actuating push-button input 015, the 9-second delay. The delay span 1 for sequence step 5 remains at two seconds.

The machine is also equipped with a simple cycle selection module to select automatic cycling with push-button input 010 and manual cycling with push-button input 011. Lamps 510 and 511 indicate the selection made. Chapter 9 gives a detailed operation explanation for such a cycle selection module.

HR 800 flip-flop is used to record the timer selection. For the short time-delay span, HR 800 is set. For the long time-delay span, HR 800 is reset (see addresses 54 to 63 in figure 8-31).

INPUT/OUTPUT ASSIGNMENT LIST

PLC inputs:

000 = start signal
001 = actuator Ⓐ retracted
002 = actuator Ⓐ extended
003 = actuator Ⓑ retracted
004 = actuator Ⓑ extended
010 = automatic cycling selection
011 = manual cycling selection
014 = short timer selection
015 = long timer selection

PLC outputs:

501 = actuator Ⓐ retract solenoid
502 = actuator Ⓐ extend solenoid
503 = actuator Ⓑ retract solenoid
504 = actuator Ⓑ extend solenoid
510 = automatic selection lamp
511 = manual cycling selection lamp

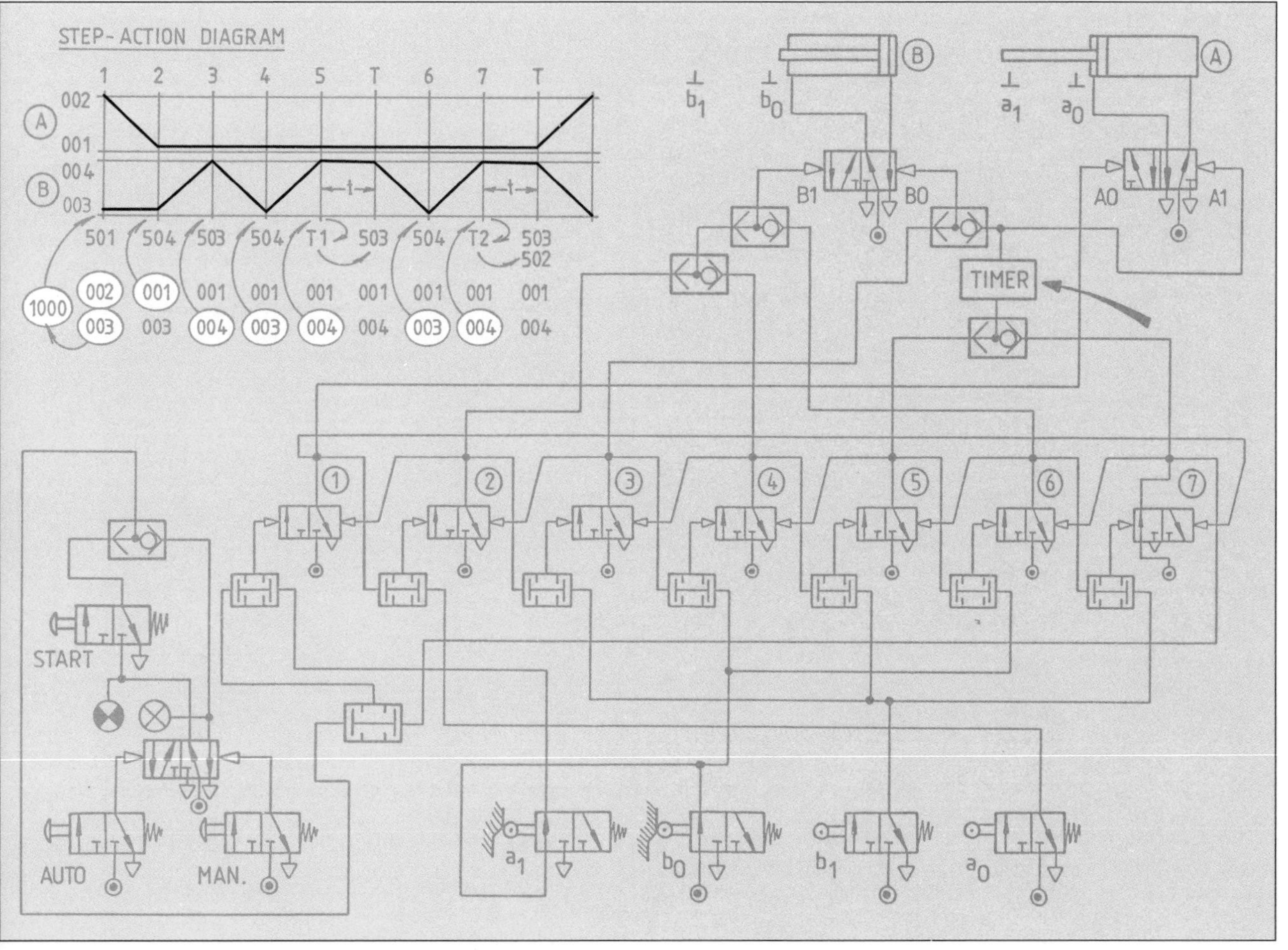

Figure 8-30. Step-action diagram and existing pneumatic control circuit for control problem 2.

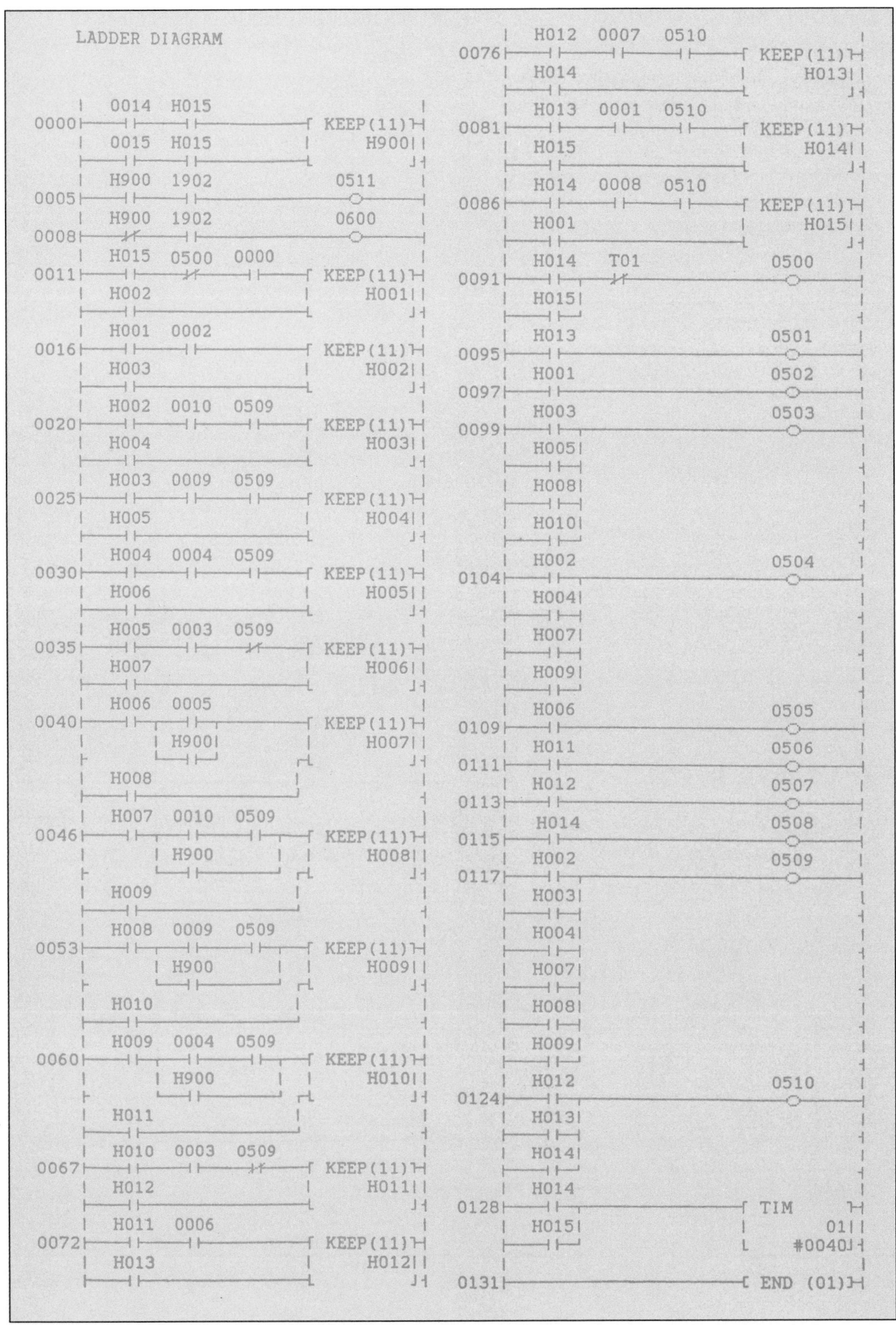

Figure 8-31. Ladder diagram and mnemonic list for control problem 2.

SEQUENTIAL PLC PROGRAMMING SUMMARY

Sequential control systems switch machine functions in precise chronological order. If well designed, they ensure that machine sequence steps are permitted to happen only when the previous step is completed and confirmed by a feedback signal from the machine. An exellent design algorithm for such controls is the well known step-counter design method, using an RS flip-flop for every sequence step. Step-action diagrams are an imperative tool for designing sequential controls. A variety of methods (e.g. bogus signal racing, step skipping) are used to cause such step-counters to skip steps, operate in parallel or repeat designated sequence sections.

9 FRINGE CONDITION MODULE PROGRAMMING

Most industrial machines that use PLCs as the heart of their control consist of some form of sequential program, as explained and illustrated in Chapter 8. However, most PLC-controlled machines are also equipped with one or more "fringe condition modules". These modules are usually attached on the fringe of the main program, hence their name. They are sometimes also called "routines".

Being of modular structure, these routines or fringe condition modules may be likened to a standardised and pretested subcircuit or circuit building block. As the given control problems in this chapter will illustrate, such fringe condition modules are absolutely essential for the proper and safe functioning of modern, automated machines.

The most commonly encountered fringe conditions in building block format are:

- emergency stop module
- machine interrupt module
- cycle selection module
- stepping/run selection module
- program selection module
- signal shunting module
- two-hand safety-start control module
- automatic actuator reversal module.

The last two of these modules have already been presented and explained in Chapter 7 as compound building blocks. One may therefore argue that fringe condition modules should also be regarded as compound building blocks, and this is exactly what they are. However, since they are often not clearly embedded in and are not an integral part of the actual sequential control circuit, and are mostly used only in conjunction with sequential step-counter controls, it may be best to present them as being distinct.

All these modules except the automatic actuator reversal module are directly controlled and influenced by a machine operator who makes decisions and presses push-buttons on the machine's control panel. These modules, therefore, are frequently equipped with control lamps, giving visible indication of the selection having been made by the operator. To illustrate: The cycle selection module, which causes a machine to run continuously in automatic mode or for a single machine cycle in manual mode by start button command for each new cycle, is usually equipped with lamps that indicate automatic or manual cycling selection in action (see figure 9-04).

Note! All fringe condition modules presented in this chapter have been rigorously tested and are applied in a number of industry controls. They may therefore be readily used as presented in this text. It is, however, advisable to check with local industrial safety authorities for their approval of such modules, since control circuit specifications vary from state to state and from country to country.

INTERRUPTING A MACHINE SEQUENCE

Most automated and PLC-controlled machines operate in exact sequential order, which means that their actuators extend or retract in accordance with precisely pre-programmed sequence steps (figures 8-08, 8-14 and 8-23). When mishaps occur, this sequence needs to be instantly, but sometimes only temporarily, interrupted, or the start of a new cycle needs to be prevented if automatic sequencing was selected. Such mishaps can include a faulty or damaged machining tool, a misaligned or altogether missing workpiece, an empty parts storage magazine, a machine structure fault, a sudden drop of fluid power supply pressure, insufficient temperature in a heat-dependent manufacturing process, or an endangered or injured machine operator.

The nature of an interruption varies with the anticipated degree of danger and the inherent safety equipment already built into the machine (safety guards, remote control, intermittent operation mode, unskilled operators on the machine during various shifts etc.). Such factors, therefore, need to be carefully considered and a decision needs to be made at the design stage of the control program on how such an interruption of machine cycling has to occur. The correct and logical nature of this interruption must prevent harm to operators or damage to machinery (figure 9-01). Such decisions may be made from the following three machine sequencing interrupt modes:

1. Stop all actuators instantly by using piston rod brakes where necessary. Piston rod brakes are available for hydraulic and pneumatic linear actuators. When actuated they will stop an actuator instantly!

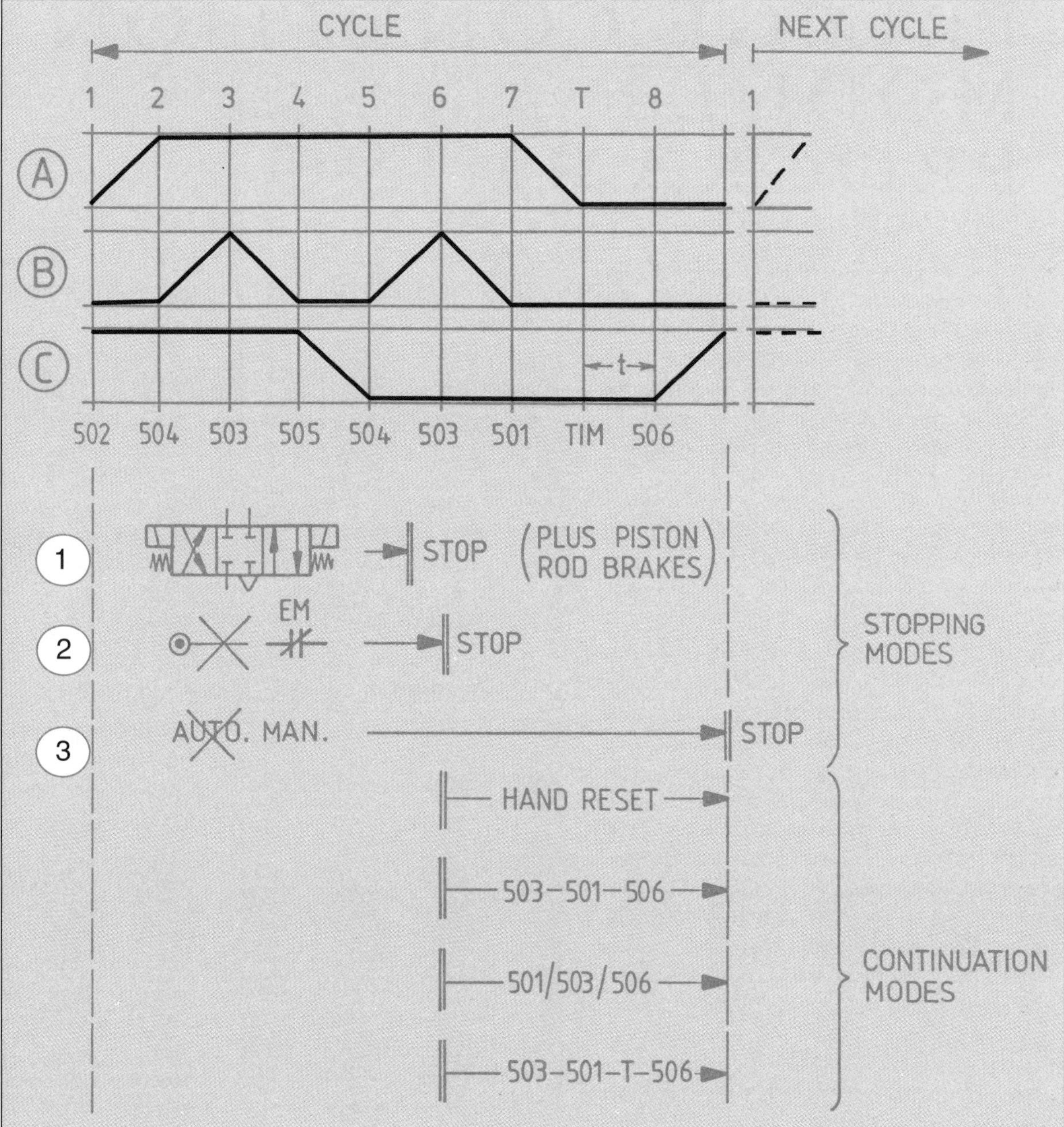

Figure 9-01. Emergency stop modes and continuation modes to interrupt and sequence a machine in a mishap.

2. Stop all actuators at the end of the sequence step in action at the time of the emergency being signalled (figure 9-01).
3. Stop the machine cycle at the end of the final sequence step (if the machine cycles automatically and does not need a start command from an operator for each cycle).

Once the machine cycle is interrupted and all actuators have stopped, the circuit designer has several options to choose from for further sequencing or machine actions. It is therefore always the circuit designer's responsibility to decide what such actions have to be. The correct choice is crucial as it can prevent serious harm to the machine operator or avoid costly damage to workpieces, tooling and machinery.

To illustrate: In an automated drilling machine, when emergency stopping or automatic interrupt followed by a machine clearing cycle is signalled, one needs to stop the drilling head from further advancing and retract it completely before the workpiece is permitted to be unclamped and the automatic workpiece ejectors are finally permitted to clear the machine. It would be disastrous if all these movements happened simultaneously, or if unclamping happened before the drill bit was retracted and disengaged. Such a continuation mode is shown in figure 9-01 and a complete control program for a similar machine is given in control problem 1 in this chapter. However, in another machine, all actuators may quite safely act in unison and move into the position required to start a new machining cycle once the emergency or interruption is cleared. In another machine application the machine cycle may be brought to a halt with either stopping mode number 1 or stopping mode number 2 (figure 9-01). When the emergency is cleared, the sequence can be permitted to continue to the end of the normal machining cycle. Such machine applications are quite common. They may be regarded as classic continuation modes for machine sequence behaviour after machine interrupt or emergency stop has been signalled.

BASIC SEQUENCING INTERRUPT MODULE

With a basic sequencing interrupt module, a temporary stop signal is stored in a holding relay (flip-flop). In this case it is essential to use a memory-retentive relay so that if power fails during machine interrupt, the condition existing before power failure is retained when power is restored. It also guarantees that the emergency condition signalled is maintained until a reset signal is given (figure 9-02). For safety reasons, it becomes advisable to place the reset switch in the control cabinet and use a key-operated reset switch. Only authorised personnel should have access to the reset switch key. Ideally, both logic states of this module's flip-flop should be displayed with a red emergency interrupt lamp and a green reset lamp, as is illustrated in figure 9-02. Control circuits showing application, function and integration of this module are presented in figures 9-12 section 1, 9-17 and 11-18. For interaction of this module with other fringe condition modules see figures 9-06, 9-12 section 1, 9-21 and 11-18.

COMPLEX SEQUENCING INTERRUPT MODULE

As shown with the basic sequencing interrupt module, the machine's sequence is permitted to continue as soon as the reset push-button switch is actuated. In some machines, this behaviour may prove to be confusing or dangerous. To eliminate this, the complex sequencing interrupt module contains an additional memory (flip-flop) that inhibits machine restart unless "reset" has been actuated and the "START" push-button is also actuated in "AND" connection. The "START" signal may come from a single start switch or from a cycle selection module (compare figures 9-02 and 9-03).

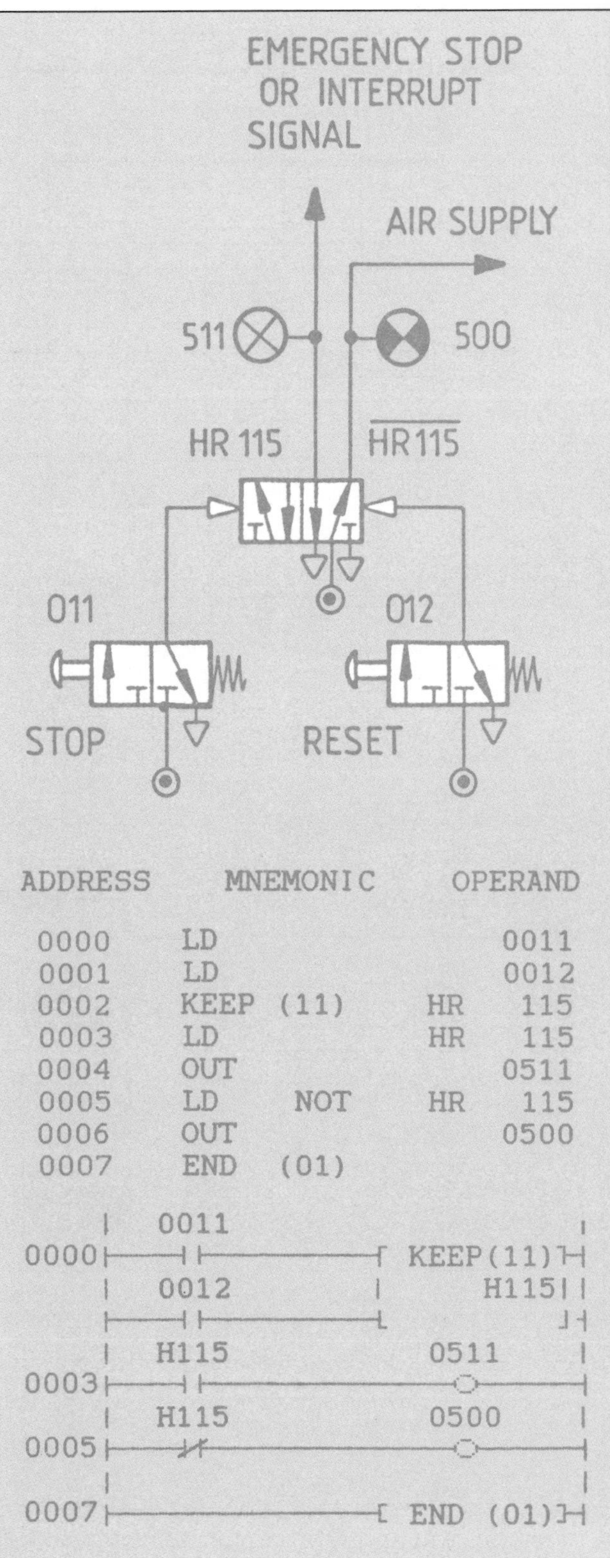

Figure 9-02. Basic sequencing interrupt module (emergency stop module) with mode selection indication lamps.

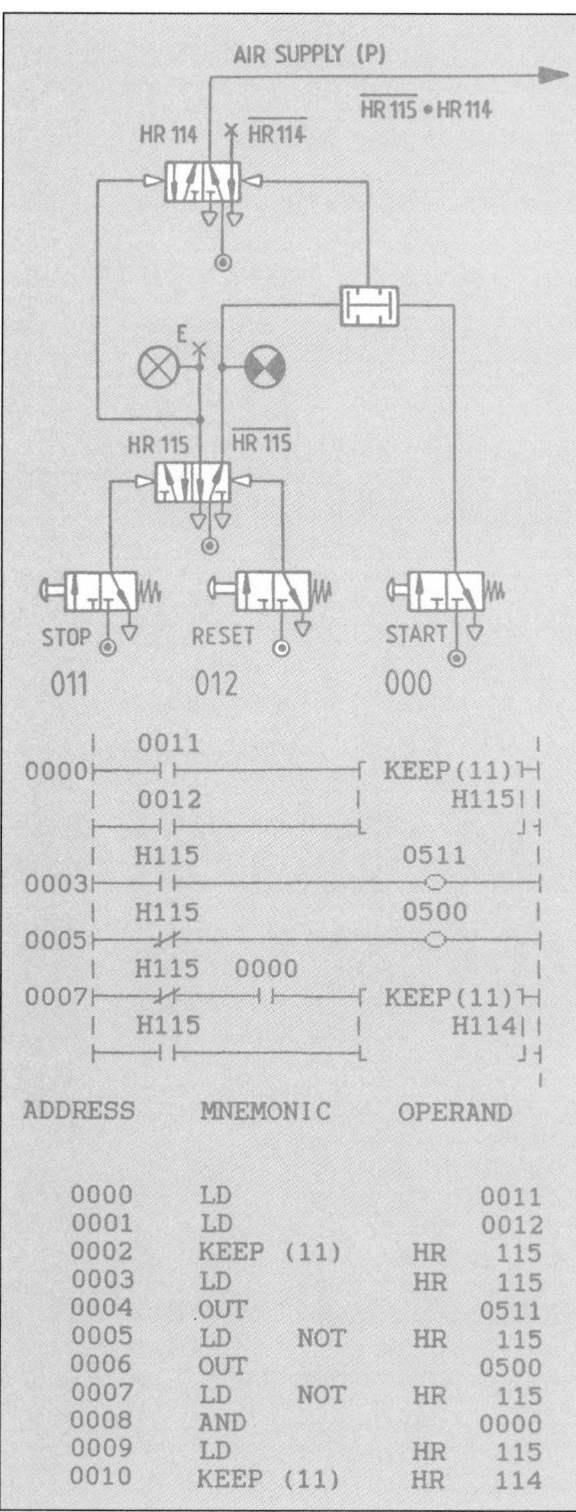

ADDRESS	MNEMONIC	OPERAND
0000	LD	0011
0001	LD	0012
0002	KEEP (11)	HR 115
0003	LD	HR 115
0004	OUT	0511
0005	LD NOT	HR 115
0006	OUT	0500
0007	LD NOT	HR 115
0008	AND	0000
0009	LD	HR 115
0010	KEEP (11)	HR 114

Figure 9-03. Complex sequencing interrupt module (complex emergency stop module).

Figure 9-04. Simple cycle selection module with mode selection indication lamps.

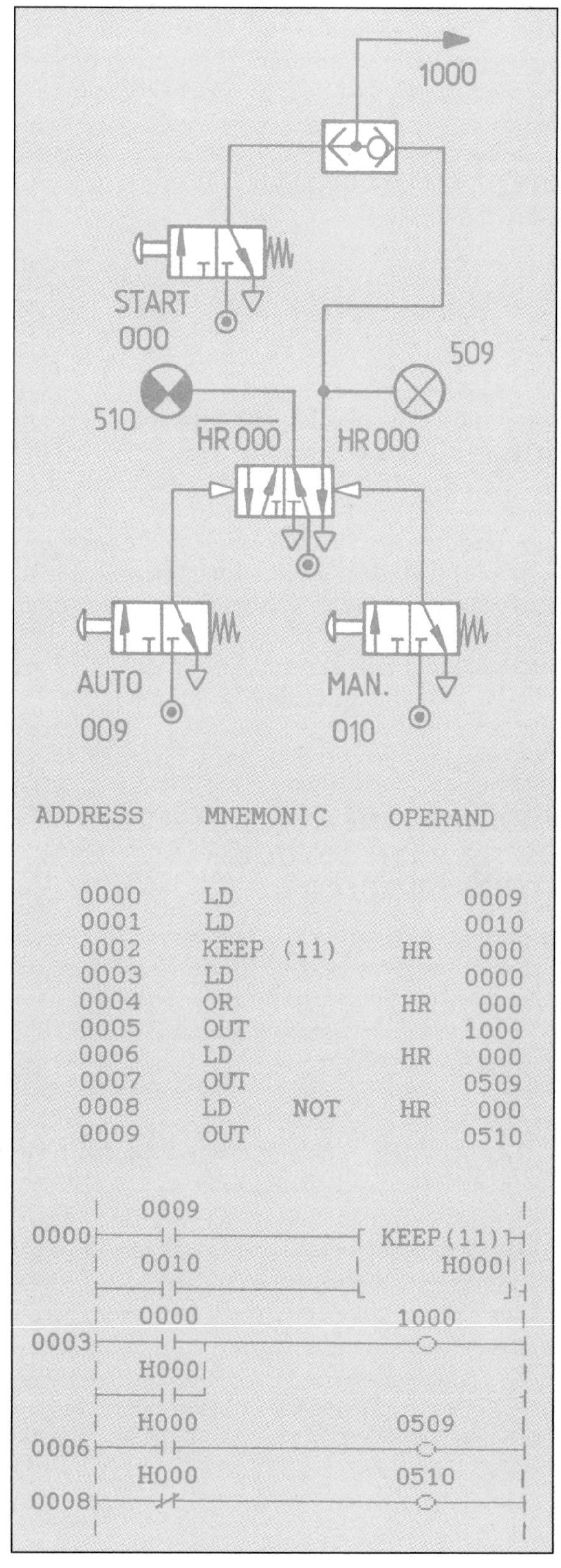

ADDRESS	MNEMONIC	OPERAND
0000	LD	0009
0001	LD	0010
0002	KEEP (11)	HR 000
0003	LD	0000
0004	OR	HR 000
0005	OUT	1000
0006	LD	HR 000
0007	OUT	0509
0008	LD NOT	HR 000
0009	OUT	0510

SIMPLE CYCLE SELECTION MODULE

Although not directly a machine interrupt or emergency stop module this simple cycle select module (figure 9-04) may be used to terminate a continuous and automatically started machine cycle. With this module, stopping mode number 3 as shown in figure

9-01 is achieved. The module also provides a simple solution to give "low danger" rated machines automatic or manual cycling selection. The disadvantage of this module is its combination of auto-selection and auto-start being accomplished by the same push-button switch. When auto-selection is made, the machine starts to operate immediately. This may prove to be confusing for some operators. The integration and application of this module is illustrated in figures 8-30 and 8-31.

EXTENDED CYCLE SELECTION MODULE

The extended cycle selection module shown in figure 9-05 eliminates the disadvantage mentioned for the simple cycle selection module given in figure 9-04. It provides separate selection for each selectable mode. Additionally, when automatic cycling is selected, the operator must also press the start push-button the first time only to initiate automatic machine operation. Thereafter, all cycles start automatically. When manual cycling is selected, the operator needs to press the start switch push-button for each cycle initiation. Output signal 1000 represents the combined outputs for each selection mode, and replaces the start signal 000, usually given to flip-flop 1 of step-counter module 1 in the step-counter chain. (Compare figure 11-17 with 11-18. For a typical integration of this module, see circuit illustration figure 9-20. Also compare this module with the module presented in figure 9-04).

COMBINING THE CYCLE SELECTION MODULE WITH A SEQUENCING INTERRUPT MODULE

When combining a cycle selection module with a sequencing interrupt module (figure 9-06), the signal linkage marked "LINK" is of great importance. This link causes the cycle selection module to be automatically forced into the manual mode when, for any reason, cycle interrupt is signalled. After machine cycling interrupt has been cancelled through the actuation of the reset input 012 push-button and automatic cycling must be resumed, a fresh automatic selection needs to be given through inputs 009 and 000 (START). The machine will then continue to cycle automatically until another mode change becomes necessary or cycle interrupt is again signalled. For an integration of these modules, see figure 9-18.

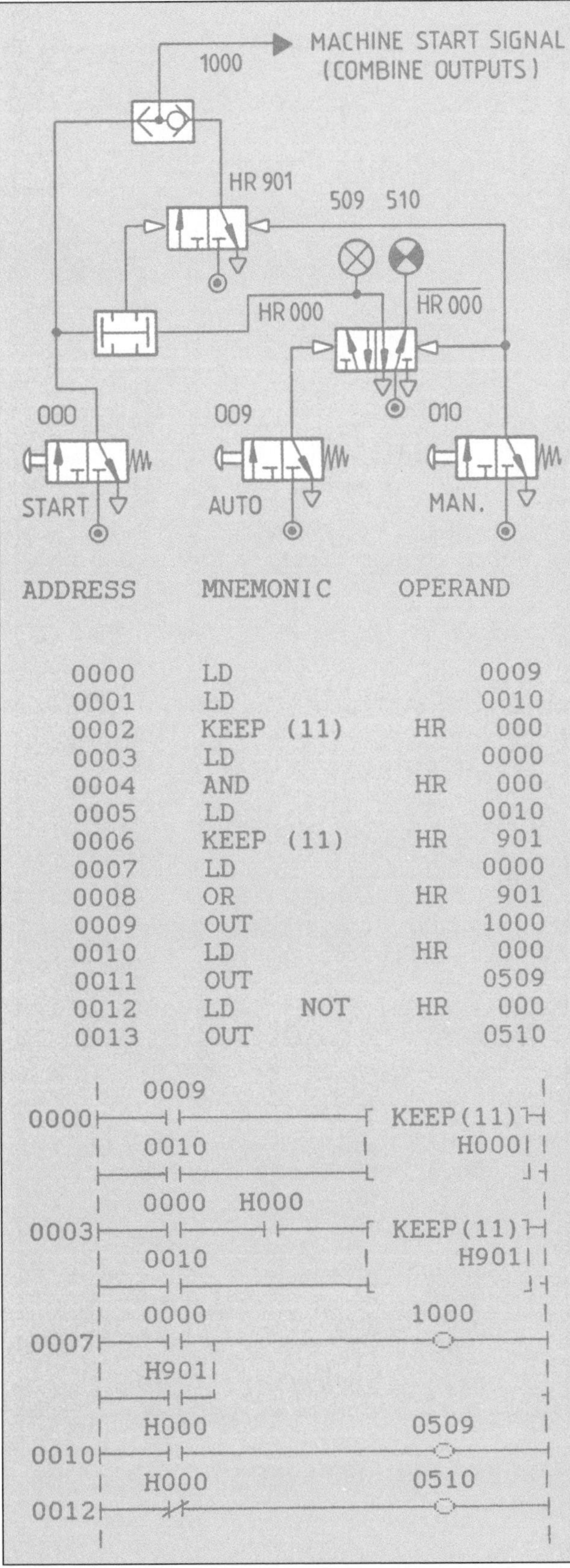

Figure 9-05. Extended cycle selection module, demanding an initial start command before automatic cycling commences.

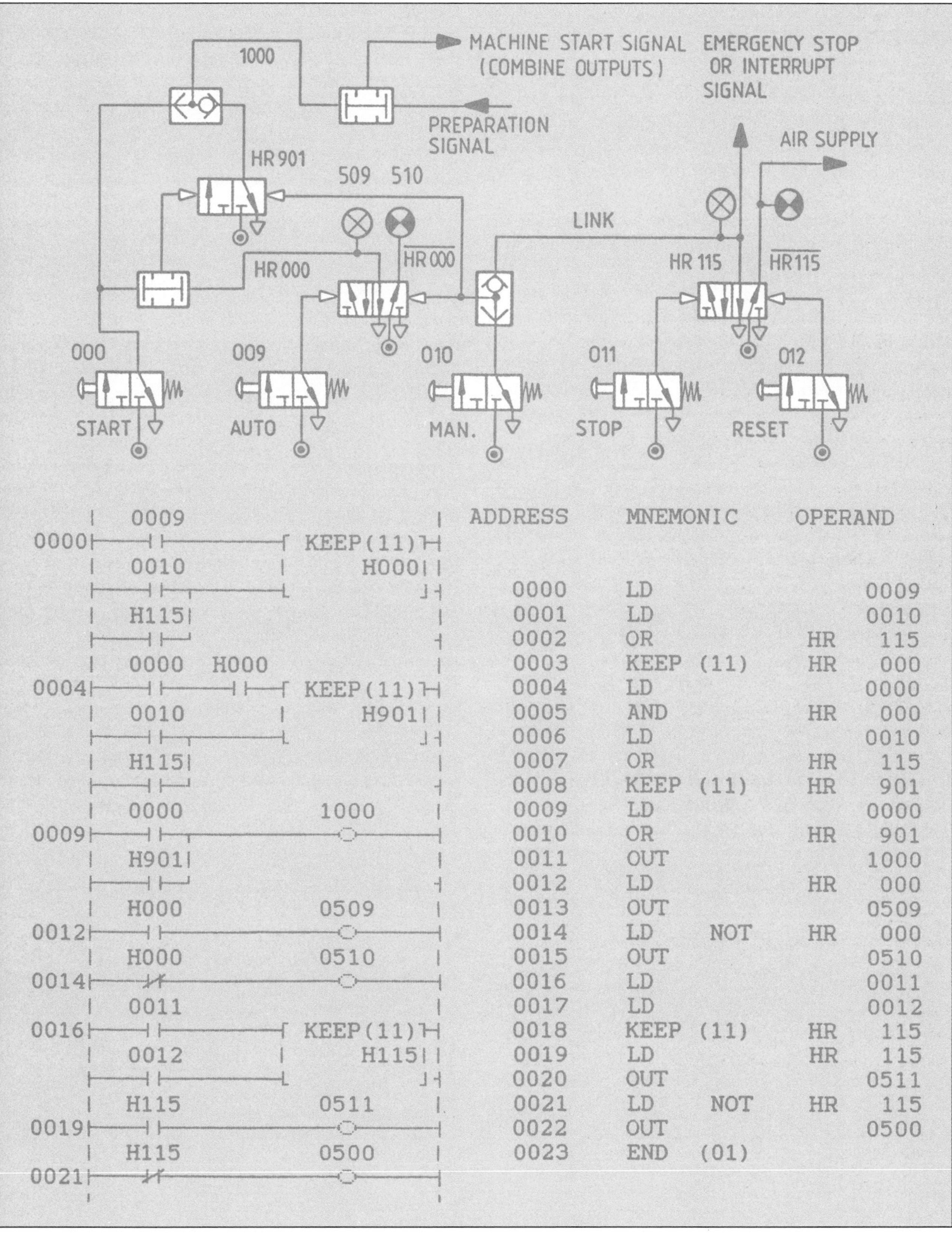

ADDRESS	MNEMONIC		OPERAND	
0000	LD			0009
0001	LD			0010
0002	OR		HR	115
0003	KEEP	(11)	HR	000
0004	LD			0000
0005	AND		HR	000
0006	LD			0010
0007	OR		HR	115
0008	KEEP	(11)	HR	901
0009	LD			0000
0010	OR		HR	901
0011	OUT			1000
0012	LD		HR	000
0013	OUT			0509
0014	LD	NOT	HR	000
0015	OUT			0510
0016	LD			0011
0017	LD			0012
0018	KEEP	(11)	HR	115
0019	LD		HR	115
0020	OUT			0511
0021	LD	NOT	HR	115
0022	OUT			0500
0023	END	(01)		

Figure 9-06. Combining a cycle selection module with a machine interrupt module. Note the link between the two modules where HR 115 resets HR 000 to manual mode (address 0002).

STEPPING MODULE FOR RUN/ STEPPING SELECTION

For the initial commissioning or setting-up of a machine, or the initial setting and adjusting of a tool or fixture on a machine, it is often mandatory to operate the machine sequence on a step-by-step basis. Once the machine is proved to function satisfactorily, the operator or tool setter can then select the "RUN MODE" (input 007), which causes the machine sequence to advance automatically through its sequence (figure 9-07). For the "STEPPING MODE" operation, the first actuation of the step signal push-button 008 causes holding relay HR 900 to change logic state, thus rendering output $\overline{\text{HR 900}}$. This first actuation, for safety reasons, will not initiate any machine sequencing action, it is merely a mode selection. Any further actuations of the stepping push-button, however, cause the step-counter to advance step-by-step.

Internal auxiliary relay output signal 1004 is the "COMMENCE CYCLE" signal. When stepping is selected, this signal requires input from 008, $\overline{\text{TIM 10}}$ and TIM 09. Hence the Boolean equation for this branch leading to 1004 in address 0009 of figure 9-07 reads:

$$(008 \bullet \overline{\text{TIM 10}} \bullet \text{TIM 09}) + (\ldots) = 1004$$

The initial actuation of input 008 causes "OFF"-delay timer 10 to time-out after 0.5 second and break its contact in address 010 of figure 9-07. At the same time "ON"-delay timer 09, having a set value of 1.0 second, and being driven by and therefore dependent on the $\overline{\text{HR 900}}$ state of the flip-flop, has not yet timed-out. It is therefore impossible to cause this branch of the logic line, leading to output 1004, to become true with an initial actuation of input 008. However, after that, when timer 09 has timed-out and input 008 is again being actuated, output 1004 will become logic 1. This causes the step-counter-controlled sequential circuit to start and switch sequence step 1.

Now to advance the step-counter beyond step 1, the same logic Boolean equation:

$$(008 \bullet \overline{\text{TIM 10}} \bullet \text{TIM 09}) = \ldots$$

will also render signal output 1003 (see address 0004 in figure 9-07). This then causes the step-counter to advance step-by-step whenever input 008 is given. For this reason, output signal 1003 is called "ADVANCE CYCLE". But since timer 10 is an "OFF"-delay timer, the operator is forced to release the stepping push-button 008 every time before initiating a new step. One could of course use two push-buttons, one to select the stepping mode and one to cause the stepping action. However, the circuit module shown in figure 9-07 combines both tasks in one single push-button.

Alternatively, if a "RUN" selection is made with input 007, the flip-flop HR 800 is caused to assume its "ON" state (or set state), thus rendering output HR 900. This causes the lower branch of the "OR" function leading to output 1003 to become true, and the lower branch to output 1004 to become true when start bit contact 000 is logic 1 (see address 0013 of figure 9-07). Where the stepping module needs to be combined with a cycle selection module, as is shown in figures 9-21 and 9-22, bit contact 000 for the start is replaced by bit contact 1000 from the combined output of the cycle selection module (see figure 9-22 and addresses 0013 and 0042).

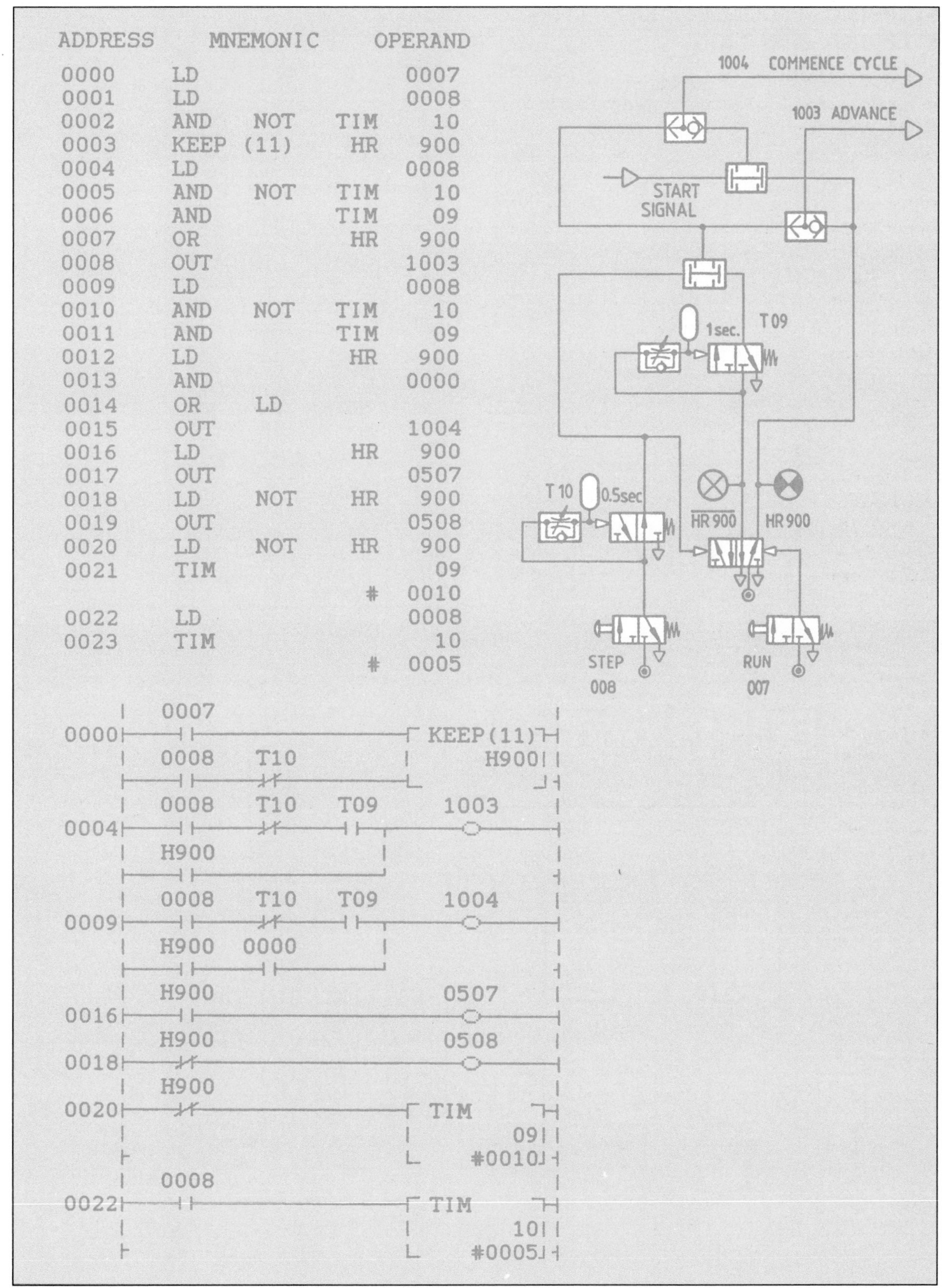

Figure 9-07. Stepping module to provide sequential machines with the option of manual stepping or automatic step advancing. Input 008 selects stepping and input 007 selects "RUN" operation (automatic step advancing). HR 900 flip-flop stores the selection made by the operator.

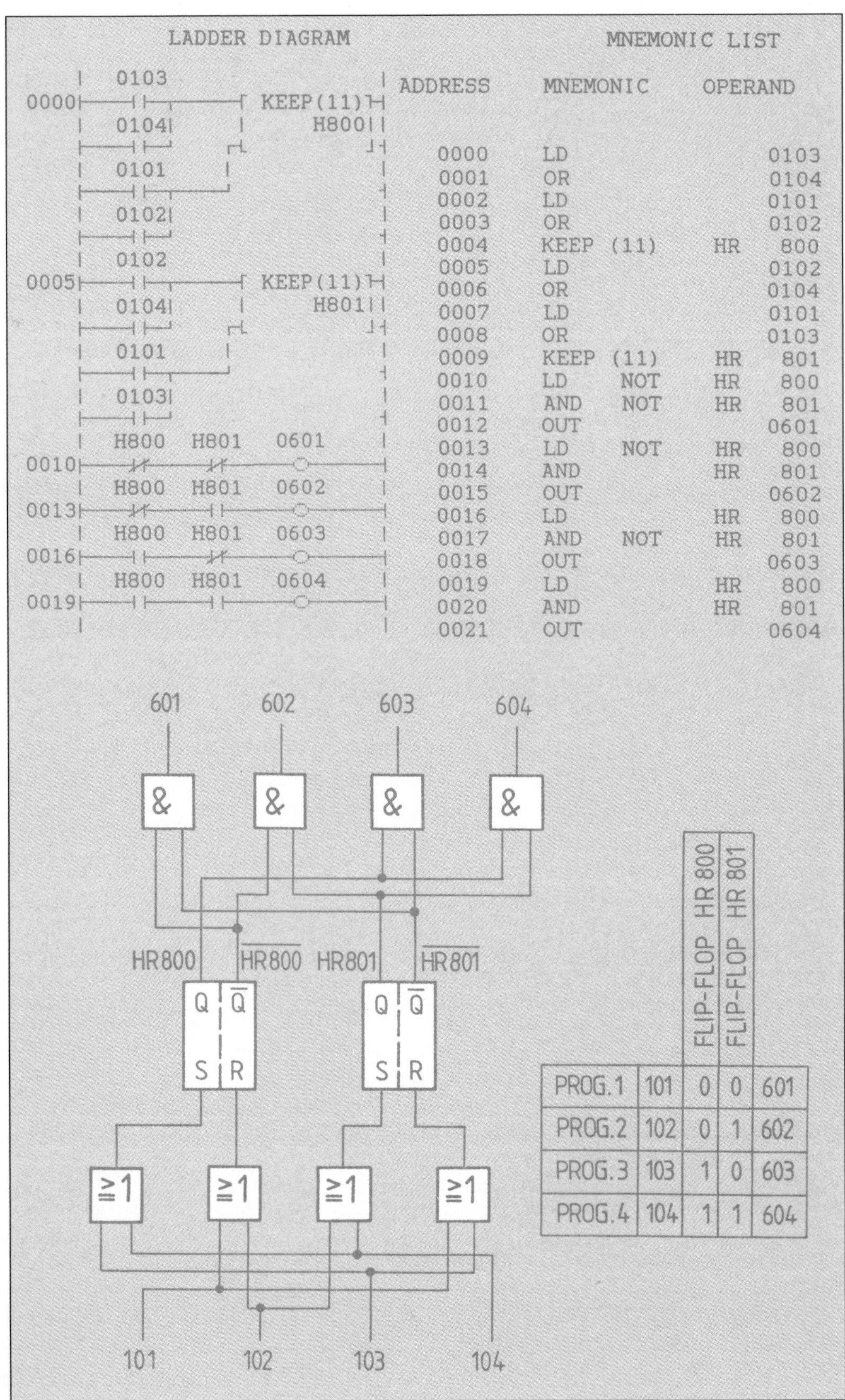

MNEMONIC LIST

ADDRESS	MNEMONIC		OPERAND	
0000	LD			0103
0001	OR			0104
0002	LD			0101
0003	OR			0102
0004	KEEP	(11)	HR	800
0005	LD			0102
0006	OR			0104
0007	LD			0101
0008	OR			0103
0009	KEEP	(11)	HR	801
0010	LD	NOT	HR	800
0011	AND	NOT	HR	801
0012	OUT			0601
0013	LD	NOT	HR	800
0014	AND		HR	801
0015	OUT			0602
0016	LD		HR	800
0017	AND	NOT	HR	801
0018	OUT			0603
0019	LD		HR	800
0020	AND		HR	801
0021	OUT			0604

		FLIP-FLOP HR 800	FLIP-FLOP HR 801	
PROG.1	101	0	0	601
PROG.2	102	0	1	602
PROG.3	103	1	0	603
PROG.4	104	1	1	604

Figure 9-08. Program selector module to select one of four possible programs. A truth table is used to design the program selector. The rows indicate the binary state of the two flip-flops for each program selection, and the columns are filled with logic 1s and logic 0s.

PROGRAM SELECTOR MODULE

Program selector modules may vary from a simple flip-flop, being set to cause program 1 to operate and reset to cause program 2 to operate, to multiple-input, binary-method-designed modules using several flip-flops. A simple, single flip-flop, program selector module is shown and applied in figure 5-13 to select program 1, reciprocating pattern of 30 seconds, or program 2, reciprocating pattern of 70 seconds. The program selection flip-flop in that circuit is HR 001, being set or reset by inputs 014 or 015 alternately. A similar program selector module application is also given in figures 8-18 and 8-19, where the program selector produces the bogus signal. Selecting flip-flop HR 900 into the "ON" logic state causes the bogus signal to appear. This then forces the step-counter to race and dictate program 2.

Such simple program selector circuits have one feature in common. They are designed to select only one or the other of two programs. If the program selector needs to cause selection and distinction between three or more programs, a program selector as is shown in figures 9-08 and 9-14 needs to be designed. This program selector is designed with the combinational binary logic method, using a truth table to list all possible selection combinations. (For detailed conceptual design of binary logic controls see Chapter 10 of the book *Pneumatic Control for Industrial Automation* by Peter Rohner and Gordon Smith.) The four input signals 101, 102, 103 and 104 stem from the four push-button selector switches (not shown in this circuit). The four output signals 601, 602, 603 and 604 are used to initiate the four step-counter circuit programs and to drive the lamps indicating the selection made. (Input 101 selects program 1 and output 601 is the lamp for program 1. The same principle applies for the other three programs.) The step-counter control circuit and the lamps are not shown in this control circuit, but an application of such a program selector module is given in control problem 4 of this chapter. If a program selection module for more than four programs needs to be designed, one must firstly establish a truth table for, say, eight input switch signals (if, for example, eight programs had to be selected). One would therefore need three flip-flops as 2 to the power of 3 (2^3) gives eight switching combination possibilities:

$$K = 2^n \qquad 8 = 2^3$$

n = number of input signals from flip-flops

K = number of truth table rows (figure 9-07)

0 = binary state of flip-flop being reset ($\overline{Q}$)

1 = binary state of flip-flop being set (Q)

For the formula shown above, 2 denotes the binary nature of flip-flops (set or reset state), 3 denotes the required number of such flip-flops, and 8 indicates the maximum possible number of switching combinations achievable with 3 flip-flops:

$$2 \times 2 \times 2 = 8$$

Following this pattern, four flip-flops could distinguish between 16 different programs and with five flip-flops one could construct a program selector for 32 programs! If one had to differentiate between only three programs, the fourth row of the truth table in figure 9-08 would not be used. This implies that the combination of both flip-flops being set would never arise. An application for such a program selection module to distinguish and select between three programs is given in control problem 2.

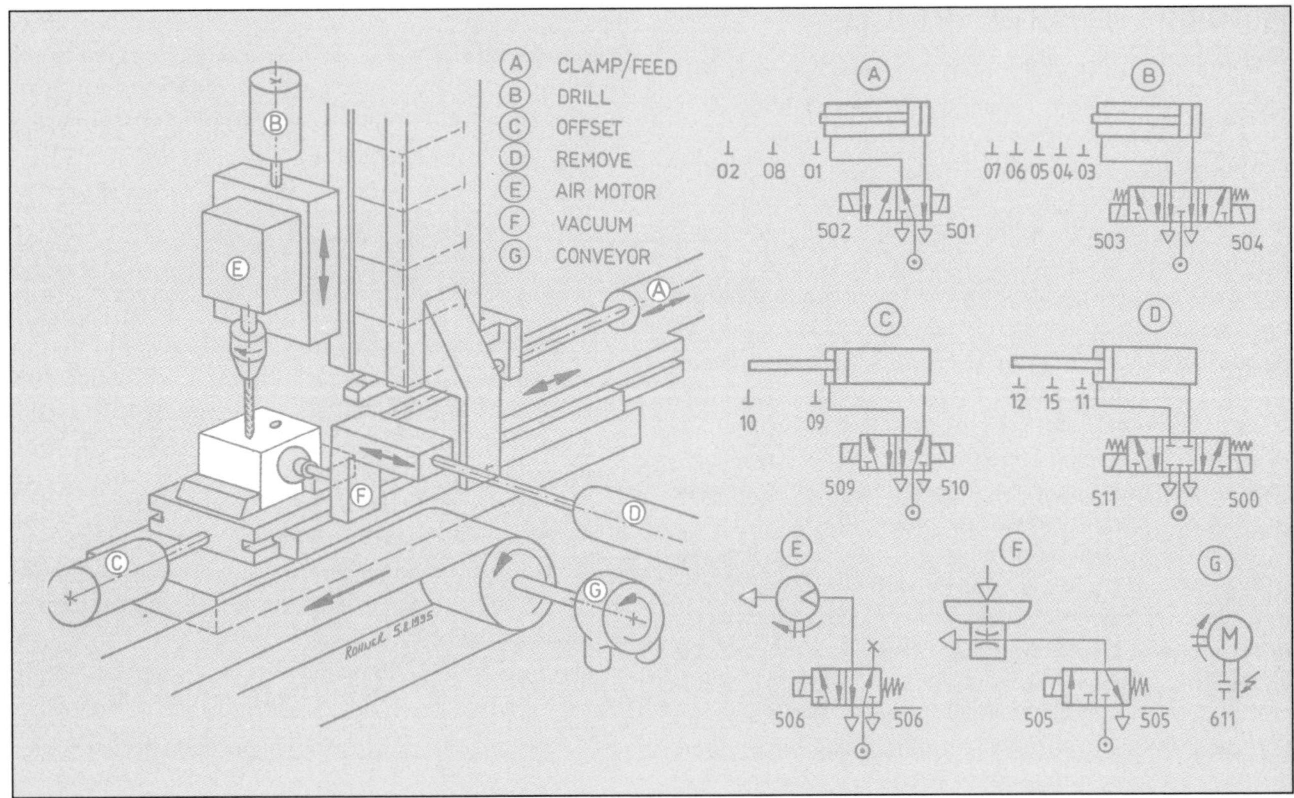

Figure 9-09. Isometric conceptual sketch and electrohydraulic power circuit for dowel-hole-drilling machine as explained and illustrated in control problem 1.

CONTROL PROBLEM 1

A PLC-controlled machine program identical to the sequence given in control problem 1 of Chapter 8, but with some different input and output data, needs to be designed (compare figure 9-09 with figure 8-27). In addition, an emergency machine-clearing sequencing program has to be integrated. This emergency machine-clearing program is designed to remove faulty components by placing them on the conveyor and then reversing its direction (switch motor rotation polarity). The emergency machine-clearing program moves the actuator according to the step-action diagram in figure 9-10.

The normal machining program is almost identical to the program shown in figures 8-28 and 8-29 (some modifications are, however, necessary).

The emergency machine-clearing program works completely independently of the normal machining program. It may be selected manually through the actuation of push-button switch input 011, or automatically by a drilling operations watchdog timer (TIM 02 in addresses 0019 and 0280). The watchdog timer springs into action when the drilling process takes longer than 6.0 seconds, which could be the case if the drill bit becomes blunt or when the drill breaks. The emergency machine-clearing program may be signalled to start in any step from 2 to 12.

The emergency stop module has a triple task, as is shown in figure 9-12, addresses 0018 to 0027:

1. It completely resets and disables the normal step-counter (the 15-step step-counter; see inputs HR 115 serving as reset commands to the step-counter flip-flops in figure 9-12).

2. At the same time it starts the emergency machine-clearing step-counter with input 107 from the stop push-button or from the watchdog timer TIM 02 (the six-step step-counter in figure 9-12 starts at address 0199).

3. It indicates the signalling of emergency machine-clearing having happened or being in motion. It does this by causing the red emergency machine-clearing indication lamp to flicker. It also actuates the green emergency reset lamp once the reset key switch is selected (see figure 9-12 addresses 0023 to 0027).

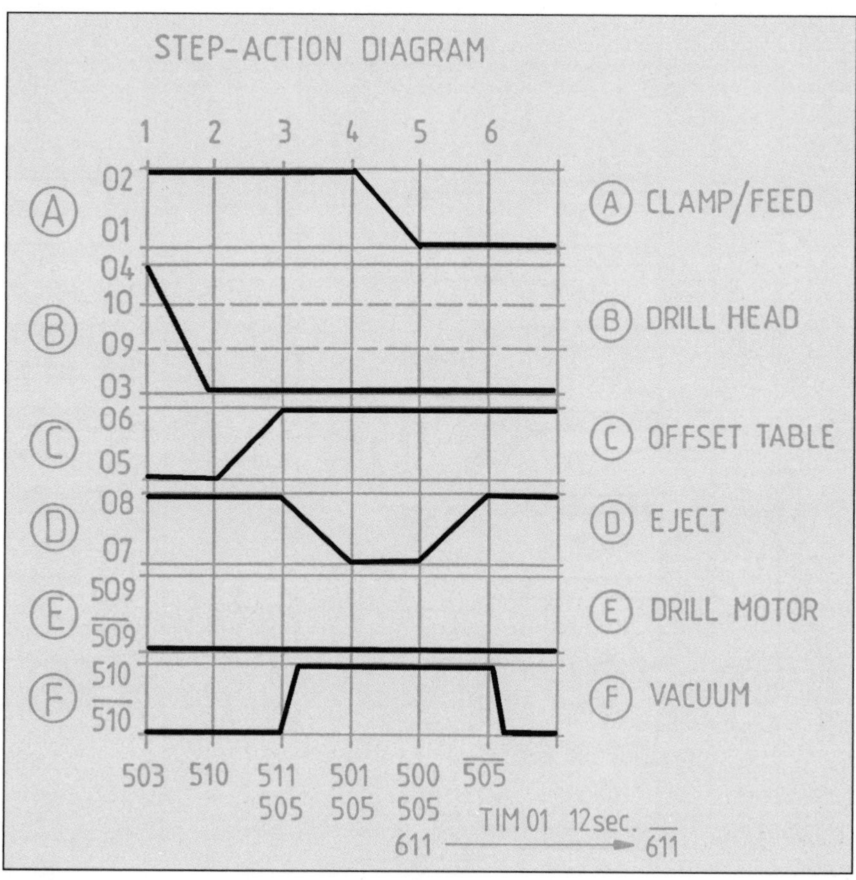

Figure 9-10. Step-action diagram for emergency machine-clearing program in control problem 1.

INPUT/OUTPUT ASSIGNMENT LIST

PLC inputs:

000 = start signal
001 = actuator Ⓐ retracted
002 = actuator Ⓐ extended
003 = actuator Ⓑ fully retracted
005 = actuator Ⓑ 1/3 extended
006 = actuator Ⓑ 2/3 extended
007 = actuator Ⓑ fully extended
009 = actuator Ⓒ retracted
010 = actuator Ⓒ extended
011 = actuator Ⓓ retracted
012 = actuator Ⓓ extended
505 = vacuum flow started Ⓕ
$\overline{505}$ = vacuum flow stopped Ⓕ
506 = actuator Ⓔ in motion
$\overline{506}$ = actuator Ⓔ not in motion
511 = conveyor motor running Ⓖ
$\overline{511}$ = conveyor motor stopped Ⓖ
100 = select automatic cycling
101 = select program 1 (1 hole drilled)
102 = select program 2 (2 holes drilled)
103 = select program 3 (no woodpecking)
104 = select run operation
105 = select stepping operation
107 = select emergency operation
108 = select emergency reset
109 = select manual cycling

PLC outputs:

500 = actuator Ⓓ extend solenoid
501 = actuator Ⓐ retract solenoid
502 = actuator Ⓐ extend solenoid
503 = actuator Ⓑ retract solenoid
504 = actuator Ⓑ extend solenoid
505 = vacuum venturi air supply solenoid
506 = drill motor run solenoid
507 = conveyor motor reversal signal
509 = actuator Ⓒ retract solenoid
510 = actuator Ⓒ extend solenoid
511 = actuator Ⓓ retract solenoid
600 = auto selection lamp
601 = program 1 lamp
602 = program 2 lamp
603 = program 3 lamp
604 = run operation lamp
605 = stepping operation lamp
607 = emergency stop lamp
608 = emergency reset lamp
609 = manual selection lamp
611 = conveyor motor run signal

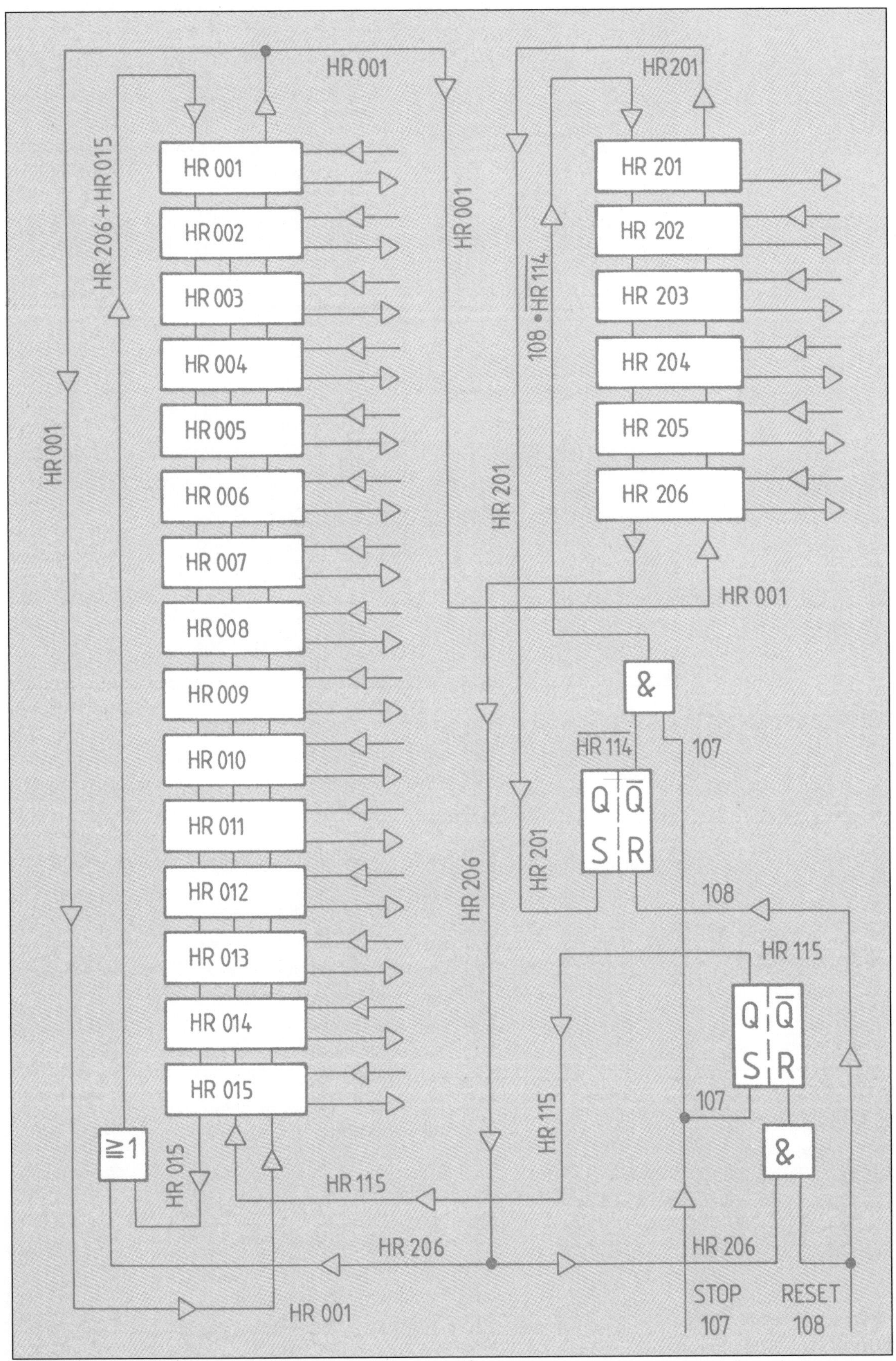

Figure 9-11. Simplified function plan for control problem 1, to show the complex interaction of the two step-counters.

- HR 001 to HR 015 flip-flops are used for the normal step-counter steps (drilling machine step-counter).
- HR 201 to HR 206 are used for step-counter steps for the emergency machine-clearing program.
- HR 115 flip-flop is used for the emergency stop module.
- HR 114 flip-flop is used to inhibit rogue input 012 signals from entering HR 201 flip-flop after initial switching of step 1 emergency step-counter module.
- HR 800 and HR 801 flip-flops are used to select one of the three programs within the program selector module.
- HR 901 and HR 000 flip-flops are used in the extended-cycle-selection module.
- HR 902 is used to reverse conveyor motion.
- TIM 01 timer is used to stop the conveyor motor after 12 seconds' running time.
- TIM 02 timer is used to supervise drilling operations (actuator Ⓑ extension movements) and to select the emergency machine-clearing program into action if such extension movements exceed 6 seconds (because of broken or blunt drills).

CIRCUIT EXPLANATIONS FOR CONTROL PROBLEM 1

A simplified function plan showing the conceptual layout for control problem 1 is given in figure 9-11 to illustrate the complexity of this circuit, the interaction of flip-flops, logic functions and the two step-counters, and the roles these play to signal the machine during normal sequencing or, if emergency is signalled, for the machine-clearing operation.

For this type of emergency machine-clearing action, the normal step-counter needs to be completely disabled (all stepping modules to be reset), and a reset signal HR 115 from the emergency stop module flip-flop renders all relevant step-counter modules inactive. The first and last step-counter modules need no emergency "OR" reset command (figures 9-11 and 9-12).

For normal cycling conditions, the preparation signal for step-counter module 1 (flip-flop HR 001) is provided by step-counter module 15 (flip-flop HR 015), as is shown in figure 8-29, address 0011. However, in an emergency, with the normal step-counter being disabled during its advancing process, the last step-counter flip-flop will not be set. This implies that flip-flop HR 015 is now no longer able to furnish step-counter module 1 with the necessary preparation signal. One therefore uses the last emergency step-counter flip-flop HR 206 to provide the necessary preparation signal. But as this preparation signal stems from a source other than the normal, it is called a "substitute preparation signal" (figures 9-11 and 9-12).

Hence the preparation signals for HR 001 are:

$$(\text{HR } 015 + \text{HR } 206) \bullet \ldots$$

(See figures 9-11 and 9-12, addresses 0074 to 0080.)

Following similar guidelines, the first step-counter module flip-flop HR 001 resets HR 015 under normal operating conditions, and resets HR 206 after the emergency condition is reset and the emergency step-counter is no longer required to produce the substitute preparation signal.

HR 114 flip-flop with its HR 114 state being "ON" ($\overline{\text{HR 114}}$ state "OFF") is used in this circuit to inhibit and therefore prevent any spurious or rogue 107 stop signals from the stop push-button to reactivate emergency step-counter module 1 while this step-counter is in motion. Once HR 206 has been reset by HR 001, signal 107 or TIM 01 is again permitted to set HR 201. It must also be noted that holding relay flip-flops in general are reset-dominant (see Chapter 4, figure 4-03 and the relevant textual explanations). Therefore, for safety reasons, it is advisable to use the reset condition output of flip-flop HR 114 to contribute to the logic 1 combination of the inhibition function to conduct signal 107, thus causing the emergency program step-counter to start. It must be noted that this first step-counter step for emergency machine-clearing has no essential confirmation in its flip-flop set line. The only two conditions are emergency being signalled (input 107 or TIM 02) and flip-flop HR 114 not inhibiting the emergency signal. This flip-flop therefore switches *unconditionally*. Unconditional starting is important and mandatory for all emergency and machine interrupt control sequences!

It is for these reasons that I suggest that the reset condition of the HR 114 flip-flop be used to conduct the

> **General rule:** Emergency stop program initiation should always be unconditional and solely signalled by actuation of the emergency stop push-button switch (or emergency signal).

107 emergency stop signal (HR 114) and not the artificial set condition. At machine commissioning stage, all HR flip-flops are reset, thus rendering $\overline{Q}$ as is shown in figures 4-02 and 4-03. If forcing of the artificial set condition ($\overline{\text{HR 114}}$ being "ON") were neglected during machine commissioning, it could prevent the operator from starting the emergency program! Hence, the starting equation for step 201 is:

$$(\text{TIM } 01 + 107) \bullet \overline{\text{HR } 114} = \text{SET HR } 201.$$

(See also ladder diagram figure 9-12 and function plan figure 9-11.) Thus, HR 114 is set by HR 201 and reset by the reset push-button signal 108.

```
                LADDER DIAGRAM    PAGE-001

     | 0100                                     |
0000 |--| |-------------------------[ KEEP(11) ]|
     | 0109                           H000      |
     |--| |--+                                  |
     | H115  |                                  |
     |--| |--|                                  |
     | H900  |                                  |
     |--|/|--+                                  |
     | 0000  H000                               |
0005 |--| |--| |--------------------[ KEEP(11) ]|
     | 0109                           H901      |
     |--| |--+                                  |
     | H115  |                                  |
     |--| |--|                                  |
     | H900  |                                  |
     |--|/|--+                                  |
     | 0000                          1000       |
0011 |--| |--+---------------------------( )----|
     | H901  |                                  |
     |--| |--+                                  |
     | H000                          0600       |
0014 |--| |------------------------------( )----|
     | H000                          0609       |
0016 |--|/|------------------------------( )----|
     | 0107                                     |
0018 |--| |--+-------------------[ KEEP(11) ]---|
     | T02   |                        H115      |
     |--| |--+                    +-            |
     | 0108  0611                 |             |
     |--| |--|/|------------------+             |
     | H115  1901                    0607       |
0023 |--| |--| |-------------------------( )----|
     | H115                          0608       |
0026 |--|/|------------------------------( )----|
     | 0104                                     |
0028 |--| |----------------------[ KEEP(11) ]---|
     | 0105  T10                      H900      |
     |--| |--|/|------------------+             |
     | 0105  T10   T09                1003      |
0032 |--| |--|/|--| |--+----------------( )----|
     | H900             |                       |
     |--| |-------------+                       |
     | 0105  T10   T09   H115         1004      |
0037 |--| |--|/|--| |--+--|/|----------( )----|
     | H900  1000      |                        |
     |--| |--| |-------+                        |
     | H900                          0604       |
0045 |--| |------------------------------( )----|
     | H900                          0605       |
0047 |--|/|------------------------------( )----|
     | H900                                     |
0049 |--|/|-------------------------[ TIM      ]|
     |                                     09   |
     |                                  #0010   |
     | 0105                                     |
0051 |--| |-------------------------[ TIM      ]|
     |                                     10   |
     |                                  #0005   |
     | 0103                                     |
0053 |--| |--+-------------------[ KEEP(11) ]---|
     | 0105  |                        H800      |
     |--| |--|                    +-            |
     | 0107  |                    |             |
     |--| |--+                    |             |
     | 0101                       |             |
     |--| |--+--------------------+             |
     | 0102  |                                  |
     |--| |--+                                  |
     | 0102                                     |
0059 |--| |----------------------[ KEEP(11) ]---|
     | 0101                           H801      |
     |--| |--+                                  |
     | 0103  |                                  |
     |--| |--|                                  |
     | 0105  |                                  |
     |--| |--|                                  |
     | 0107  |                                  |
     |--| |--+                                  |
```

MNEMONIC LIST

ADDRESS	MNEMONIC			OPERAND
0000	LD			0100
0001	LD			0109
0002	OR		HR	115
0003	OR	NOT	HR	900
0004	KEEP	(11)	HR	000
0005	LD			0000
0006	AND		HR	000
0007	LD			0109
0008	OR		HR	115
0009	OR	NOT	HR	900
0010	KEEP	(11)	HR	901
0011	LD			0000
0012	OR		HR	901
0013	OUT			1000
0014	LD		HR	000
0015	OUT			0600
0016	LD	NOT	HR	000
0017	OUT			0609
0018	LD			0107
0019	OR		TIM	02
0020	LD			0108
0021	AND	NOT		0611
0022	KEEP	(11)	HR	115
0023	LD		HR	115
0024	AND			1901
0025	OUT			0607
0026	LD	NOT	HR	115
0027	OUT			0608
0028	LD			0104
0029	LD			0105
0030	AND	NOT	TIM	10
0031	KEEP	(11)	HR	900
0032	LD			0105
0033	AND	NOT	TIM	10
0034	AND		TIM	09
0035	OR		HR	900
0036	OUT			1003
0037	LD			0105
0038	AND	NOT	TIM	10
0039	AND		TIM	09
0040	LD		HR	900
0041	AND			1000
0042	OR	LD		
0043	AND	NOT	HR	115
0044	OUT			1004
0045	LD		HR	900
0046	OUT			0604
0047	LD	NOT	HR	900
0048	OUT			0605
0049	LD	NOT	HR	900
0050	TIM			09
			#	0010
0051	LD			0105
0052	TIM			10
			#	0005
0053	LD			0103
0054	OR			0105
0055	OR			0107
0056	LD			0101
0057	OR			0102
0058	KEEP	(11)	HR	800
0059	LD			0102
0060	LD			0101
0061	OR			0103
0062	OR			0105
0063	OR			0107
0064	KEEP	(11)	HR	801
0065	LD	NOT	HR	800
0066	AND	NOT	HR	801
0067	OUT			0601
0068	LD	NOT	HR	800
0069	AND		HR	801
0070	OUT			0602

Figure 9-12 Ladder diagram for control problem (section 1 of 4)

Figure 9-12. Ladder diagram for control problem 1 (section 2 of 4).

LADDER DIAGRAM PAGE-002

```
     H800  H801                         0601
0065
     H800  H801                         0602
0068
     H800  H801                         0603
0071
     H015  0505  0611  1004
0074                             KEEP(11)
     H206                           H001
     H002
     H001  0002  1003
0081                             KEEP(11)
     H003                           H002
     H115
     H002  0006  0506  1003
0087                             KEEP(11)
           0603  0007               H003
     H004
     H115
     H003  0005  0506  1003
0098                             KEEP(11)
           0603                     H004
     H005
     H115
     H004  0007  0506  1003
0107                             KEEP(11)
     H006                           H005
     H115
     H005  0003  0506  1003
0114                             KEEP(11)
     H007                           H006
     H115
     H006  0009  1003
0121                             KEEP(11)
           0601                     H007
     H008
     H115
     H007  0602  0006  0506  1003
0129                             KEEP(11)
           0603  0007  0506         H008
           0601
     H009
     H115
     H008  0602  0005  0506  1003
0143                             KEEP(11)
           0603  0007  0506         H008
           0601
     H009
     H115
     H008  0602  0005  0506  1003
0143                             KEEP(11)
           0603                     H009
           0601
     H010
     H115
```

MNEMONIC LIST

ADDRESS	MNEMONIC		OPERAND	
0071	LD		HR	800
0072	AND	NOT	HR	801
0073	OUT			0603
0074	LD		HR	015
0075	OR		HR	206
0076	AND	NOT		0505
0077	AND	NOT		0611
0078	AND			1004
0079	LD		HR	002
0080	KEEP	(11)	HR	001
0081	LD		HR	001
0082	AND			0002
0083	AND			1003
0084	LD		HR	003
0085	OR		HR	115
0086	KEEP	(11)	HR	002
0087	LD		HR	002
0088	LD			0006
0089	AND			0506
0090	LD			0603
0091	AND			0007
0092	OR	LD		
0093	AND	LD		
0094	AND			1003
0095	LD		HR	004
0096	OR		HR	115
0097	KEEP	(11)	HR	003
0098	LD		HR	003
0099	LD			0005
0100	AND			0506
0101	OR			0603
0102	AND	LD		
0103	AND			1003
0104	LD		HR	005
0105	OR		HR	115
0106	KEEP	(11)	HR	004
0107	LD		HR	004
0108	AND			0007
0109	AND			0506
0110	AND			1003
0111	LD		HR	006
0112	OR		HR	115
0113	KEEP	(11)	HR	005
0114	LD		HR	005
0115	AND			0003
0116	AND	NOT		0506
0117	AND			1003
0118	LD		HR	007
0119	OR		HR	115
0120	KEEP	(11)	HR	006
0121	LD		HR	006
0122	LD			0009
0123	OR			0601
0124	AND	LD		
0125	AND			1003
0126	LD		HR	008
0127	OR		HR	115
0128	KEEP	(11)	HR	007
0129	LD		HR	007
0130	LD			0602
0131	AND			0006
0132	AND			0506
0133	LD			0603
0134	AND			0007
0125	AND			1003
0126	LD		HR	008
0127	OR		HR	115
0128	KEEP	(11)	HR	007
0129	LD		HR	007
0130	LD			0602
0131	AND			0006
0132	AND			0506
0133	LD			0603
0134	AND			0007
0135	AND			0506
0136	OR	LD		
0137	OR			0601
0138	AND	LD		
0139	AND			1003
0140	LD		HR	009
0141	OR		HR	115
0142	KEEP	(11)	HR	008

```
LADDER DIAGRAM      PAGE-003

     | H009  0007  0506  1003                  |
0154 |--| |--+--| |----| |--+--| |--------[ KEEP(11)
     |       | 0601        |                 |  H010
     |       +--| |--------+                 |
     | H011                                  |
     |--| |-----------------------------------
     | H115
     |--| |--
     | H010  0003  0506  1003
0163 |--| |----| |----|/|----| |-----------[ KEEP(11)
     | H012                                     H011
     |--| |--+---------------------------
     | H115
     |--| |--
     | H011  0010  1003
0170 |--| |----| |----| |-----------------[ KEEP(11)
     | H013                                     H012
     |--| |--+---------------------------
     | H115
     |--| |--
     | H012  0011  0505  1003
0176 |--| |----| |----| |----| |----------[ KEEP(11)
     | H014                                     H013
     |--| |--+---------------------------
     | H115
     |--| |--
     | H013  0001  0505  1003
0183 |--| |----| |----| |----| |----------[ KEEP(11)
     | H015                                     H014
     |--| |--+---------------------------
     | H115
     |--| |--
     | H014  0012  0505  1003
0190 |--| |----| |----| |----| |----------[ KEEP(11)
     | H001                                     H015
     |--| |------------------------------
     | H201
0196 |--| |-------------------------------[ KEEP(11)
     | 0108                                     H114
     |--| |------------------------------
     | 0107  H114
0199 |--| |--+--|/|-----------------------[ KEEP(11)
     | T02                                      H201
     |--| |--
     | H202
     |--| |------------------------------
     | H201  0003
0204 |--| |----| |------------------------[ KEEP(11)
     | H203                                     H202
     |--| |------------------------------
     | H202  0010
0208 |--| |----| |------------------------[ KEEP(11)
     | H204                                     H203
     |--| |------------------------------
     | H203  0011  0505
0212 |--| |----| |----| |-----------------[ KEEP(11)
     | H205                                     H204
     |--| |------------------------------
     | H204  0001  0505
0217 |--| |----| |----| |-----------------[ KEEP(11)
     | H206                                     H205
     |--| |------------------------------
     | H205  0012  0505  0611
0222 |--| |----| |----| |----| |----------[ KEEP(11)
     | H001                                     H206
     |--| |------------------------------
     | H014
0228 |--| |--+----------------------------[ KEEP(11)
     | H205                                     H902
     |--| |--
     | T01
     |--| |------------------------------
     | H014                          0500
0232 |--| |--+--------------------------( )--
     | H205
     |--| |--
```

MNEMONIC LIST

ADDRESS	MNEMONIC		OPERAND	
0143	LD		HR	008
0144	LD			0602
0145	AND			0005
0146	AND			0506
0147	OR			0603
0148	OR			0601
0149	AND	LD		
0150	AND			1003
0151	LD		HR	010
0152	OR		HR	115
0153	KEEP	(11)	HR	009
0154	LD		HR	009
0155	LD			0007
0156	AND			0506
0157	OR			0601
0158	AND	LD		
0159	AND			1003
0160	LD		HR	011
0161	OR		HR	115
0162	KEEP	(11)	HR	010
0163	LD		HR	010
0164	AND			0003
0165	AND	NOT		0506
0166	AND			1003
0167	LD		HR	012
0168	OR		HR	115
0169	KEEP	(11)	HR	011
0170	LD		HR	011
0171	AND			0010
0172	AND			1003
0173	LD		HR	013
0174	OR		HR	115
0175	KEEP	(11)	HR	012
0176	LD		HR	012
0177	AND			0011
0178	AND			0505
0179	AND			1003
0180	LD		HR	014
0181	OR		HR	115
0182	KEEP	(11)	HR	013
0183	LD		HR	013
0184	AND			0001
0185	AND			0505
0186	AND			1003
0187	LD		HR	015
0188	OR		HR	115
0189	KEEP	(11)	HR	014
0190	LD		HR	014
0191	AND			0012
0192	AND			0505
0193	AND			1003
0194	LD		HR	001
0195	KEEP	(11)	HR	015
0196	LD		HR	201
0197	LD			0108
0198	KEEP	(11)	HR	114
0199	LD			0107
0200	OR		TIM	02
0201	AND	NOT	HR	114
0202	LD		HR	202
0203	KEEP	(11)	HR	201
0204	LD		HR	201
0205	AND			0003
0206	LD		HR	203
0207	KEEP	(11)	HR	202
0208	LD		HR	202
0209	AND			0010
0210	LD		HR	204
0211	KEEP	(11)	HR	203
0212	LD		HR	203
0213	AND			0011
0214	AND			0505
0215	LD		HR	205
0216	KEEP	(11)	HR	204

Figure 9-12. Ladder diagram for control problem 1 (section 3 of 4).

Figure 9-12. Ladder diagram for control problem 1 (section 4 of 4).

LADDER DIAGRAM PAGE-004

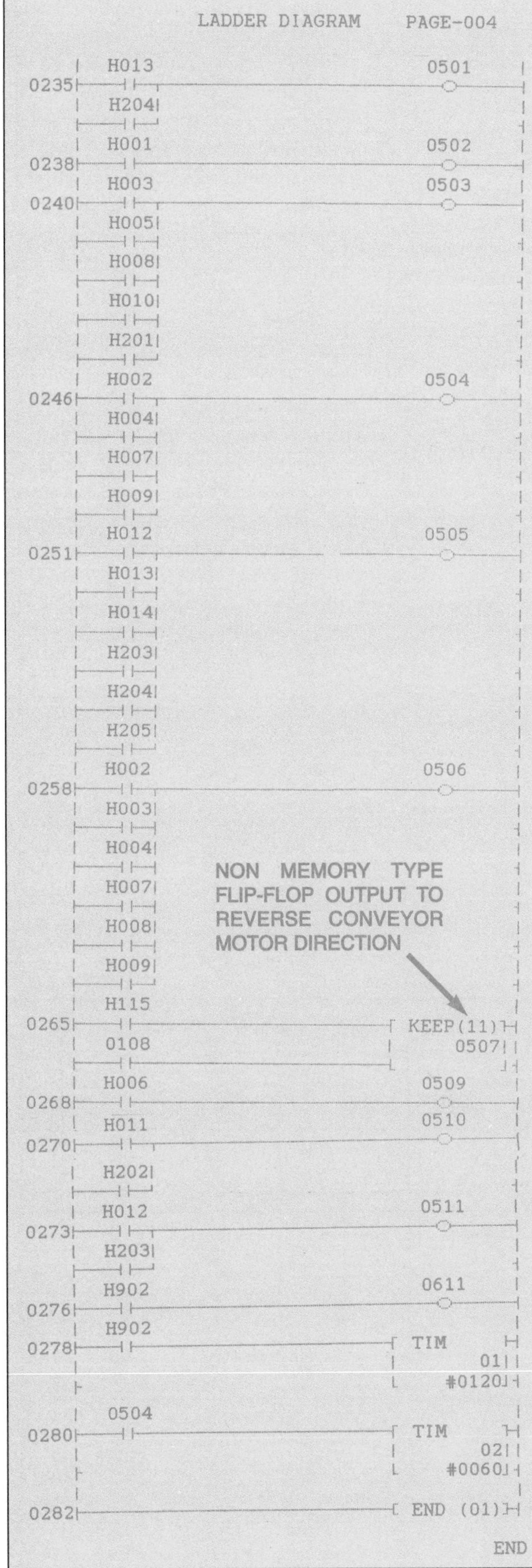

```
      | H013                                   0501    |
0235  |--| |--+--------------------------------( )-----|
      | H204  |                                        |
      |--| |--+                                        |
      | H001                                   0502    |
0238  |--| |-----------------------------------( )-----|
      | H003                                   0503    |
0240  |--| |--+--------------------------------( )-----|
      | H005  |                                        |
      |--| |--+                                        |
      | H008  |                                        |
      |--| |--+                                        |
      | H010  |                                        |
      |--| |--+                                        |
      | H201  |                                        |
      |--| |--+                                        |
      | H002                                   0504    |
0246  |--| |--+--------------------------------( )-----|
      | H004  |                                        |
      |--| |--+                                        |
      | H007  |                                        |
      |--| |--+                                        |
      | H009  |                                        |
      |--| |--+                                        |
      | H012                                   0505    |
0251  |--| |--+--------------------------------( )-----|
      | H013  |                                        |
      |--| |--+                                        |
      | H014  |                                        |
      |--| |--+                                        |
      | H203  |                                        |
      |--| |--+                                        |
      | H204  |                                        |
      |--| |--+                                        |
      | H205  |                                        |
      |--| |--+                                        |
      | H002                                   0506    |
0258  |--| |--+--------------------------------( )-----|
      | H003  |                                        |
      |--| |--+                                        |
      | H004  |                                        |
      |--| |--+                                        |
      | H007  |                                        |
      |--| |--+                                        |
      | H008  |                                        |
      |--| |--+                                        |
      | H009  |                                        |
      |--| |--+                                        |
      | H115                                           |
0265  |--| |-----------------------[ KEEP(11) ]-------|
      | 0108                       |     0507 |        |
      |--| |-----------------------+----------+        |
      | H006                                   0509    |
0268  |--| |-----------------------------------( )-----|
      | H011                                   0510    |
0270  |--| |--+--------------------------------( )-----|
      | H202  |                                        |
      |--| |--+                                        |
      | H012                                   0511    |
0273  |--| |--+--------------------------------( )-----|
      | H203  |                                        |
      |--| |--+                                        |
      | H902                                   0611    |
0276  |--| |-----------------------------------( )-----|
      | H902                                           |
0278  |--| |-----------------------[ TIM          ]----|
      |                            |          01 |     |
      |                            |      #0120  |     |
      | 0504                                           |
0280  |--| |-----------------------[ TIM          ]----|
      |                            |          02 |     |
      |                            |      #0060  |     |
      |                                                |
0282  |----------------------------[ END (01)    ]----|
                                                   END
```

NON MEMORY TYPE FLIP-FLOP OUTPUT TO REVERSE CONVEYOR MOTOR DIRECTION

MNEMONIC LIST

ADDRESS	MNEMONIC	OPERAND
0217	LD	HR 204
0218	AND	0001
0219	AND	0505
0220	LD	HR 206
0221	KEEP (11)	HR 205
0222	LD	HR 205
0223	AND	0012
0224	AND	0505
0225	AND	0611
0226	LD	HR 001
0227	KEEP (11)	HR 206
0228	LD	HR 014
0229	OR	HR 205
0230	LD	TIM 01
0231	KEEP (11)	HR 902
0232	LD	HR 014
0233	OR	HR 205
0234	OUT	0500
0235	LD	HR 013
0236	OR	HR 204
0237	OUT	0501
0238	LD	HR 001
0239	OUT	0502
0240	LD	HR 003
0241	OR	HR 005
0242	OR	HR 008
0243	OR	HR 010
0244	OR	HR 201
0245	OUT	0503
0246	LD	HR 002
0247	OR	HR 004
0248	OR	HR 007
0249	OR	HR 009
0250	OUT	0504
0251	LD	HR 012
0252	OR	HR 013
0253	OR	HR 014
0254	OR	HR 203
0255	OR	HR 204
0256	OR	HR 205
0257	OUT	0505
0258	LD	HR 002
0259	OR	HR 003
0260	OR	HR 004
0261	OR	HR 007
0262	OR	HR 008
0263	OR	HR 009
0264	OUT	0506
0265	LD	HR 115
0266	LD	0108
0267	KEEP (11)	0507
0268	LD	HR 006
0269	OUT	0509
0270	LD	HR 011
0271	OR	HR 202
0272	OUT	0510
0273	LD	HR 012
0274	OR	HR 203
0275	OUT	0511
0276	LD	HR 902
0277	OUT	0611
0278	LD	HR 902
0279	TIM	01
		# 0120
0280	LD	0504
0281	TIM	02
		# 0060
0282	END (01)	

CONTROL PROBLEM 2

This control problem illustrates a possibility of merging three independent step-counter circuits. All three step-counters controlling the three programs send output signals to the same solenoid valves and therefore control the same fluid power actuators. They also receive input signals from common limit switches on the machine (see step-action diagrams in figure 9-13). The design concept applied in this control problem is called "selective parallel program control". The program selector module calls up the program chosen by the machine operator and activates the corresponding step-counter (figure 9-14). The other two step-counters are prevented from starting. It must be noted that the last flip-flop of the two inactive step-counters remains set while the selected step-counter is in operation. But to prevent these last step-counter flip-flops from causing opposing signals on the fluid power directional control valve for actuator Ⓐ:

$$(101 \Rightarrow \Leftarrow 102),$$

their enabling signal is "AND" connected to the program selector output signal (see addresses 0095 to 0107 in figure 9-14).

The program selector circuit shown in figure 9-08, when used for parallel program selection, needs a minor modification so that changing of programs may occur only if all three step-counters are at their last sequence step (see addresses 0012 to 0040). Note also that the output circuit gathers (connects with logic "OR") the step-counter output signals of all three step-counters and merges them to common output relays 101 and 102 (see addresses 0095 to 0107).

If the last step of the active step-counter also had to be confirmed, so that the newly selected program could start only when the previously selected program is at its last sequence step and that last sequence step is completed, one could build additional safety into the program selector by including the essential confirmation signal of this last step in all three programs. Such a modification to the program selector module is shown at the bottom of figure 9-14. There, step-counter 1 completion is denoted with signal x, step-counter 2 completion with signal y, and step-counter 3 completion with signal z. Relay 1002 is "ON" when all three step-counters are at their last step and their last step action is completed and confirmed (see figure 9-14 section 2).

For control problem 2, such a modification is not required, as all three programs finish their last step with the retraction motion of actuator Ⓐ to limit switch 0001. Hence, none of the three programs is permitted to start if input 0001 limit switch signal is "OFF" (see addresses 0042, 0060 and 0078 in the ladder diagram and the mnemonic list of figure 9-14).

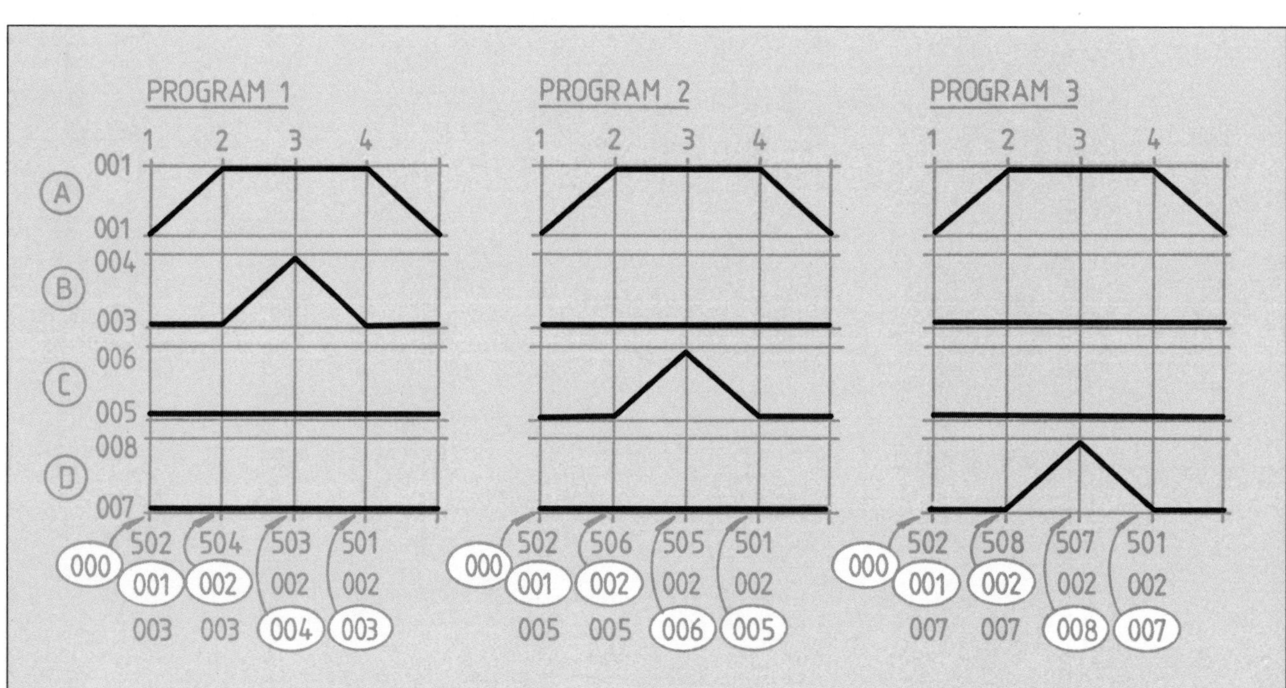

Figure 9-13. Step-action diagram for control problem 2.

Figure 9-14. Ladder diagram and mnemonic list for control problem 2 (section 1 of 2, for section 2 see page 138).

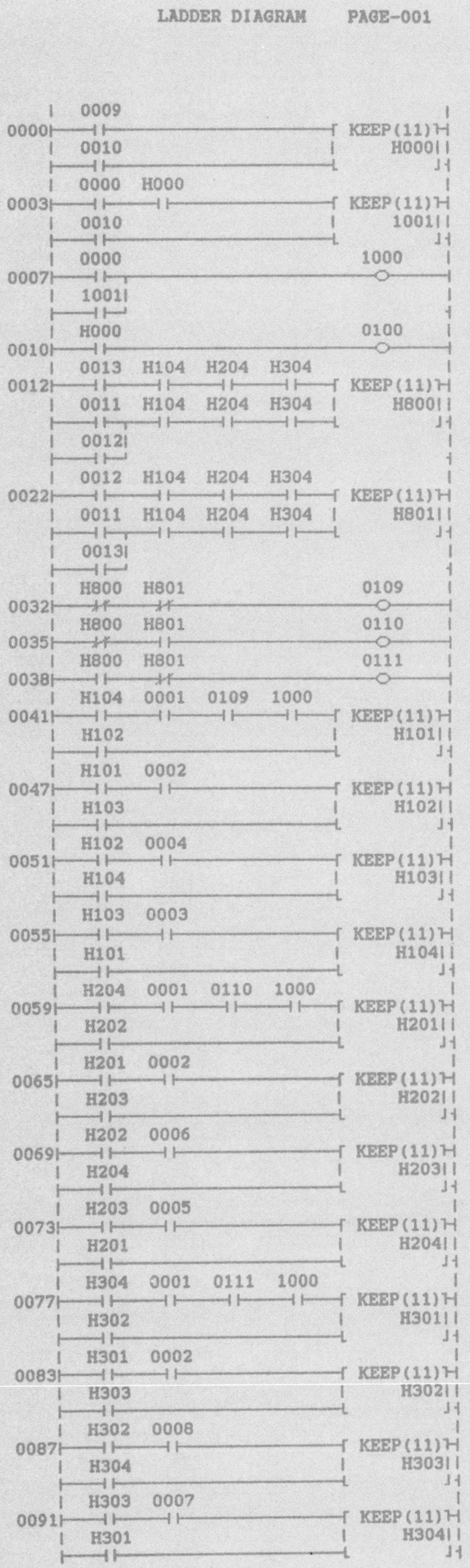

```
            LADDER DIAGRAM    PAGE-001

     | 0009                                |
0000 |--| |-------------------+ KEEP(11)|--
     | 0010                   |    H000||
     |--| |-------------------+         |
     | 0000  H000                      |
0003 |--| |---| |-------------+ KEEP(11)|--
     | 0010                   |    1001||
     |--| |-------------------+         |
     | 0000                    1000    |
0007 |--| |-+--------------------( )----|
     | 1001|                           |
     |--| |-+                          |
     | H000                    0100    |
0010 |--| |----------------------( )----|
     | 0013  H104  H204  H304          |
0012 |--| |---| |---| |---| |-+ KEEP(11)|--
     | 0011  H104  H204  H304 |    H800||
     |--| |-+-| |---| |---| |-+         |
     | 0012|                           |
     |--| |-+                          |
     | 0012  H104  H204  H304          |
0022 |--| |---| |---| |---| |-+ KEEP(11)|--
     | 0011  H104  H204  H304 |    H801||
     |--| |-+-| |---| |---| |-+         |
     | 0013|                           |
     |--| |-+                          |
     | H800  H801              0109    |
0032 |--|/|---|/|----------------( )----|
     | H800  H801              0110    |
0035 |--|/|---| |----------------( )----|
     | H800  H801              0111    |
0038 |--| |---|/|----------------( )----|
     | H104  0001  0109  1000          |
0041 |--| |---| |---| |---| |-+ KEEP(11)|--
     | H102                   |    H101||
     |--| |-------------------+         |
     | H101  0002                      |
0047 |--| |---| |-------------+ KEEP(11)|--
     | H103                   |    H102||
     |--| |-------------------+         |
     | H102  0004                      |
0051 |--| |---| |-------------+ KEEP(11)|--
     | H104                   |    H103||
     |--| |-------------------+         |
     | H103  0003                      |
0055 |--| |---| |-------------+ KEEP(11)|--
     | H101                   |    H104||
     |--| |-------------------+         |
     | H204  0001  0110  1000          |
0059 |--| |---| |---| |---| |-+ KEEP(11)|--
     | H202                   |    H201||
     |--| |-------------------+         |
     | H201  0002                      |
0065 |--| |---| |-------------+ KEEP(11)|--
     | H203                   |    H202||
     |--| |-------------------+         |
     | H202  0006                      |
0069 |--| |---| |-------------+ KEEP(11)|--
     | H204                   |    H203||
     |--| |-------------------+         |
     | H203  0005                      |
0073 |--| |---| |-------------+ KEEP(11)|--
     | H201                   |    H204||
     |--| |-------------------+         |
     | H304  0001  0111  1000          |
0077 |--| |---| |---| |---| |-+ KEEP(11)|--
     | H302                   |    H301||
     |--| |-------------------+         |
     | H301  0002                      |
0083 |--| |---| |-------------+ KEEP(11)|--
     | H303                   |    H302||
     |--| |-------------------+         |
     | H302  0008                      |
0087 |--| |---| |-------------+ KEEP(11)|--
     | H304                   |    H303||
     |--| |-------------------+         |
     | H303  0007                      |
0091 |--| |---| |-------------+ KEEP(11)|--
     | H301                   |    H304||
     |--| |-------------------+         |
```

ADDRESS	MNEMONIC		OPERAND
0000	LD		0009
0001	LD		0010
0002	KEEP (11)	HR	000
0003	LD		0000
0004	AND	HR	000
0005	LD		0010
0006	KEEP (11)		1001
0007	LD		0000
0008	OR		1001
0009	OUT		1000
0010	LD	HR	000
0011	OUT		0100
0012	LD		0013
0013	AND	HR	104
0014	AND	HR	204
0015	AND	HR	304
0016	LD		0011
0017	OR		0012
0018	AND	HR	104
0019	AND	HR	204
0020	AND	HR	304
0021	KEEP (11)	HR	800
0022	LD		0012
0023	AND	HR	104
0024	AND	HR	204
0025	AND	HR	304
0026	LD		0011
0027	OR		0013
0028	AND	HR	104
0029	AND	HR	204
0030	AND	HR	304
0031	KEEP (11)	HR	801
0032	LD NOT	HR	800
0033	AND NOT	HR	801
0034	OUT		0109
0035	LD NOT	HR	800
0036	AND	HR	801
0037	OUT		0110
0038	LD	HR	800
0039	AND NOT	HR	801
0040	OUT		0111
0041	LD	HR	104
0042	AND		0001
0043	AND		0109
0044	AND		1000
0045	LD	HR	102
0046	KEEP (11)	HR	101
0047	LD	HR	101
0048	AND		0002
0049	LD	HR	103
0050	KEEP (11)	HR	102
0051	LD	HR	102
0052	AND		0004
0053	LD	HR	104
0054	KEEP (11)	HR	103
0055	LD	HR	103
0056	AND		0003
0057	LD	HR	101
0058	KEEP (11)	HR	104
0059	LD	HR	204
0060	AND		0001
0061	AND		0110
0062	AND		1000
0063	LD	HR	202
0064	KEEP (11)	HR	201
0065	LD	HR	201
0066	AND		0002
0067	LD	HR	203
0068	KEEP (11)	HR	202
0069	LD	HR	202
0070	AND		0006
0071	LD	HR	204
0072	KEEP (11)	HR	203
0073	LD	HR	203
0074	AND		0005
0075	LD	HR	201
0076	KEEP (11)	HR	204
0077	LD	HR	304
0078	AND		0001
0079	AND		0111
0080	AND		1000
0081	LD	HR	302
0082	KEEP (11)	HR	301

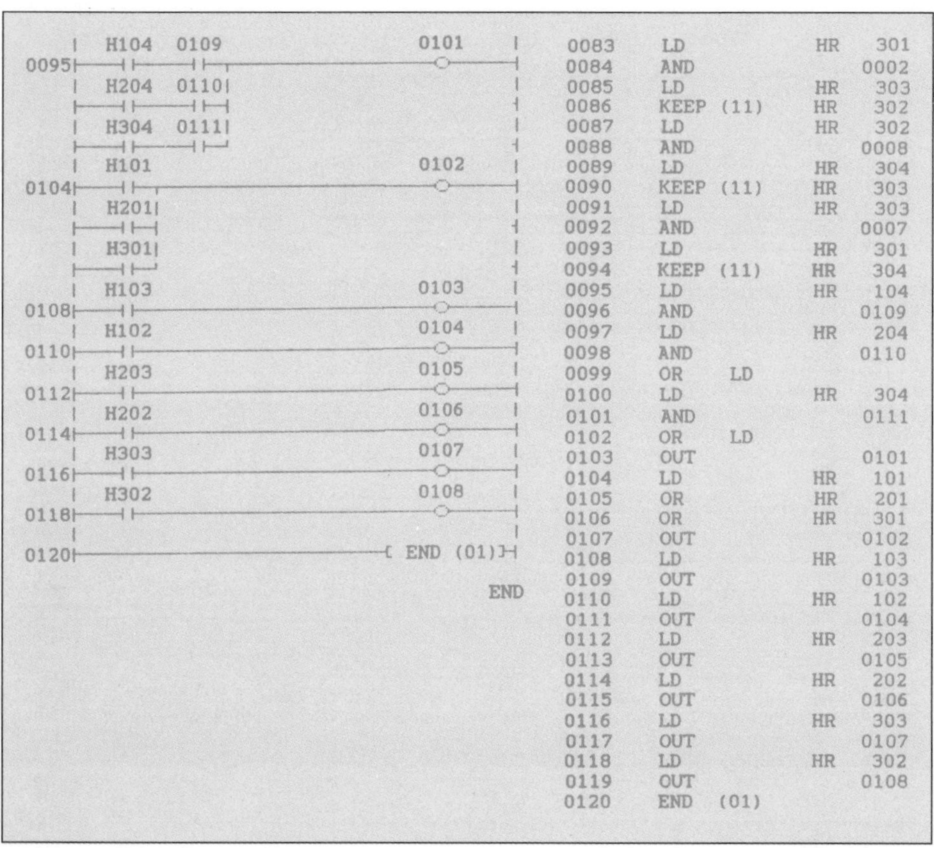

0083	LD	HR	301
0084	AND		0002
0085	LD	HR	303
0086	KEEP (11)	HR	302
0087	LD	HR	302
0088	AND		0008
0089	LD	HR	304
0090	KEEP (11)	HR	303
0091	LD	HR	303
0092	AND		0007
0093	LD	HR	301
0094	KEEP (11)	HR	304
0095	LD	HR	104
0096	AND		0109
0097	LD	HR	204
0098	AND		0110
0099	OR LD		
0100	LD	HR	304
0101	AND		0111
0102	OR LD		
0103	OUT		0101
0104	LD	HR	101
0105	OR	HR	201
0106	OR	HR	301
0107	OUT		0102
0108	LD	HR	103
0109	OUT		0103
0110	LD	HR	102
0111	OUT		0104
0112	LD	HR	203
0113	OUT		0105
0114	LD	HR	202
0115	OUT		0106
0116	LD	HR	303
0117	OUT		0107
0118	LD	HR	302
0119	OUT		0108
0120	END (01)		

Figure 9-14. Ladder diagram and mnemonic list for control problem 2. (section 2 of 2, for section 1 see page 137).

INPUT/OUTPUT ASSIGNMENT LIST

PLC inputs:

000 = manual start signal
001 = actuator Ⓐ retracted
002 = actuator Ⓐ extended
003 = actuator Ⓑ retracted
004 = actuator Ⓑ extended
005 = actuator Ⓒ retracted
006 = actuator Ⓒ extended
007 = actuator Ⓓ retracted
008 = actuator Ⓓ extended
009 = select automatic cycling
010 = select manual cycling
011 = select program 1
012 = select program 2
013 = select program 3

PLC outputs:

100 = auto selection lamp
101 = actuator Ⓐ retract solenoid
102 = actuator Ⓐ extend solenoid
103 = actuator Ⓑ retract solenoid
104 = actuator Ⓑ extend solenoid
105 = actuator Ⓒ retract solenoid
106 = actuator Ⓒ extend solenoid
107 = actuator Ⓓ retract solenoid
108 = actuator Ⓓ extend solenoid
109 = program 1 lamp
110 = program 2 lamp
111 = program 3 lamp

HR 00 = auto/manual selector flip-flop
HR 101 to HR 104 = program 1 flip-flops
HR 201 to HR 204 = program 2 flip-flops
HR 301 to HR 304 = program 3 flip-flops
HR 800 and HR 801 = program selector flip-flops
1000 and HR 1001 = auto/manual selector flip-flop.

Note: For an Omron C28K PLC, the solenoid output signals in the step-action diagram of figure 9-13 would be taken from channel 01 instead of channel 05 (example: 502 becomes 102).

PLC TO CATER FOR THE PECULIARITY OF THREE-POSITION HYDRAULIC VALVES

Electrically signalled, three-position, hydraulic solenoid valves (usually featuring closed centre, open centre or tandem centre valve middle position) with a spring-selected centre position will automatically select the centre position (spring action) when neither of the two electrical solenoid signals is present. Electrical solenoid signals from PLC controllers leading to such hydraulic directional control valves must therefore be carefully timed and sustained for a precise duration to match the stringent requirements of the hydraulic machine sequence (figures 9-15 and 9-18).

Figure 9-15. Step-action diagram for control problem 3.

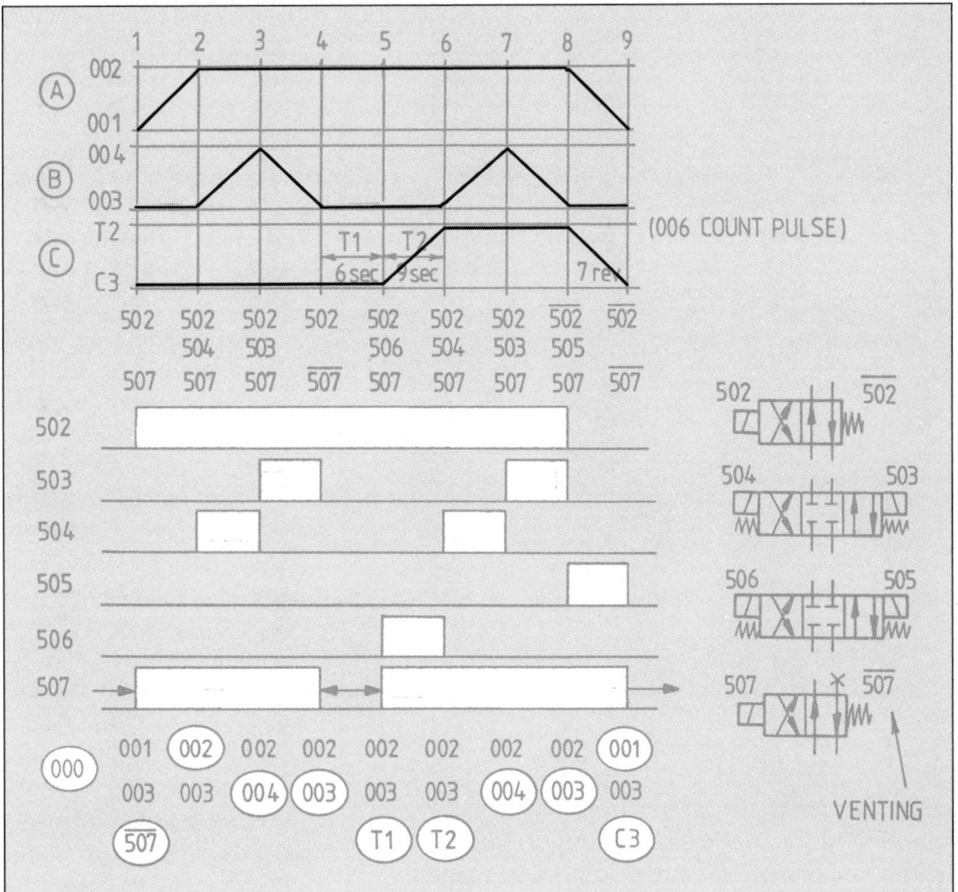

PLC sequence control systems designed with the step-counter algorithm automatically cancel the output signal of the previous sequence step flip-flop whenever the present step is actuated (figures 8-05, 8-06, 8-07 and 8-12). Therefore, wherever step-counter-based PLC control systems have to control and operate three-position hydraulic or pneumatic valves or spring-returned two-position valves, the circuit designer may choose one of three options to sustain (or prolong) step-counter output signals. The options available are as follows:

- The monoflop type output relay leading to the non-memory type hydraulic or pneumatic valve may be converted into an RS flip-flop type set/reset relay, which requires specific set and reset signalling instructions (figures 4-03, 4-04, 4-02 and 8-17 A)
- Additional holding relays (flip-flops) may be used to maintain the PLC's output signal and thus the position of such non-memory type pneumatic or hydraulic directional control valves (see figure 8-17 B)
- Step-counter output signals may be "OR" connected in the ladder diagram output circuit so that each consecutive step-counter flip-flop sustains the output signal leading to a mono-flop type directional control valve until cancellation is required (see output signals 509 and 510 in figure 8-29).

The last of these three suggested methods is the simplest and best for designing circuits and sustaining step-counter output signals leading to non-memory type hydraulic or pneumatic valves (see also the section "PLC Step-Counter Programming for Non-Memory Type Directional Control Valves" in Chapter 8).

CONTROL PROBLEM 3

Control problem 3 illustrates the design for a hydraulically powered machine consisting of two linear hydraulic actuators (actuators Ⓐ and Ⓑ) and a hydraulic motor (actuator Ⓒ). The sequence of this machine is illustrated in the step-action diagram of figure 9-15. The electrohydraulic control circuit displaying all relevant PLC input and output signals is given in figure 9-16.

Hydraulic actuator Ⓐ is controlled by a spring-return type directional control valve (DCV) (figure 9-16). Actuators Ⓑ and Ⓒ require a closed-centre, three-position DCV. A fixed displacement pump recirculates its flow back to tank via the vented compound relief valve during inactive stages of the machine sequence. This venting valve is also a monoflop type spring-return valve. Solenoid actuation of this venting valve causes the pump to come on-stream to supply pump flow into the system. The absence of the PLC output signal 507 causes venting and flow recirculation back to the tank (see electrohydraulic circuit in figure 9-16).

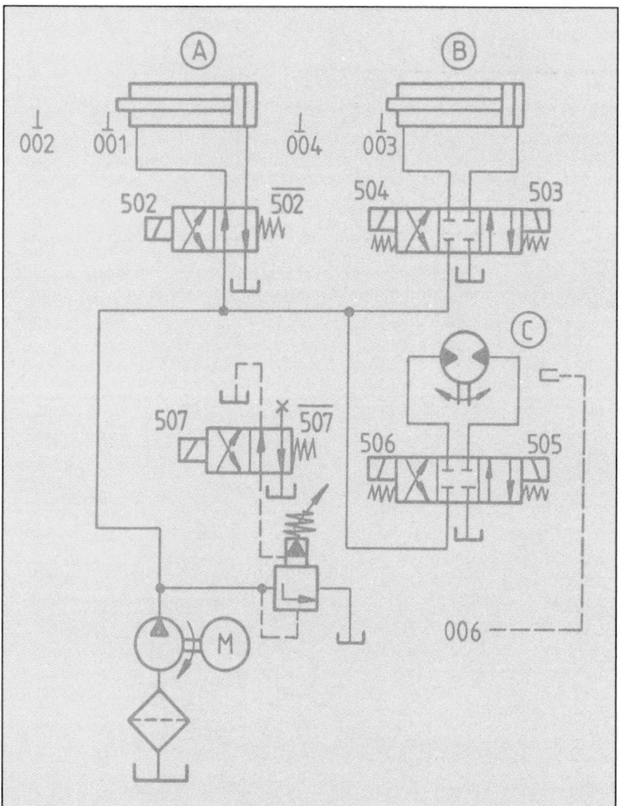

Figure 9-16. Electrohydraulic control circuit for control problem 3. This circuit applies the vented compound relief method for pump off-loading.

The sequencing action for the two hydraulic actuators Ⓐ and Ⓑ is according to the step-action diagram in figure 9-15. The motions of the hydraulic motor Ⓒ are also given in that diagram. The motor starts to rotate anticlockwise at sequence step 5 for 9 seconds. The anticlockwise rotations are terminated and confirmed by timer 2 (TIM 02). At sequence step 8, the linear actuator Ⓐ and the motor Ⓒ operate simultaneously. The motor, however, is now rotating clockwise, and its rotations are counted by counter 3 and must amount to seven full rotations. Motor rotations at step 8 are required to stop as soon as seven rotations are completed, regardless of the position of actuator Ⓐ. To accomplish this, the PLC output signal 505 connected to solenoid 505 of the DCV that controls actuator Ⓒ must be terminated as soon as the counter flag has switched to "ON" (seven rotations completed), and its termination signal must not be dependent on the completion signal 001 from the retraction stroke of linear actuator Ⓐ (figure 9-17).

To achieve minimum power consumption and no system overheating during inactive stages of the machine, the compound relief valve must be vented during the time delay period, starting with sequence step 4 and at the completion of sequence step 8, when the motion cycle is completed. Step-counter steps 4 and 9 (flip-flops HR 004 and HR 009) signal this venting action. Venting must also occur when the sequence is interrupted by a machine sequence interrupt module, as is shown in figure 9-01. When sequence interrupt is signalled (HR 115), flip-flop 114 renders $\overline{\text{HR 114}}$. This breaks contact HR 114 in the logic line to output 507 (address 0081) and thus automatically activates the venting action. The venting action recirculates pump flow back to tank for as long as interrupt is signalled.

The bar chart below the step-action diagram for this machine (figure 9-15) shows at what times the various solenoid signals from the PLC need to be switched "ON" or "OFF". Solenoid output signal 502 from the PLC controller must be deliberately held (sustained) during sequence steps 1 to 7, because actuator Ⓐ (spring-return valve) has to remain in the extended position during those steps. To achieve this, a multiple "OR" function is used: holding relay 1 (HR 001) switches output signal 502 to the "ON" position and holding relays HR 002 to HR 007 are used to sustain this output signal (see ladder diagram figure 9-17, addresses 0056 to 0063). The same principle is applied for the venting valve, which also needs its signal sustained from steps 1 to 7 (see ladder diagram figure 9-17, addresses 0074 to 0082). Alternatively, one could use an auxiliary relay, as is shown in figure 8-17, to sustain the output during the required period.

CONTROL PROBLEM 4

Control problem 4 follows the same machine sequence given in control problem 3 (figure 9-18). But in this application, to achieve minimum power consumption or no power wastage during machine inaction, pump off-loading is achieved with a tandem centre, directional control valve allocated for actuator Ⓑ. (For the concept of pump off-loading see the book *Industrial Hydraulic Control* by Peter Rohner.) To achieve this pump off-loading, the PLC must not send any signals to the directional control valve for actuator Ⓑ during the time-delay period starting at the end of sequence step 3, or during the machine "OFF" period starting at the end of sequence step 7 (see electrohydraulic control circuit figure 9-19). With no PLC output signal on the solenoids of actuator Ⓑ, the springs cause the valve to centre and pump flow is recirculated back to the tank (see figure 9-19). This is called "tandem centre pump offloading".

Whenever flow from the pump is required, the directional control valve for actuator Ⓑ must receive either a signal 504 or a signal 503 from the PLC to switch the valve's spool out of its centre position. For this reason, signal 503 is given at sequence steps 1, 4 and 7, even though actuator Ⓑ is already in the retracted position.

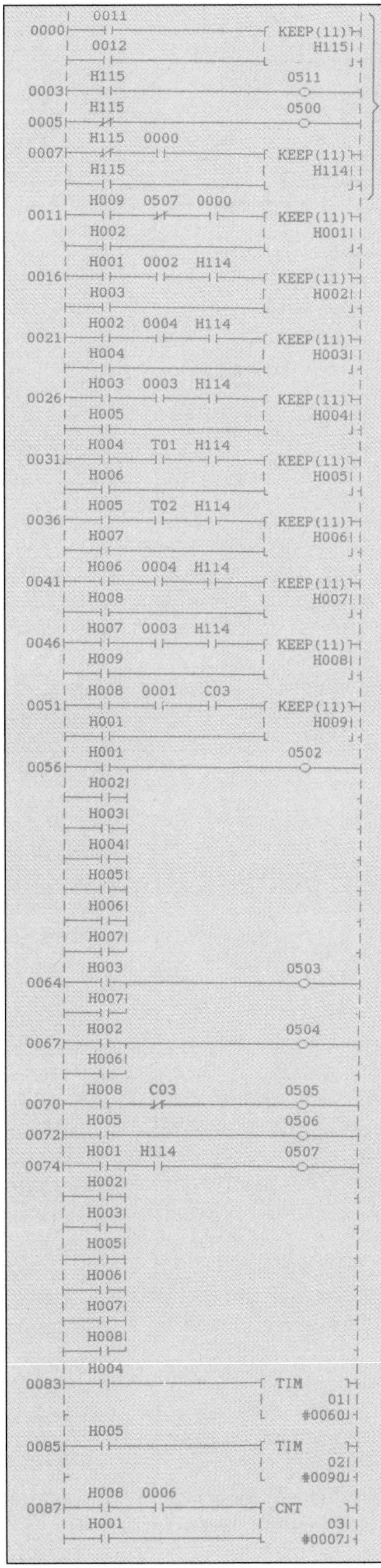

Figure 9-17.
Ladder diagram and mnemonic list for control problem 3.

```
INTERRUPT MODULE
     | 0011                                   |
0000 |--| |-------------------------+ KEEP(11)|
     | 0012                         |   H115  |
     |--| |-------------------------+         |
     | H115                           0511    |
0003 |--| |---------------------------( )-----|
     | H115                           0500    |
0005 |--|/|---------------------------( )-----|
     | H115  0000                             |
0007 |--|/|---| |-------------------+ KEEP(11)|
     | H115                         |   H114  |
     |--| |-------------------------+         |
     | H009  0507  0000                       |
0011 |--| |---|/|---| |-------------+ KEEP(11)|
     | H002                         |   H001  |
     |--| |-------------------------+         |
     | H001  0002  H114                       |
0016 |--| |---| |---| |-------------+ KEEP(11)|
     | H003                         |   H002  |
     |--| |-------------------------+         |
     | H002  0004  H114                       |
0021 |--| |---| |---| |-------------+ KEEP(11)|
     | H004                         |   H003  |
     |--| |-------------------------+         |
     | H003  0003  H114                       |
0026 |--| |---| |---| |-------------+ KEEP(11)|
     | H005                         |   H004  |
     |--| |-------------------------+         |
     | H004  T01   H114                       |
0031 |--| |---| |---| |-------------+ KEEP(11)|
     | H006                         |   H005  |
     |--| |-------------------------+         |
     | H005  T02   H114                       |
0036 |--| |---| |---| |-------------+ KEEP(11)|
     | H007                         |   H006  |
     |--| |-------------------------+         |
     | H006  0004  H114                       |
0041 |--| |---| |---| |-------------+ KEEP(11)|
     | H008                         |   H007  |
     |--| |-------------------------+         |
     | H007  0003  H114                       |
0046 |--| |---| |---| |-------------+ KEEP(11)|
     | H009                         |   H008  |
     |--| |-------------------------+         |
     | H008  0001  C03                        |
0051 |--| |---| |---| |-------------+ KEEP(11)|
     | H001                         |   H009  |
     |--| |-------------------------+         |
     | H001                           0502    |
0056 |--| |-+-------------------------( )-----|
     | H002|                                  |
     |--| |-+                                 |
     | H003|                                  |
     |--| |-+                                 |
     | H004|                                  |
     |--| |-+                                 |
     | H005|                                  |
     |--| |-+                                 |
     | H006|                                  |
     |--| |-+                                 |
     | H007|                                  |
     |--| |-+                                 |
     | H003                           0503    |
0064 |--| |-+-------------------------( )-----|
     | H007|                                  |
     |--| |-+                                 |
     | H002                           0504    |
0067 |--| |-+-------------------------( )-----|
     | H006|                                  |
     |--| |-+                                 |
     | H008  C03                      0505    |
0070 |--| |---|/|-----------------------( )-----|
     | H005                           0506    |
0072 |--| |---------------------------( )-----|
     | H001  H114                     0507    |
0074 |--| |-+-| |---------------------( )-----|
     | H002|                                  |
     |--| |-+                                 |
     | H003|                                  |
     |--| |-+                                 |
     | H005|                                  |
     |--| |-+                                 |
     | H006|                                  |
     |--| |-+                                 |
     | H007|                                  |
     |--| |-+                                 |
     | H008|                                  |
     |--| |-+                                 |
     | H004                                   |
0083 |--| |-------------------------+ TIM     |
     |                              |    01   |
     |                              +  #0060  |
     | H005                                   |
0085 |--| |-------------------------+ TIM     |
     |                              |    02   |
     |                              +  #0090  |
     | H008  0006                             |
0087 |--| |---| |-------------------+ CNT     |
     | H001                         |    03   |
     |--| |-------------------------+  #0007  |
```

ADDRESS	MNEMONIC		OPERAND	
0000	LD			0011
0001	LD			0012
0002	KEEP	(11)	HR	115
0003	LD		HR	115
0004	OUT			0511
0005	LD	NOT	HR	115
0006	OUT			0500
0007	LD	NOT	HR	115
0008	AND			0000
0009	LD		HR	115
0010	KEEP	(11)	HR	114
0011	LD		HR	009
0012	AND	NOT		0507
0013	AND			0000
0014	LD		HR	002
0015	KEEP	(11)	HR	001
0016	LD		HR	001
0017	AND			0002
0018	AND		HR	114
0019	LD		HR	003
0020	KEEP	(11)	HR	002
0021	LD		HR	002
0022	AND			0004
0023	AND		HR	114
0024	LD		HR	004
0025	KEEP	(11)	HR	003
0026	LD		HR	003
0027	AND			0003
0028	AND		HR	114
0029	LD		HR	005
0030	KEEP	(11)	HR	004
0031	LD		HR	004
0032	AND		TIM	01
0033	AND		HR	114
0034	LD		HR	006
0035	KEEP	(11)	HR	005
0036	LD		HR	005
0037	AND		TIM	02
0038	AND		HR	114
0039	LD		HR	007
0040	KEEP	(11)	HR	006
0041	LD		HR	006
0042	AND			0004
0043	AND		HR	114
0044	LD		HR	008
0045	KEEP	(11)	HR	007
0046	LD		HR	007
0047	AND			0003
0048	AND		HR	114
0049	LD		HR	009
0050	KEEP	(11)	HR	008
0051	LD		HR	008
0052	AND			0001
0053	AND		CNT	03
0054	LD		HR	001
0055	KEEP	(11)	HR	009
0056	LD		HR	001
0057	OR		HR	002
0058	OR		HR	003
0059	OR		HR	004
0060	OR		HR	005
0061	OR		HR	006
0062	OR		HR	007
0063	OUT			0502
0064	LD		HR	003
0065	OR		HR	007
0066	OUT			0503
0067	LD		HR	002
0068	OR		HR	006
0069	OUT			0504
0070	LD		HR	008
0071	OUT			0505
0072	LD		HR	005
0073	OUT			0506
0074	LD		HR	001
0075	OR		HR	002
0076	OR		HR	003
0077	OR		HR	005
0078	OR		HR	006
0079	OR		HR	007
0080	OR		HR	008
0081	AND		HR	114
0082	OUT			0507
0083	LD		HR	004
0084	TIM			01
			#	0060
0085	LD		HR	005
0086	TIM			02
			#	0090
0087	LD		HR	008
0088	AND			0006
0089	LD		HR	001
0090	CNT			03
			#	0007
0091	END	(01)		

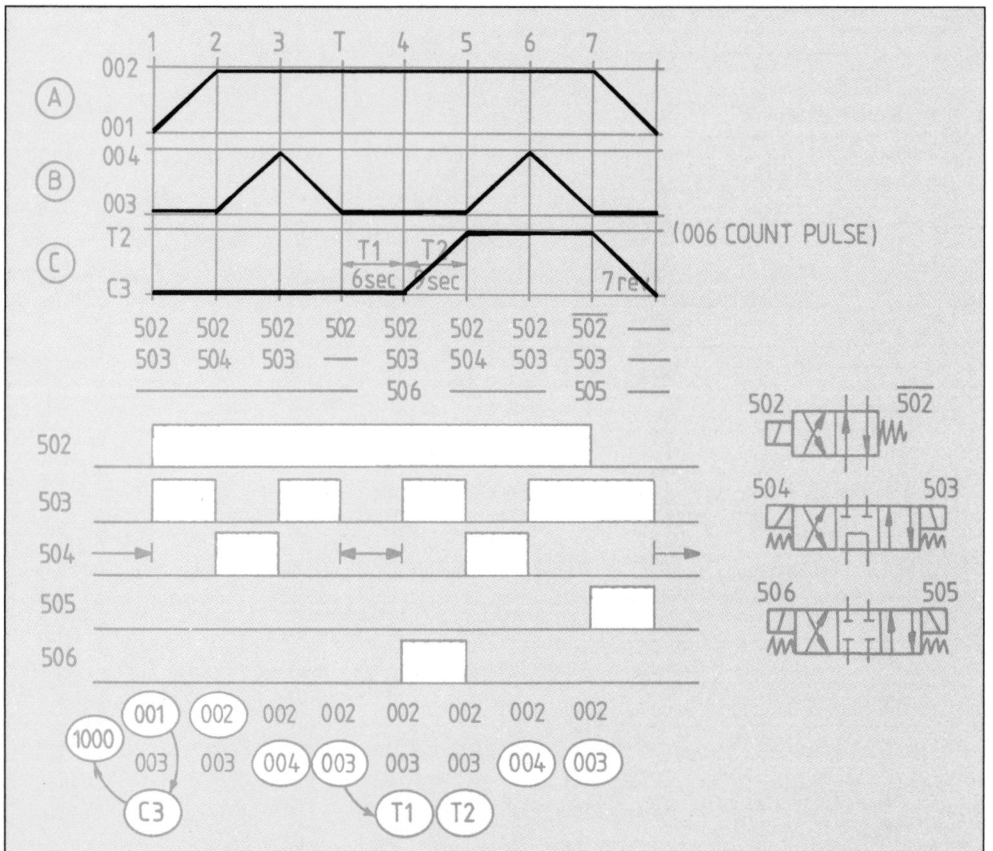

Figure 9-18. Step-action diagram for control problem 4.

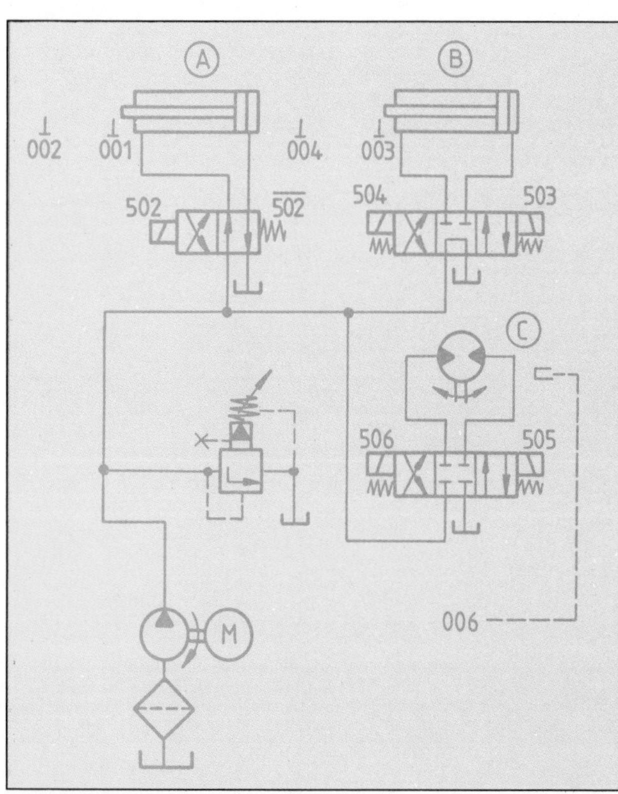

Figure 9-19. Electrohydraulic control circuit for control problem 4. This circuit applies the tandem centre pump off-load method.

Step-counter module 4 obviously resets step-counter module 3 when limit switch 003 is actuated at the end of step 3 (figure 9-20). Pump off-loading (flow recirculation back to the tank via the tandem centre of the DCV), however, therefore occurs as soon as actuator Ⓑ is fully retrated. Similarly, pump offloading is applied at the completion of sequence step 7. Signal 503, however, is not cancelled until counter contact C03 signals "motor rotation" completed (figure 9-20, address 0052), and limit switch 001 signals "actuator Ⓐ fully retracted" (figure 9-20, addresses 0056 and 0057). If one of the two actuators finishes its task ahead of the other, the PLC's output signal 503 must then still be maintained to cause pump flow to the system (see addresses 0050 to 0062). This means that the directional control valve used here to recirculate pump flow back to the tank is not permitted to move into the centre position until all actuator motions are completed. However, as soon as both tasks are completed (figure 9-18), pump flow must stop. Hence PLC output signal 503 must then disappear and the directional control valve Ⓑ must assume its centre position. Here, De Morgan's theorem of switching theory, as described and illustrated in Chapter 3, figures 3-10, 3-11 and 3-12, is applied and yields:

$$\overline{0001 \bullet C03} = \overline{0001} + \overline{C03}$$

As an incomplete Boolean equation, one could write:

$$\ldots + (HR007 \bullet \overline{0001}) + (HR007 \bullet \overline{C03}) = 503$$

Figure 9-20. Ladder diagram and mnemonic list for control problem 4.

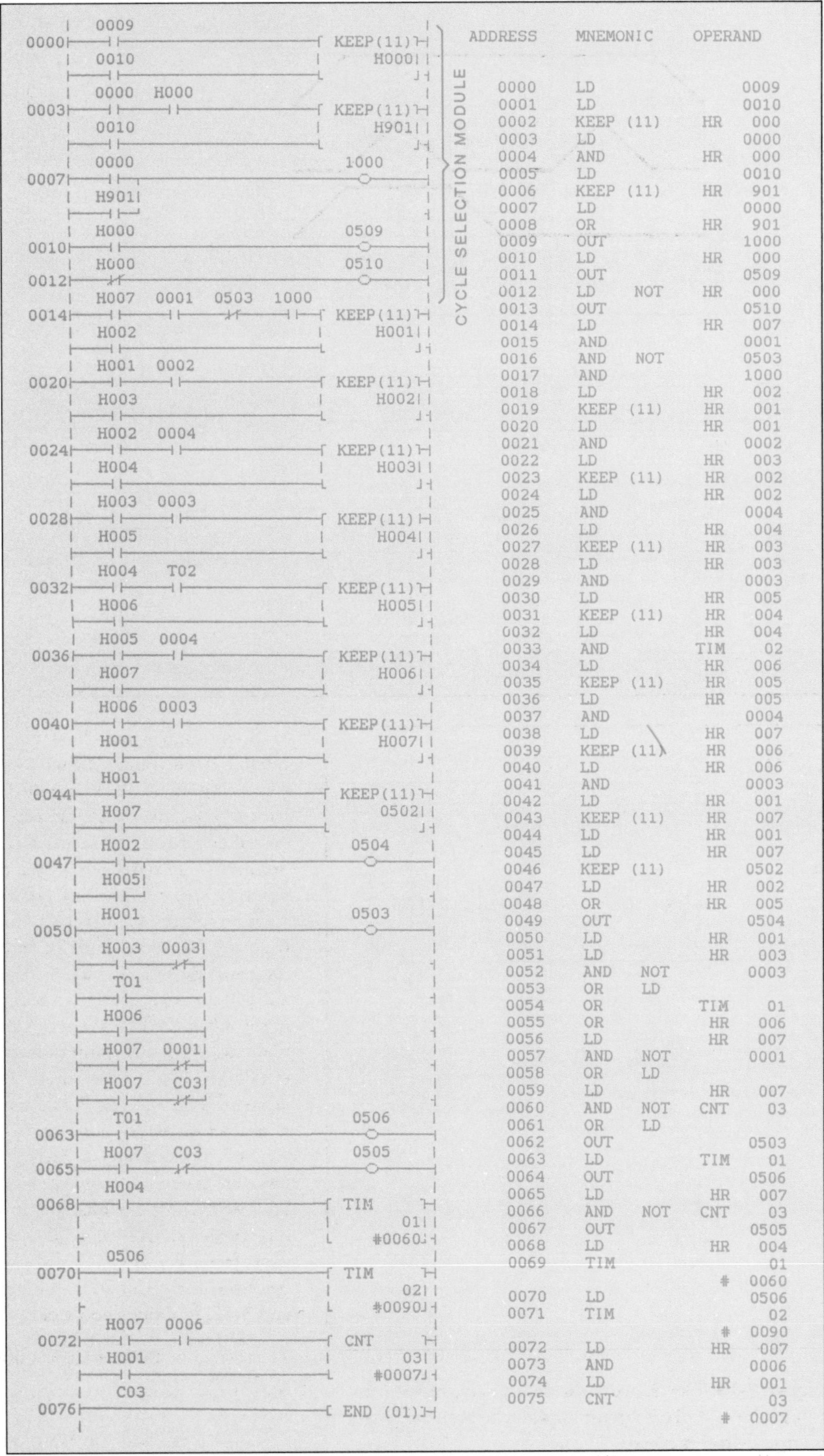

ADDRESS	MNEMONIC	OPERAND
0000	LD	0009
0001	LD	0010
0002	KEEP (11)	HR 000
0003	LD	0000
0004	AND	HR 000
0005	LD	0010
0006	KEEP (11)	HR 901
0007	LD	0000
0008	OR	HR 901
0009	OUT	1000
0010	LD	HR 000
0011	OUT	0509
0012	LD NOT	HR 000
0013	OUT	0510
0014	LD	HR 007
0015	AND	0001
0016	AND NOT	0503
0017	AND	1000
0018	LD	HR 002
0019	KEEP (11)	HR 001
0020	LD	HR 001
0021	AND	0002
0022	LD	HR 003
0023	KEEP (11)	HR 002
0024	LD	HR 002
0025	AND	0004
0026	LD	HR 004
0027	KEEP (11)	HR 003
0028	LD	HR 003
0029	AND	0003
0030	LD	HR 005
0031	KEEP (11)	HR 004
0032	LD	HR 004
0033	AND	TIM 02
0034	LD	HR 006
0035	KEEP (11)	HR 005
0036	LD	HR 005
0037	AND	0004
0038	LD	HR 007
0039	KEEP (11)	HR 006
0040	LD	HR 006
0041	AND	0003
0042	LD	HR 001
0043	KEEP (11)	HR 007
0044	LD	HR 001
0045	LD	HR 007
0046	KEEP (11)	0502
0047	LD	HR 002
0048	OR	HR 005
0049	OUT	0504
0050	LD	HR 001
0051	LD	HR 003
0052	AND NOT	0003
0053	OR LD	
0054	OR	TIM 01
0055	OR	HR 006
0056	LD	HR 007
0057	AND NOT	0001
0058	OR LD	
0059	LD	HR 007
0060	AND NOT	CNT 03
0061	OR LD	
0062	OUT	0503
0063	LD	TIM 01
0064	OUT	0506
0065	LD	HR 007
0066	AND NOT	CNT 03
0067	OUT	0505
0068	LD	HR 004
0069	TIM	01
		# 0060
0070	LD	0506
0071	TIM	02
		# 0090
0072	LD	HR 007
0073	AND	0006
0074	LD	HR 001
0075	CNT	03
		# 0007

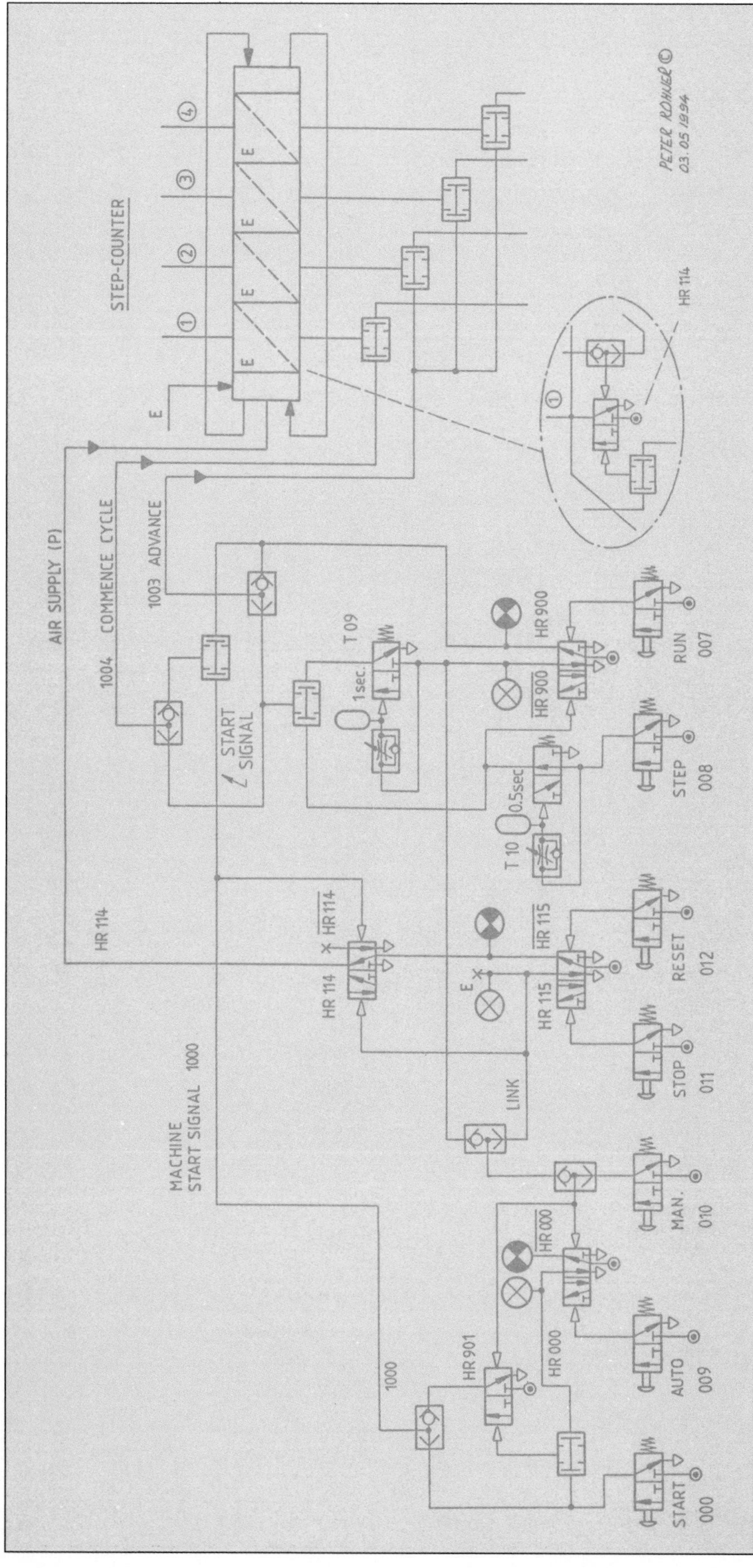

Figure 9-21. Fluid power circuit with three fringe condition modules integrated with a 4-step step-counter. The modules illustrated are : complex cycle selection module (left), sequencing interrupt module (middle) and stepping module (right).

Figure 9-22. Ladder diagram and mnemonic list for the pneumatic circuit in figure 9-21.

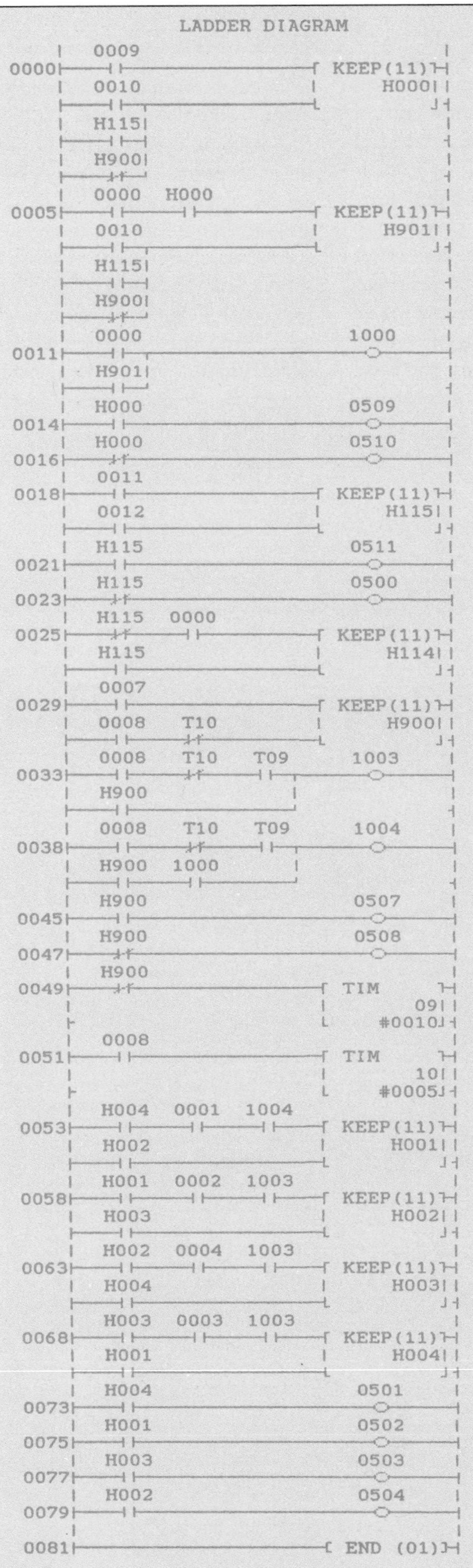

LADDER DIAGRAM

```
     | 0009                              |
0000 |--| |-------------------+ KEEP(11)|-
     | 0010                   |    H000 |
     |--| |-------------------+         |
     | H115 |                            |
     |--| |-|                            |
     | H900 |                            |
     |--|/|-|                            |
     | 0000    H000                      |
0005 |--| |-----| |-----------+ KEEP(11)|-
     | 0010                   |    H901 |
     |--| |-------------------+         |
     | H115 |                            |
     |--| |-|                            |
     | H900 |                            |
     |--|/|-|                            |
     | 0000                    1000      |
0011 |--| |-+-------------------( )------|
     | H901 |                            |
     |--| |-|                            |
     | H000                    0509      |
0014 |--| |---------------------( )------|
     | H000                    0510      |
0016 |--|/|---------------------( )------|
     | 0011                              |
0018 |--| |-------------------+ KEEP(11)|-
     | 0012                   |    H115 |
     |--| |-------------------+         |
     | H115                    0511      |
0021 |--| |---------------------( )------|
     | H115                    0500      |
0023 |--|/|---------------------( )------|
     | H115    0000                      |
0025 |--|/|-----| |-----------+ KEEP(11)|-
     | H115                   |    H114 |
     |--| |-------------------+         |
     | 0007                              |
0029 |--| |-------------------+ KEEP(11)|-
     | 0008    T10            |    H900 |
     |--| |-----|/|-----------+         |
     | 0008    T10     T09     1003      |
0033 |--| |-----|/|-----| |-+--( )------|
     | H900                 |            |
     |--| |-----------------+            |
     | 0008    T10     T09     1004      |
0038 |--| |-----|/|-----| |-+--( )------|
     | H900    1000         |            |
     |--| |-----| |---------+            |
     | H900                    0507      |
0045 |--| |---------------------( )------|
     | H900                    0508      |
0047 |--|/|---------------------( )------|
     | H900                              |
0049 |--|/|-------------------+ TIM     |-
     |                        |      09 |
     |                        +  #0010  |
     | 0008                              |
0051 |--| |-------------------+ TIM     |-
     |                        |      10 |
     |                        +  #0005  |
     | H004    0001    1004              |
0053 |--| |-----| |-----| |---+ KEEP(11)|-
     | H002                   |    H001 |
     |--| |-------------------+         |
     | H001    0002    1003              |
0058 |--| |-----| |-----| |---+ KEEP(11)|-
     | H003                   |    H002 |
     |--| |-------------------+         |
     | H002    0004    1003              |
0063 |--| |-----| |-----| |---+ KEEP(11)|-
     | H004                   |    H003 |
     |--| |-------------------+         |
     | H003    0003    1003              |
0068 |--| |-----| |-----| |---+ KEEP(11)|-
     | H001                   |    H004 |
     |--| |-------------------+         |
     | H004                    0501      |
0073 |--| |---------------------( )------|
     | H001                    0502      |
0075 |--| |---------------------( )------|
     | H003                    0503      |
0077 |--| |---------------------( )------|
     | H002                    0504      |
0079 |--| |---------------------( )------|
     |                                   |
0081 |------------------------[ END (01)]-
```

ADDRESS	MNEMONIC		OPERAND	
0000	LD			0009
0001	LD			0010
0002	OR		HR	115
0003	OR	NOT	HR	900
0004	KEEP	(11)	HR	000
0005	LD			0000
0006	AND		HR	000
0007	LD			0010
0008	OR		HR	115
0009	OR	NOT	HR	900
0010	KEEP	(11)	HR	901
0011	LD			0000
0012	OR		HR	901
0013	OUT			1000
0014	LD		HR	000
0015	OUT			0509
0016	LD	NOT	HR	000
0017	OUT			0510
0018	LD			0011
0019	LD			0012
0020	KEEP	(11)	HR	115
0021	LD		HR	115
0022	OUT			0511
0023	LD	NOT	HR	115
0024	OUT			0500
0025	LD	NOT	HR	115
0026	AND			0000
0027	LD		HR	115
0028	KEEP	(11)	HR	114
0029	LD			0007
0030	LD			0008
0031	AND	NOT	TIM	10
0032	KEEP	(11)	HR	900
0033	LD			0008
0034	AND	NOT	TIM	10
0035	AND		TIM	09
0036	OR		HR	900
0037	OUT			1003
0038	LD			0008
0039	AND	NOT	TIM	10
0040	AND		TIM	09
0041	LD		HR	900
0042	AND			1000
0043	OR	LD		
0044	OUT			1004
0045	LD		HR	900
0046	OUT			0507
0047	LD	NOT	HR	900
0048	OUT			0508
0049	LD	NOT	HR	900
0050	TIM			09
			#	0010
0051	LD			0008
0052	TIM			10
			#	0005
0053	LD		HR	004
0054	AND			0001
0055	AND			1004
0056	LD		HR	002
0057	KEEP	(11)	HR	001
0058	LD		HR	001
0059	AND			0002
0060	AND			1003
0061	LD		HR	003
0062	KEEP	(11)	HR	002
0063	LD		HR	002
0064	AND			0004
0065	AND			1003
0066	LD		HR	004
0067	KEEP	(11)	HR	003
0068	LD		HR	003
0069	AND			0003
0070	AND			1003
0071	LD		HR	001
0072	KEEP	(11)	HR	004
0073	LD		HR	004
0074	OUT			0501
0075	LD		HR	001
0076	OUT			0502
0077	LD		HR	003
0078	OUT			0503
0079	LD		HR	002
0080	OUT			0504
0081	END	(01)		

CONCLUSION

The complexity of hydraulic and pneumatic control systems with applications of three-position, non-memory-type valves, and the sometimes elaborate pump control circuits of hydraulics, necessitates that programmers and designers of PLC control systems have intricate knowledge of the peculiarities of industrial, mining and agricultural hydraulic controls. Only then will they have a chance to successfully design functional and economical PLC control circuitry to do justice to both PLC electronic control design and hydraulic control design. It is therefore desirable, if not mandatory, that university engineering schools include PLC programming and fluid power control subjects under the umbrella of mechatronics in their engineering course curriculum.

FRINGE CONDITION MODULES SUMMARY

Fringe condition modules are frequently used to complement sequential controls, with such features as interrupt the machine cycle, select automatic or manual cycling, select stepping or run mode, select a particular machine program, cause a machine to abort the normal program and continue with an emergency program. Such fringe condition modules are often standardised and pretested subcircuits that are attached to the fringe of the sequential control. Some of these modules may be subject to stringent local design regulations. Check such regulations before designing fringe condition modules!

10 ARITHMETIC FUNCTION PROGRAMMING ON PLCs

Programmable logic controllers can also execute arithmetic functions, such as ADD, SUBTRACT, MULTIPLY, DIVIDE and COMPARE as well as take SQUARE ROOT. Closely related to these arithmetic functions are the special PLC functions for converting binary data to binary-coded decimal (BCD) data, for converting BCD data to binary and hexadecimal data, and for data movement from one data area or word (channel) to another. All these complex and exotic functions are listed in the Appendix. This chapter describes the more common functions used for everyday PLC programming applications.

Unfortunately, none of these arithmetic functions can easily be understood and programmed without at least a basic command of numbering systems used for PLCs.

NUMBERING SYSTEMS FOR PLCs

We human beings are most familiar with the decimal numbering system, where a particular number, for example 106 324, automatically tells us that the quantities displayed by this number have four units (on the far right of the number), two tens, three hundreds, six thousands, zero ten thousands and one hundred thousand. We also learn from infancy onwards what these figures and their compiled total number mean. The decimal numbering system obviously has ten digits, which can display any given quantity, no matter how small or large. These ten digits are represented by the numerals 0, 1, 2, 3, 4, 5, 6, 7, 8 and 9. Historically, the decimal system may be traced back to biblical times, where the Bible describes its use in Genesis 18:32, Exodus 34:28 and Revelation 2:10. However, some attribute its origin to human beings having ten fingers and ten toes. Whatever the origins of the decimal system may be, it is the system we are most comfortable and familiar with.

PLCs and, for the same reason, computers, are digital binary devices that operate and respond to only two discrete states. These two discrete states are:

Signal "ON"
Signal "OFF"

The numerical digit "1" stands for the "ON" state and the numerical digit "0" stands for the "OFF" state. Hence, to express large or small numbers with only two digits is not easy. For this reason, concepts had to be found to compile these digits into groups that could represent decimal numbers.

Five numbering systems are commonly used with PLCs. These numbering systems are:

- decimal
- octal
- binary
- hexadecimal
- binary-coded decimal (BCD).

The first, third, fourth and fifth are described in this chapter, as they are the most commonly used numbering systems in modern PLCs. For the better comprehension of the three last listed systems, they are always compared with the decimal system. Many books are also available on the subject of numbering systems.

THE DECIMAL NUMBERING SYSTEM

The decimal numbering system is the most commonly used system in our everyday life. To identify value or magnitude expressed with that system, it uses ten digits ranging from zero (0) to nine (9). The decimal system has a radix (root) of 10, since it employs ten digits or symbols. In the decimal system, digits to the left of the decimal point represent units, tens, hundreds, thousands etc. (see figure 10-01). These digits are also called integers. The positional values or weights can be expressed as powers of ten: 10^0, 10^1, 10^2, 10^3 etc. Digits to the right of the decimal point represent tenths, hundredths, thousandths etc. and therefore have positional values of 10^{-1}, 10^{-2}, 10^{-3} etc. The decimal number 952.68 could therefore be written as:

$$(9 \times 10^2) + (5 \times 10^1) + (2 \times 10^0) + (6 \times 10^{-1}) + (8 \times 10^{-2})$$

Here it must be remembered that any radix with zero (0) exponent equals 1.
Thus, $2^0 = 1$, $10^0 = 1$, $16^0 = 1$, $23^0 = 1$.

When reading such a decimal number, we automatically assess its whole number value, as our eyes scan from left to right before the decimal point; and its fractional value after the decimal point.

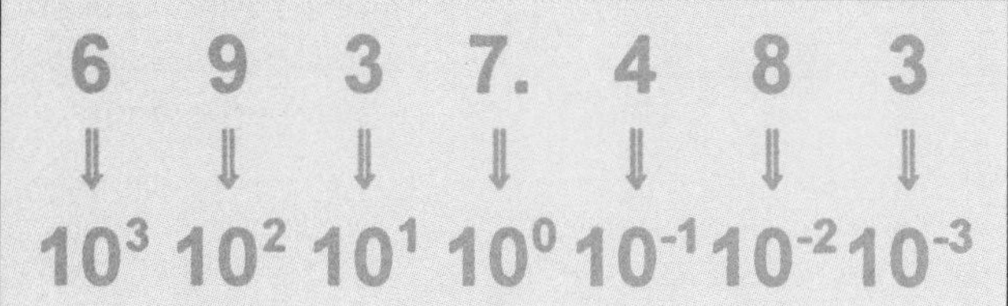

Figure 10-01. Decimal numbering system, using a radix of 10 and positional values, indicated by the exponent next to the radix, to express a large number.

THE BINARY NUMBERING SYSTEM

Whereas the decimal system has a radix of 10, the binary system has a radix of 2 (compare figures 10-01 and 10-02). All numbers are constructed on this radix basis of 2. The binary system, as its name "binary" implies, uses only two digits, 0 (zero) and 1 (one). All assembled digits are weighted with powers, ranging from 0 up (often from 0 to 7). This is shown in figure 10-02.

Two important terms are used to define positional value in binary-coded numbers:

- "Rightmost bit" (RMB), sometimes also called "least significant bit" (LSB).
- "Leftmost bit" (LMB), sometimes also called "most significant bit" (MSB).

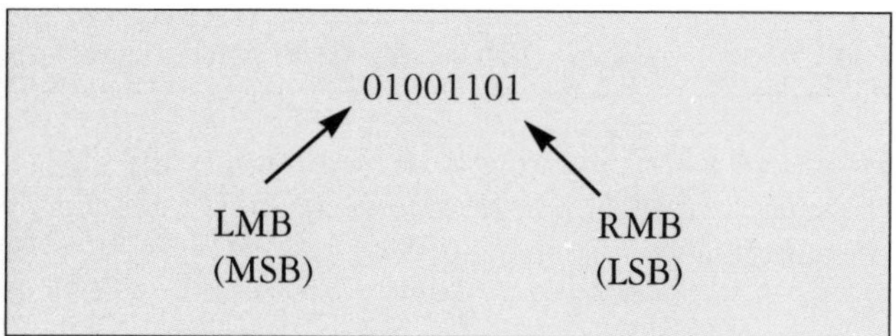

To convert the binary number given in figure 10-02 to decimal, one would work as is shown in figure 10-03.

A number expressed in the decimal system can, of course, also be converted to binary by successive division by two (see figure 10-04). The remainder of each division is retained as a bit of the binary number. The first

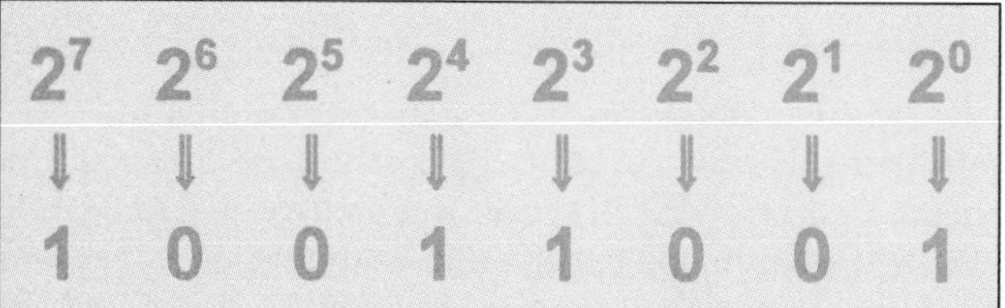

Figure 10-02. Representation of the binary numbering system. It has a radix of 2 and positional values are indicated by the exponent next to that radix.

remainder obtained when dividing the decimal number by two becomes the least significant bit (rightmost bit) in that converted binary number.

Conversion of complex data numbers to other numbering systems is very tedious and often time-consuming. Errors may also easily creep in during the conversion. This could have disastrous effects on actions taken by a PLC using such converted data. For this reason, most modern PLCs possess numerous data conversion functions so that the PLC takes over the task of converting entered data. These data converting functions are explained and illustrated in the following pages.

THE BINARY-CODED DECIMAL CONCEPT (BCD)

To evaluate or display the binary value of a large decimal number, one can perform complex calculations or use conversion tables (see figure 10-09). A much more practical method is the use of a numbering presentation form in which a binary word is constructed so that decimal numbers can be read out directly from their corresponding binary grouping presentation. This modification is called binary-coded decimal, or BCD. In all reality, BCD is not a true numbering system, but rather a clever adaptation or merging of two numbering systems. It retains characteristics of both binary coding and decimal presentation. This coding is sometimes called the BCD-8421 code (see figure 10-05).

The ground rules for BCD quantity representation are as follows:

- Four bits of binary logic state equal 1 digit. The digit's maximum decimal value must never exceed 9.
- Each such BCD digit group represents a decimal base 10 value (radix) with increasing exponent numbers starting from the rightmost bit and moving to the leftmost bit (RMB to LMB).

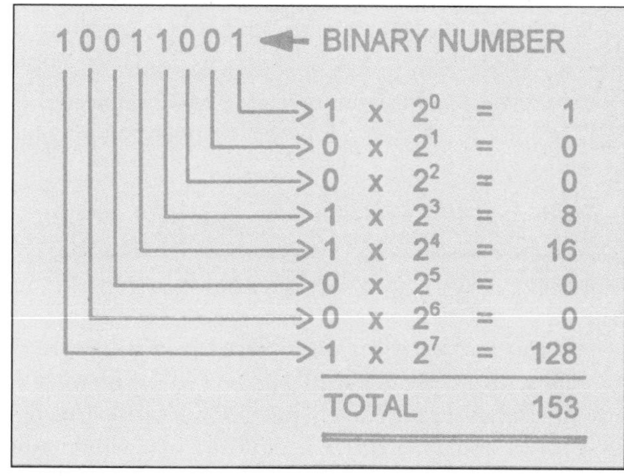

Figure 10-03. Manual binary to decimal conversion procedure.

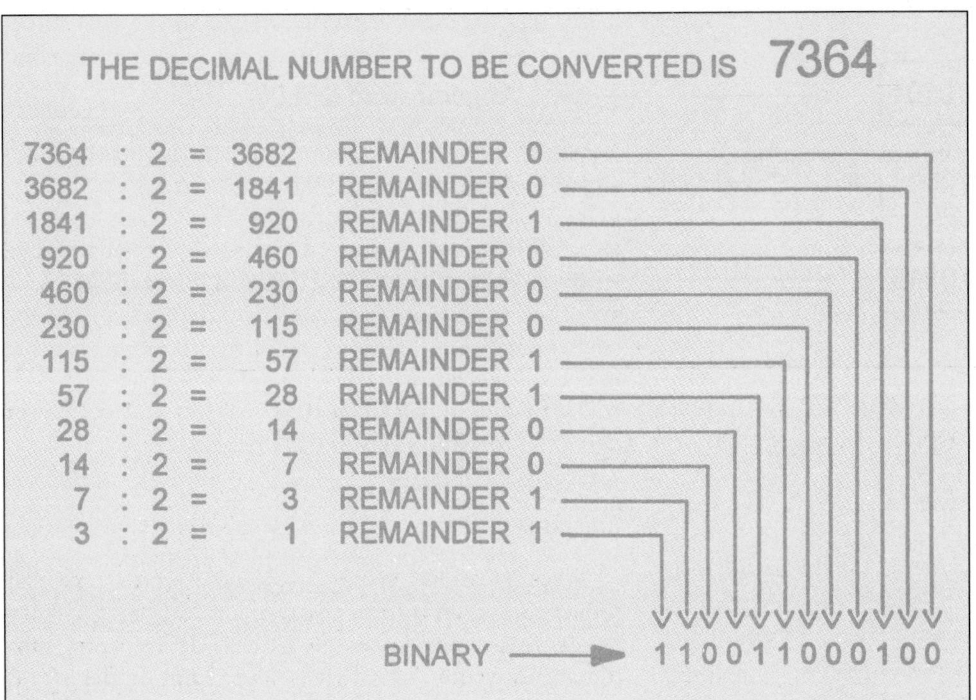

Figure 10-04. Manual decimal to binary conversion procedure.

Converting a large decimal number such as **8937** to BCD would use the procedure shown in figure 10-05.

8	9	3	7	DECIMAL QUANTITY
1000	1001	0011	0111	BCD EXPRESSION
8421	8421	8421	8421	BCD - 8421 CODE
10^3	10^2	10^1	10^0	WEIGHTS

Figure 10-05. Converting a decimal number (8937) manually to a BCD number, using the 8421 code for the conversion.

If one wishes to convert the decimal value of 9999 to BCD the BCD word would then be:

1001 1001 1001 1001.

This is the maximum decimal value, in binary representation, one can allocate to timers and counters (PV = #9999; see applications in Chapters 5 and 6). There are 16 digits. This is equivalent to a channel or word (see figures 1-08, 1-10, 1-12 and 1-14). Hence, a value with four decimal digits needs a storage space in the PLC's data memory area of 16 bits grouped into one word or channel. The same way as an overflow in the decimal system is carried to the next digit to the left, an overflow in the BCD system is carried to the next digit to the left, so that no digit ever exceeds the maximum of 9999. Therefore, the decimal number of 10 000 expressed in BCD would need two words or channels and would appear as:

word 01	word 00
0000 0000 0000 0001	1001 1001 1001 1001

For this reason, when multiplying, adding and subtracting with FUN 54 and FUN 55, it is always mandatory to allocate two channels to store the result (see figures 10-26 and 10-29). BCD addition and subtraction use a carry flag (see figures 10-06, 10-14 and 10-15). This carry flag indicates overflow to the next higher word (channel). Hence, an addition in BCD of an augend decimal figure of say **8365** with an addend decimal figure of 3741 would give a decimal result of **12 106**. In this instance, carrying the leftmost digit (1, representing 10 000) causes the carry flag 1904 to turn "ON" (see also figure 10-20).

THE HEXADECIMAL NUMBERING SYSTEM

The hexadecimal numbering system is another compact numbering system used in PLC arithmetic that allows the programmer and PLC user to obtain even greater values than with BCD. The hexadecimal numbering system has a base or radix of 16. Its numerals range from 0 to 9 and are then continued with alphabetical letters, ranging from A to F. Their equivalent decimal values are shown below:

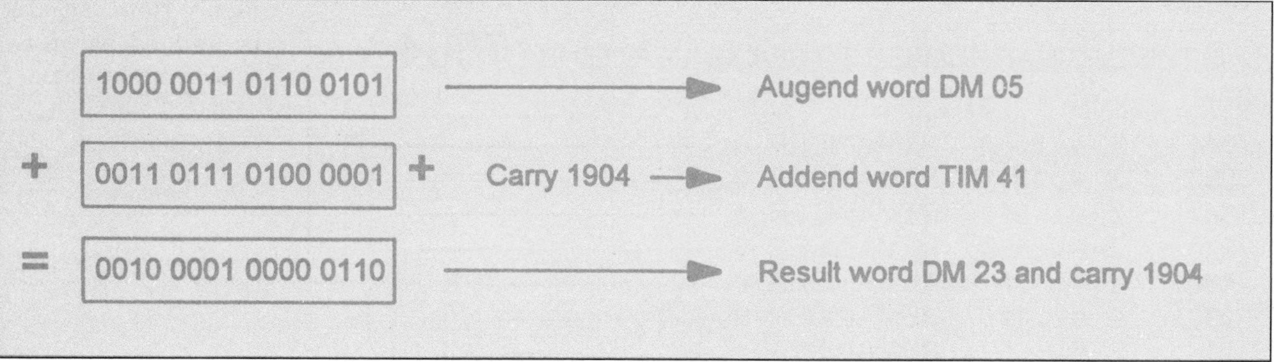

Figure 10-06. The stored present value (PV) of timer 41 is added to data in word DM 05.The result is deposited in DM word 23 and a carry of 10000 is created, which turns "ON" the carry flag 1904.

DECIMAL	HEXADECIMAL
0	0
1	1
2	2
3	3
4	4
5	5
6	6
7	7
8	8
9	9
10	A
11	B
12	C
13	D
14	E
15	F

A3F0 is an example of a hexadecimal number. To convert it, one may use a table such as figure 10-10 and then do as is shown in figure 10-07.

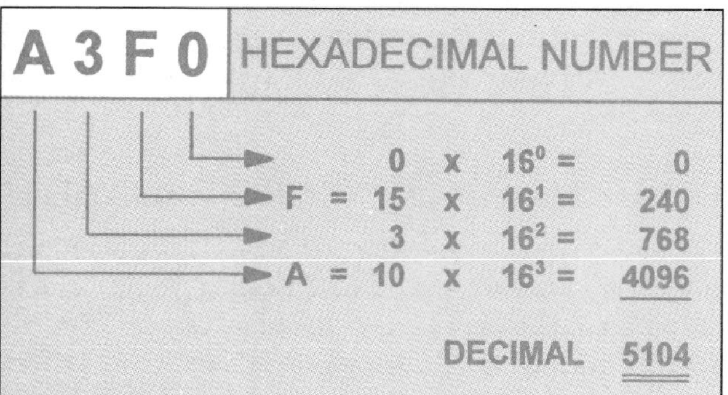

Figure 10-07. Manually converting hexadecimal to decimal channel data. The hexadecimal number A3F0 equals 5104 decimal.

Since most PLCs use 16-bit data words, one can numerically monitor the status of all 16 bits. A display on the programming console or PLC desktop programming unit would show what is depicted in figures 10-08 or 10-09. For monitoring of channels see figures 11-11 and 11-16.

The following conversion illustrates how one can obtain a decimal number from a hexadecimal number. Use figure 10-10 (table) to make this conversion. Hence, a large hexadecimal figure of say FFFF would have a binary equivalent of:

1111 1111 1111 1111

Figure 10-08 shows a conversion from binary to hexadecimal. Use figure 10-10 to obtain the converted hexadecimal value.

Figure 10-09 shows the reverse procedure of converting a hexadecimal number to a binary number.

DATA MEMORY STRUCTURE

The data memory area (DM) is accessible only in 16-bit channel units. The DM area retains data during power failure.

1001	1010	0101	1110	BINARY
9	A	5	E	HEXADECIMAL

Figure 10-08. Manual binary to hexadecimal conversion using table in figure 10-10.

7	F	B	2	HEXADECIMAL
0111	1111	1011	0010	BINARY

Figure 10-09. Manual hexadecimal to binary conversion using table in figure 10-10.

Decimal	BCD	HEX	Binary
00	00000000	00	00000000
01	00000001	01	00000001
02	00000010	02	00000010
03	00000011	03	00000011
04	00000100	04	00000100
05	00000101	05	00000101
06	00000110	06	00000110
07	00000111	07	00000111
08	00001000	08	00001000
09	00001001	09	00001001
10	00010000	0A	00001010
11	00010001	0B	00001011
12	00010010	0C	00001100
13	00010011	0D	00001101
14	00010100	0E	00001110
15	00010101	0F	00001111
16	00010110	10	00010000
17	00010111	11	00010001
18	00011000	12	00010010
19	00011001	13	00010011
20	00100000	14	00010100
21	00100001	15	00010101
22	00100010	16	00010110
23	00100011	17	00010111
24	00100100	18	00011000
25	00100101	19	00011001
26	00100110	1A	00011010
27	00100111	1B	00011011
28	00101000	1C	00011100
29	00101001	1D	00011101
30	00110000	1E	00011110
31	00110001	1F	00011111
32	00110010	20	00100000

Figure 10-10. Table for conversion between decimal, BCD, hexadecimal and binary.

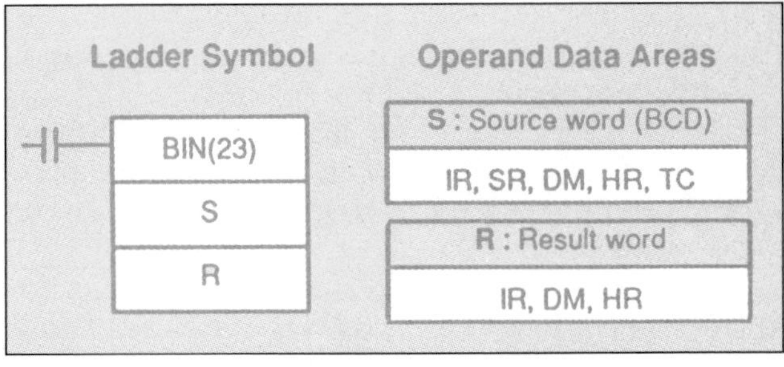

Figure 10-12. FUN 23 for automatic BCD to binary data conversion.

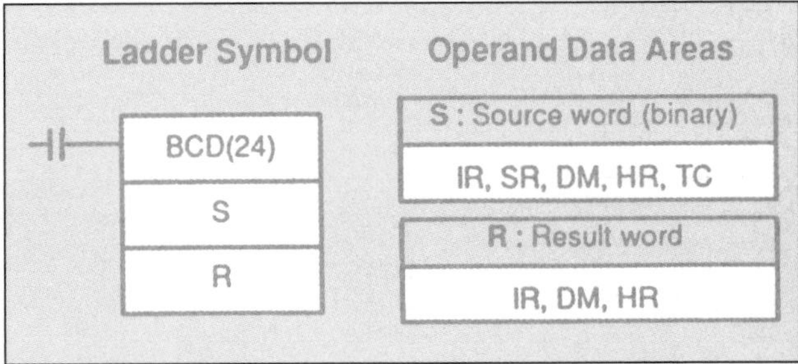

Figure 10-13. FUN 24 for automatic binary to BCD data conversion.

DM channel numbers for an Omron C200H PLC range from 0000 to 1999, but only DM channels 0000 to 0999 can be written to by the operator. This area cannot be used by instructions with bit-size operands, such as "LD", "OUT", "AND" and "OR" (see figure 10-11).

AUTOMATIC DATA CONVERSION WITH PLC

The binary to BCD conversion function for PLCs is essential when the PLC needs to communicate with outside devices for data gathering and storage. Analogue to digital and digital to analogue translation

Data Memory											
Read/Write						Read Only					
DM0000	DM0010	DM0020	~	DM0980	DM0990	DM1000	DM1010	~	DM1970	DM1980	DM1990
DM0001	DM0011	DM0021		DM0981	DM0991	DM1001	DM1011		DM1971	DM1981	DM1991
DM0002	DM0012	DM0022		DM0982	DM0992	DM1002	DM1012		DM1972	DM1982	DM1992
DM0003	DM0013	DM0023		DM0983	DM0993	DM1003	DM1013		DM1973	DM1983	DM1993
DM0004	DM0014	DM0024		DM0984	DM0994	DM1004	DM1014		DM1974	DM1984	DM1994
DM0005	DM0015	DM0025		DM0985	DM0995	DM1005	DM1015		DM1975	DM1985	DM1995
DM0006	DM0016	DM0026	~	DM0986	DM0996	DM1006	DM1016	~	DM1976	DM1986	DM1996
DM0007	DM0017	DM0027		DM0987	DM0997	DM1007	DM1017		DM1977	DM1987	DM1997
DM0008	DM0018	DM0028		DM0988	DM0998	DM1008	DM1018		DM1978	DM1988	DM1998
DM0009	DM0019	DM0029		DM0989	DM0999	DM1009	DM1019		DM1979	DM1989	DM1999

Figure 10-11. Data memory structure for an Omron C200H PLC. For other PLC types check with manufacturer's manual.

modules require such data conversion functions, as these modules translate and communicate the incoming and outgoing signals in binary only. All arithmetic functions within the PLC, on the other hand, operate only in BCD. For this reason, an "interface" function such as BCD to binary (FUN 23) and binary to BCD (FUN 24) is essential.

Function 23, shown in figure 10-12, converts a four-digit decimal number held in the source channel into a 16-bit binary (hexadecimal) number, placed into the result channel. After conversion, the decimal number held in the source channel is still available in decimal.

Function 24, shown in figure 10-13, converts a 16-bit binary number held in the source channel to a four-digit decimal number in the result channel. After conversion, the binary number in the source channel is still available in binary. An application and experimental circuit for automatic continuous data conversion from decimal to binary (hexadecimal) is given in control problem 3 at the end of this chapter.

DATA SHIFTING AND DATA CLEARING

To clear data from a single digit, several digits, a word or several words, one may use the data shift instruction, or "word shift". For Omron PLCs this is WSFT or FUN 16. Function 16 shifts data contained in digit 1 of the start word to the left into digit 2 (see figure 10-14). At the same time it shifts four zeros (0) into digit 1 of the start word. The previous content of digit 1 is now in digit 2, the content of digit 2 is in digit 3, the content of digit 3 is in digit 4, and the original content of digit 4 is lost. For as long as the word-shift enabling condition is "ON", with every scan the PLC shifts the data digit by digit to the left, until the entire starting word contains only zeros. If data shifting needs to be done on several words (channels), one nominates the starting word, for example DM 03, and the end word, for example DM 07. The PLC would then need 20 scans to replace data in these five words with zeros. Word shift may of course not only be used for data clearing, but also for data shifting from one word or digit to another word or digit. If data are to be cleared from one word only, the end word is then the same as the start word.

MANUAL DATA ENTRY

For circuit testing and diagnostic purposes, the PLC programmer can enter numbers into the PLC manually. There are three ways of entering such data:

- data input with a reversible counter.
- data input with the PLC's data change facility.
- data input with a thumb-wheel switch.

For data input with a reversible counter, a reversible counter is temporarily added at the very end of an existing program (see figure 10-38). The reversible counter's present value (PV) is then used to simulate a decimal number. To set this decimal number, the counter is incremented or decremented with a pulse bit (e.g. 1900 on an Omron C28K PLC) until the desired decimal value

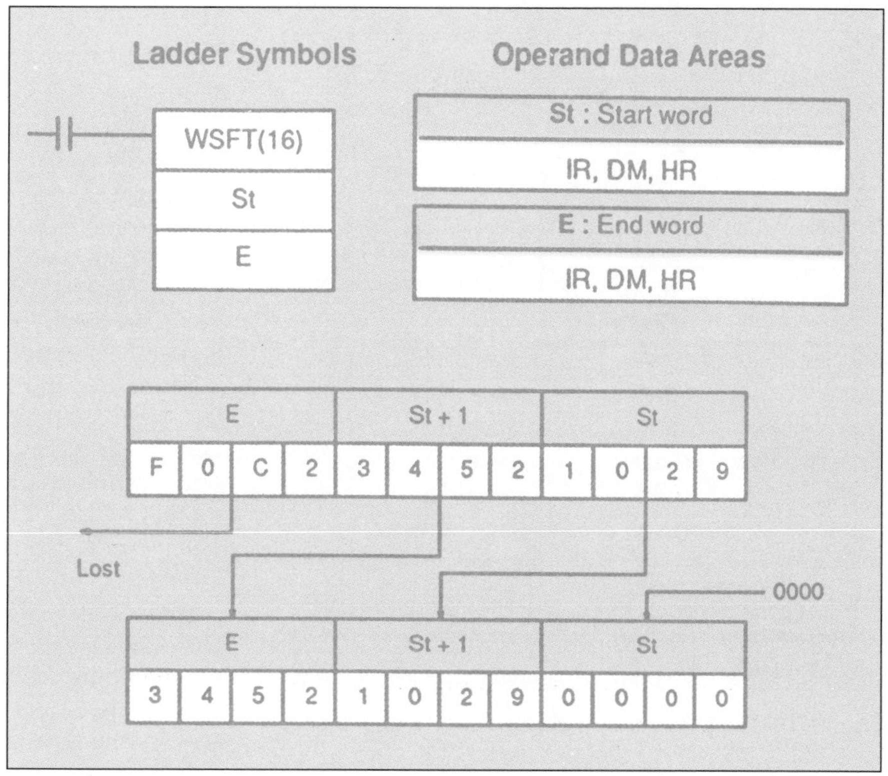

Figure 10-14. Digit by digit word shift. This may be used for clearing data memory.

Figure 10-15. Procedural steps to change data memory contents in a channel (word).

is reached. For this, one may use the monitor and data change procedure shown in figure 11-10. However, a simpler way is to call up the counter in the monitor mode. For example, counter 47 in figure 10-33 is called up by pressing the keys:

CLR, CNT, 4, 7, MONTR.

Now, one may use input 0013 to increment the counter to 1013 present value, to simulate, say, an overweight

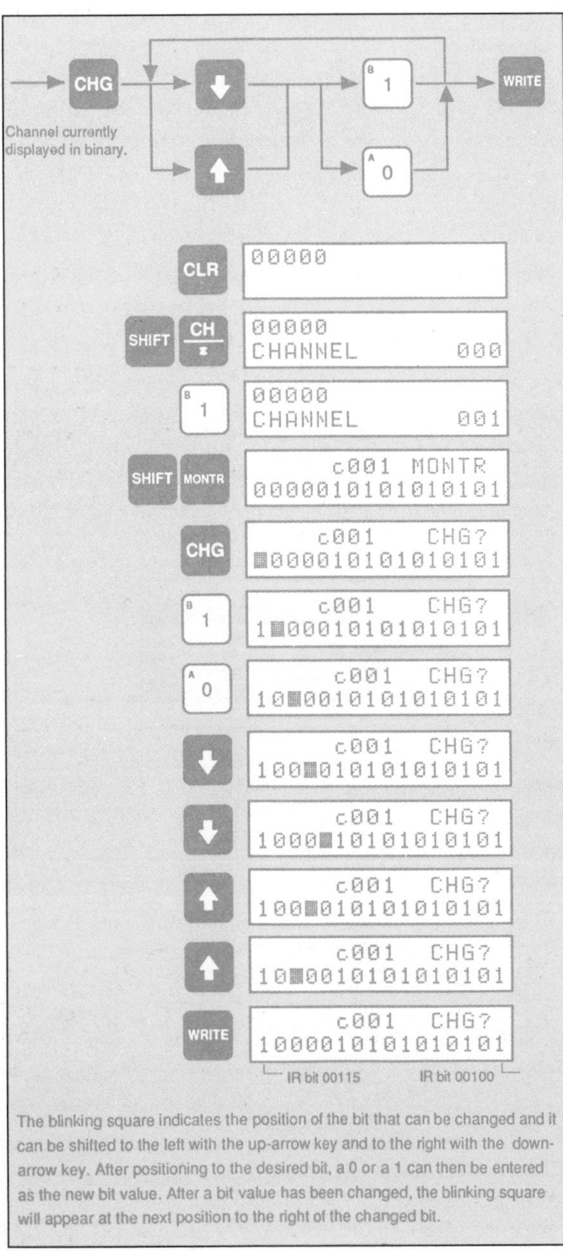

Figure 10-16. Examples for changing data in a channel bit by bit.

condition; or decrement the counter with input 0011 to 0989 present value, to simulate, say, an underweight condition (underweight and overweight conditions are used in control problem 2 and figures 10-37 and 10-38 at the end of this chapter). The programming console will then constantly show the counter's incrementing or decrementing PV.

Now, one may move the counter's PV to a data memory word (DM) by using the move instruction (FUN 21), as is shown in figures 10-19, 10-22 and 10-23, or one may use the counter directly, to simulate data for an arithmetic operation (see figure 10-38). Data entered in such a way may then be used as an augend, addend, minuend, subtrahend, dividend, divisor, multiplicand or multiplier, or as data for floating point and square root computations.

Entering data with the PLC's data change facility is similar. For this, the programmer may use the steps outlined in figures 10-15 and 10-16. By pressing the first four keys shown in figure 10-15, the data memory area is located. The targeted channel (word) is then called up. On pressing monitor (MONTR), the targeted channel's content in decimal value is displayed. Then, by pressing the change key (CHG), the programmer is given an option to enter different decimal data. The process is sealed by pressing the write key (WRITE).

Experimental circuit

The reader is now encouraged to practise this procedure with the control circuit in figure 10-17. After the PLC is programmed as shown in figure 10-17, the counter is incremented to 0080 decimal. For this, the programming console is switched to monitor mode (black switch), and counter 47 is monitored by pressing CLR, CNT, 4, 7, MONTR. Input 0013 is now activated until 0080 decimal present value (PV) is displayed. Remember, input 0011 decreases the counter. Now, the subtrahend of this practice circuit (being the counter) is set. Next comes the setting of the minuend, which is DM 01. At this stage DM 01 is zero, and only zeros are stored in this word. To enter data manually, one presses the keys CLR, SHIFT, CH, DM, 1, MONTR, CHG. The desired data in decimal value may be entered. To subtract the PV of counter 47 (0080) from the contents of DM 01 (9750), one then enters 9750 into DM 01. This is accomplished by pressing the white programming console keys 9, 7, 5, 0, in that order. The newly entered data option 9750 decimal is now displayed, and when the write key (WRITE) is pressed, the optional data are entered and sealed (see figure 10-15). Now press DM, 9, MONTR, CNT, 4, 7, MONTR. The display on the programming console should now show the following:

C47	D09	D01
0080	0000	9750

By activating input 0000, the subtraction is executed and DM 09 should now display 9670. Why? Because:

```
    9750
 –  0080
 =  9670
```

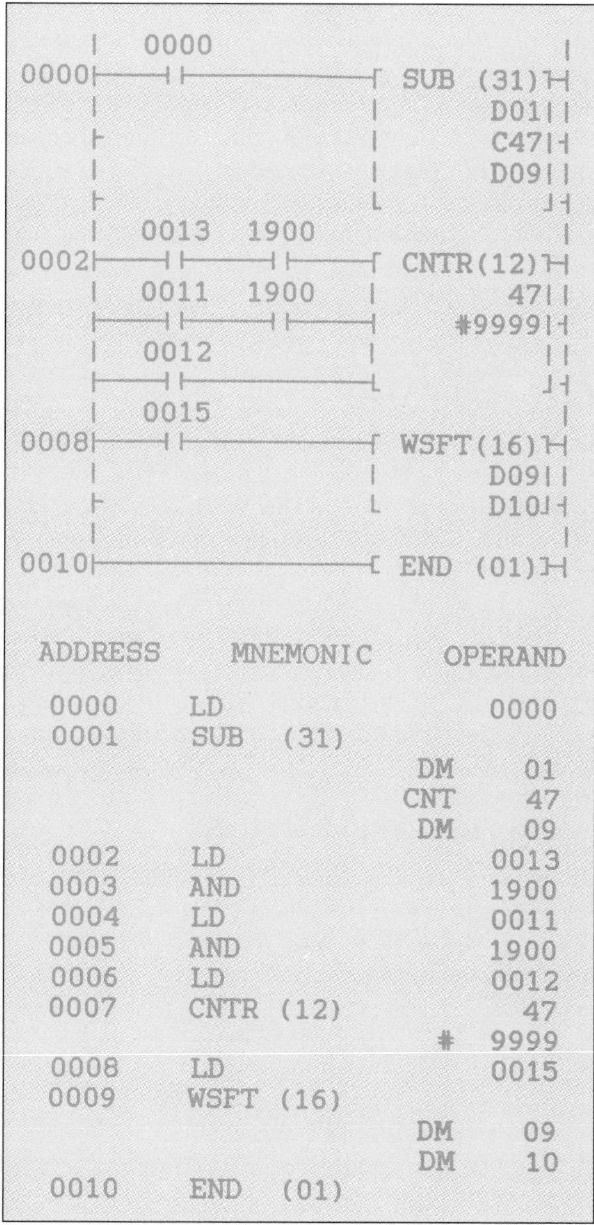

ADDRESS	MNEMONIC		OPERAND
0000	LD		0000
0001	SUB (31)		
		DM	01
		CNT	47
		DM	09
0002	LD		0013
0003	AND		1900
0004	LD		0011
0005	AND		1900
0006	LD		0012
0007	CNTR (12)		47
		#	9999
0008	LD		0015
0009	WSFT (16)		
		DM	09
		DM	10
0010	END (01)		

Figure 10-17. Experimental circuit.

To clear the data out of DM 09, a word shift function 16 has been added. When input 0015 is activated, the word shift clears the value from DM 09 and replaces it with 0000.

This practice exercise has now demonstrated how data may be entered with a reversible counter, and changed manually with the data change procedure. Both procedures were explained in the previous pages. The exercise has also shown how data may be cleared with the word shift instruction (see figures 10-14 and 10-17).

THE THUMBWHEEL SWITCH

The thumbwheel switch is a digital device frequently used to enter BCD data manually into a PLC (see figure 10-18). As four bits are necessary to represent each BCD digit and the thumbwheel switch has a four-digit, wheel-driven display, sixteen bits of PLC input data are required for its connection (see also figure 10-05). One would therefore need all 16 input points on a PLC input module to connect such a four-digit selector switch. The switch can now be adjusted by rotating the switch wheel with one's thumb. With these four wheels and their mechanically interlinked display numerals engraved on a drum, one can display and enter decimal data ranging from 0000 to 9999. Some thumbwheel switches do not have a wheel to rotate the numerals, but instead use a push-button-operated ratchet mechanism to rotate the numerals. Such selector switches perform the same function as the thumbwheel switch, but the selection process is somewhat easier.

Figure 10-18. Thumbwheel switch to enter four-digit BCD data into a PLC word.

AUTOMATIC DATA ENTRY

Whereas manual data entry relies on data from programmed decimal constants (for example, #9358), or data being entered from a BCD thumbwheel switch or originating from a manually changed counter or timer PV as explained in control problem 1, data may also enter the PLC automatically with no direct human influence. Such data may come from an automatically incremented or decremented timer or counter, or through an analogue to digital conversion module, which converts analogue data from a pressure, flow or temperature switch or a strain gauge. Data may also come from a host computer or networked PLCs through link relays. Whatever the data source may be, automatically entered data are treated the same way as manually entered data. An application for automatic input of constantly changing data is given in control problem 3.

DATA MOVEMENT

Data movement may become necessary if one wishes to fully use all of the PLC's internal data areas. A data move instruction transfers data (either the data in a specified channel, or a four-digit hexadecimal constant) to a destination channel. Because only a small percentage of the PLC's I/O registers have access to the real PLC input and output words, move instructions become increasingly more important for advanced PLC applications. It must be noted that data in the source word are not lost after data move has occurred. Typical data move applications are given in figures 10-22 and 10-23.

For Omron PLCs, the data move instruction is function 21 (FUN 21). This is illustrated with figure 10-19.

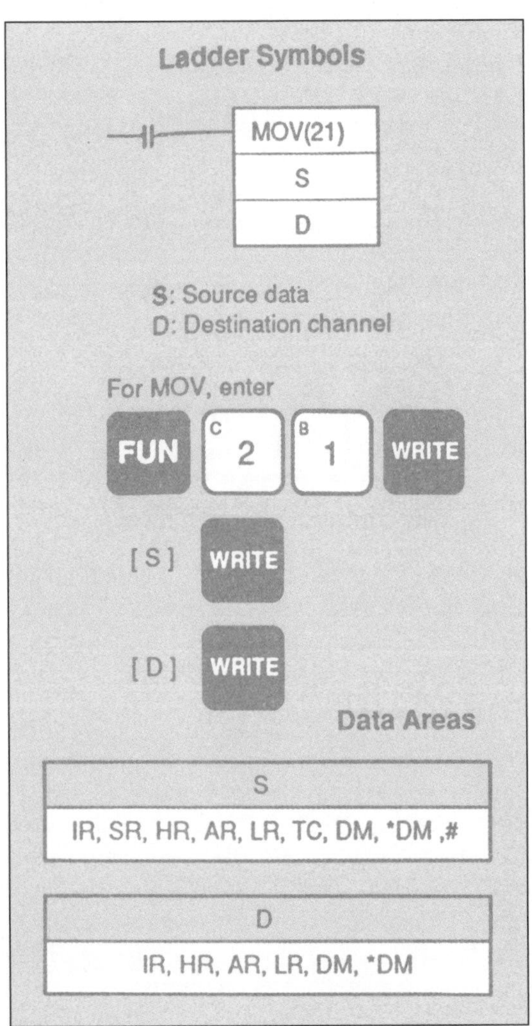

Figure 10-19. Data move instruction (FUN 21) to move data between I/O words or DM words.

DIFFERENTIATE FUNCTION

The differentiate function is a special instruction used to ensure that the operand bit to which the differentiate function is attached is never turned "ON" for more than one scan length after the enabling condition goes from "OFF" to "ON" for differentiate-up. Differentiate-down is used to actuate the operand for only one scan length if the enabling condition goes from "ON" to "OFF". These two functions were applied and explained in Chapter 5. Larger or more modern PLCs make extensive use of the differentiate function (also called edge flag or edge pulse function). Where the PLC is used to perform arithmetic instructions, this function is of great importance. If enabling signals for such arithmetic functions were not converted to an edge pulse, they could cause accumulative multiplication, division, subtraction or addition, and therefore a false result (see figures 10-22 and 10-23). With Omron PLCs, wherever the "at" sign "@" is attached to a PLC instruction, the function will then be executed once only, even if the enabling condition remains "ON". This is sometimes also referred to as transitional execution or transitional enabling. To program it into the PLC, press the function key, nominate the function number and press the "NOT" key (see figure 10-24). Pressing the "NOT" key makes the function execution transitional (only "on " at leading edge of signal).

BCD CALCULATIONS

The PLC executes all calculations on BCD data. The BCD calculation instructions—ADD (FUN 30), SUB (FUN 31), MUL (FUN 32), DIV (FUN 33), FDIV (FUN 79) and ROOT (FUN 72)—all perform arithmetic operations on BCD data.

STC (FUN 40) and CLC (FUN 41), which set and clear the carry flag, are included in this chapter because most of the BCD arithmetic operations make use of the carry flag (CY) in their results. Binary calculations and shift operations also use the carry flag (see figures 10-20 and 10-21).

CARRY FLAG APPLICATIONS

A carry is generated whenever an addition function (FUN 30 or FUN 54) produces a result exceeding the capacity of the result word(s). For example, if FUN 30 adds the following augend and addend:

	9574	⇔ DM 00
+	8912	⇔ DM 01
=	18486	⇔ DM 02

the result, 18 486, will then exceed the holding capacity of the result word, which has a maximum capacity of 9999 decimal. The carry flag is turned "ON", representing 10 000. The carry flag is bit 25 504 for an Omron C200 H PLC and bit 1904 for an Omron C28K or C20 PLC. If this ADD function (FUN 30) were now followed by a SUB function (FUN 31), and the ADD result of 18486 in DM 02 word were used as the minuend word, and a constant of #9743 were subtracted as subtrahend, the result would then be as follows:

carry (10 000)=	(1)8486	⇔	DM 02
–	9743	⇔	#
=	8743	⇔	DM 09

If, for the same experiment, the constant in the subtract function were altered to # 7000, the calculation would then be as follows:

	9574		18486
+	8912	–	7000
=	18486	=	11486

The overflow in the addition function will then generate a carry of 10 000 decimal, which is carried into the subtraction function. The subtraction function will also generate a carry of 10000, which is not carried any further, but will also turn "ON" the carry flag 1904.

To obtain an accurate result, a "clear carry" function (FUN 41) must be placed in front of the subtract function (FUN 31). Without such a clear carry, the result

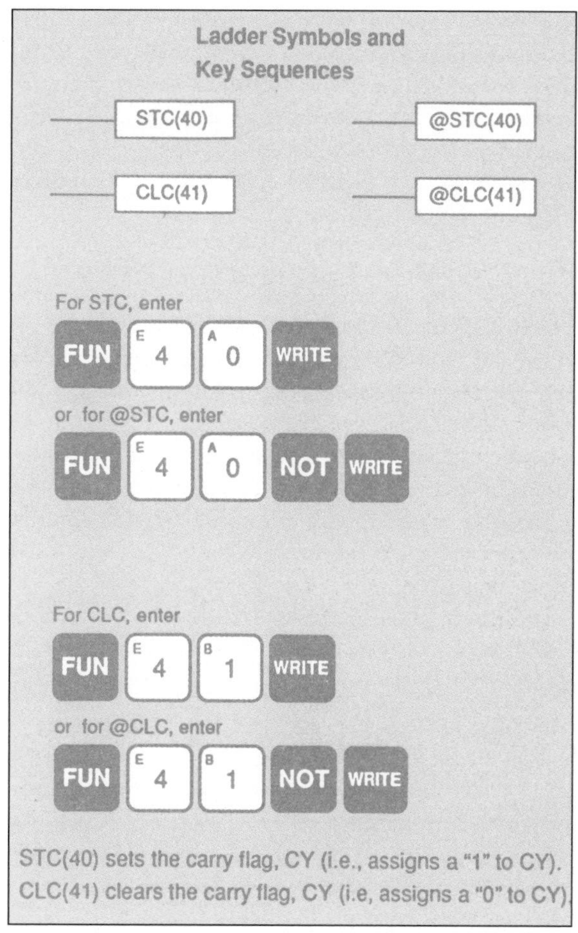

Figure 10-20. Programming procedures to clear carry and set carry.

The carry flag is affected by the following instructions:

Instruction	FUN	Meaning of Carry Flag Value	
		1	0
ADD/@ADD	30	There was an overflow in the result of an addition operation.	No overflow occurred.
ADDL/@ADDL	54		
ADB/@ADB	50		
SUB/@SUB	31	Subtraction result < 0	Subtraction result is correct.
SUBL/@SUBL	55		
SBB/@SBB	51		
ASL/@ASL	25	Before shifting, Bit 15 was ON.	Before shifting, Bit 15 was OFF.
ROL/@ROL	27		
ASR/@ASR	26	Before shifting, Bit 00 was ON.	Before shifting, Bit 00 was OFF.
ROR/@ROR	28		
SFTR/@SFTR	84	If right-shifting: Bit 00 was ON	Bit 00 was OFF.
		If left-shifting: Bit 15 was ON	Bit 15 was OFF.
STC/@STC	40	STC/@STC execution	——
CLC/@CLC	41	——	CLC execution
END	01	——	END execution.

Note: You should execute CLC before any addition, subtraction or shift operation to ensure a correct result. Simply connect CLC to the same input as the instruction it precedes.

Figure 10-21. PLC functions affected by the carry flag.

would be 1485 instead of 1486 in word 09 (DM 09).

The addition and subtraction instructions include carry (CY) in the calculation as well as in the result. One must therefore make sure to clear the carry if its previous status is not required in the calculation, or to use the result placed in carry, if required, before it is changed by the execution of any other instructions.

PLC ADDITION FUNCTION (BCD)

When the addition function (FUN 30) is enabled, the addend will be added to the augend (plus any existing carry). The obtained result is stored in the area specified by the result word. If further carry is generated, it will also turn on the carry flag 1904. Figure 10-22 shows the various data areas the programmer can select to nominate the data "A" word for the augend, the data "B" word for the addend, and the data "C" word for the result to be stored in.

The circuit given in figure 10-22 has a number of special testing features from which a student of this book can profit.

- Starting on top is a reversible counter, which may be incremented or decremented to set its PV. Input 0013 increments the counter and input 0011 decrements the counter. Pulse bits 1900 are "AND" connected to speed up the setting process. Input 0012 resets the counter to SV (set value).
- Once the counter is set, its PV can be moved into DM 00 to "load" the augend. Data movement (FUN 21) is enabled with input 0015 (see also figure 10-19). For this experiment, set the counter to #0684.
- The addition function (FUN 30) adds constant #1000, entered into the addend word during programming, and places the result, stored in DM 00, back into the augend word DM 00 (the augend word is the same as the result word).
- To slow down the continuous addition process, the enabling condition for the addition function (FUN 30) is shortened to a single scan pulse with the edge flag function DIFU (FUN 13). Hence, every addition forces the operator to actuate and deactuate input 0000.
- Through a repeated actuation of input 0000, the preset counter value, having been moved into DM 00, is being increased by value #1000 until a carry is generated.
- A carry, when generated, sets HR 000, which then drives output 0100.
- After completion of the testing experiment, a short actuation of input 0001 will clear all data from DM 00 and replace it with zeros.

This experimental circuit shows and proves the functions and features of:

- setting a reversible counter to PV (FUN 12)
- moving data (FUN 21)
- clearing a carry (FUN 41)
- ordinary addition (FUN 30)
- continuous accumulating (result word = augend)
- word shift for data clearing (FUN 16)
- storing a carry with a flip-flop (FUN 11).

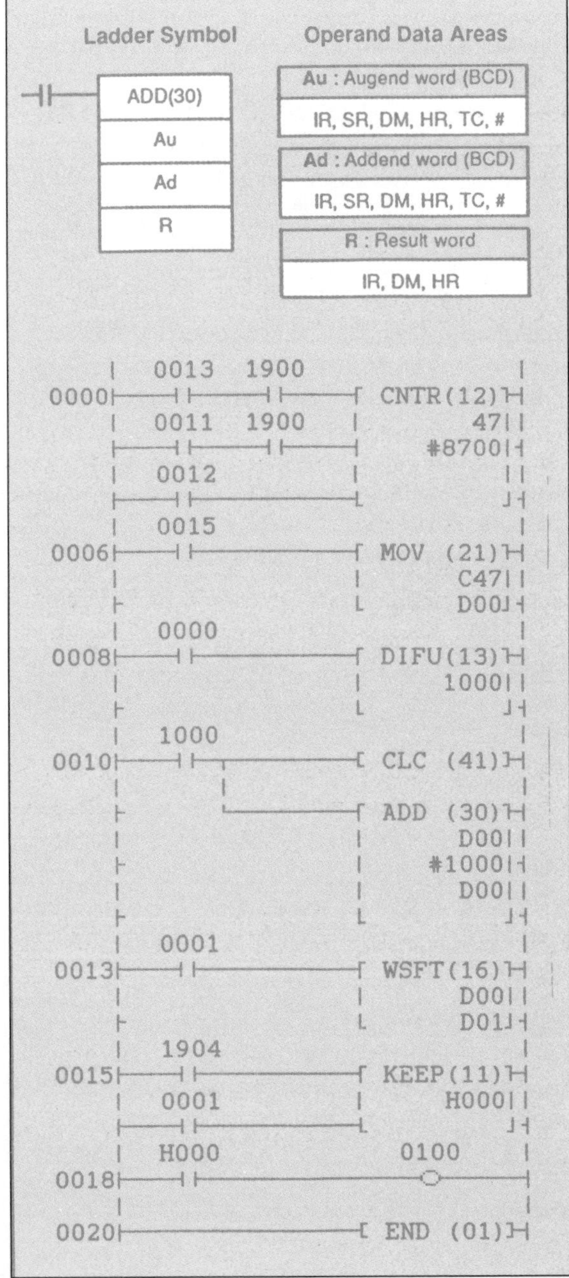

Figure 10-22. BCD addition function with application circuit.

PLC SUBTRACTION FUNCTION (BCD)

When the subtraction function (FUN 31) is enabled, the subtrahend will be subtracted from the minuend. The obtained result is stored in the area specified by the result word. The subtraction function operates on BCD data. Figure 10-23 shows the various data areas the programmer can select to nominate the data "A" word for the minuend, the data "B" word for the subtrahend, and the data "C" word to specify the result word.

The circuit given in figure 10-23 has a number of special testing features from which a student of this book can profit.

- Starting on top is a reversible counter, which may be incremented or decremented to set its PV. Input 0013 increments the counter and input 0011 decrements the counter. Pulse bits 1900 are "AND" connected to speed up the setting process. Input 0012 resets the counter to SV (set value).
- Once the counter is set, its PV can be moved into DM 00 to "load" the minuend. Data movement (FUN 21) is enabled with input 0015 (see also figure 10-19). For this experiment set the counter to PV #0999.
- The subtraction function (FUN 31) subtracts #0100, entered into the subtrahend word during programming, and places the result, stored in DM 00, back into the minuend word DM 00 (the minuend word is the same as the result word).
- To slow down the continuous subtraction process, the enabling condition for the subtraction function (FUN 31) is shortened to a single scan pulse with DIFU (FUN 13), the edge flag function. Hence, every subtraction forces the operator to actuate and deactuate input 0000.
- Through a repeated actuation of input 0000, the preset counter value, having been moved into DM 00, is being reduced by value #0100 until a carry is generated.
- A carry is generated when the result is less than 0 (<0). A carry, when generated sets HR 000, which drives output 0100.
- After completion of the testing experiment, a short actuation of input 0001 will clear all data from DM 00 and replace it with zeros.

This experimental circuit shows and proves the functions and features of:

- setting a reversible counter to PV (FUN 12)
- moving data (FUN 21)
- clearing a carry (FUN 41)
- ordinary subtraction (FUN 31)
- continuous subtraction (result word = minuend)
- word shift for data clearing (FUN 16)
- storing a carry with a flip-flop (FUN 11).

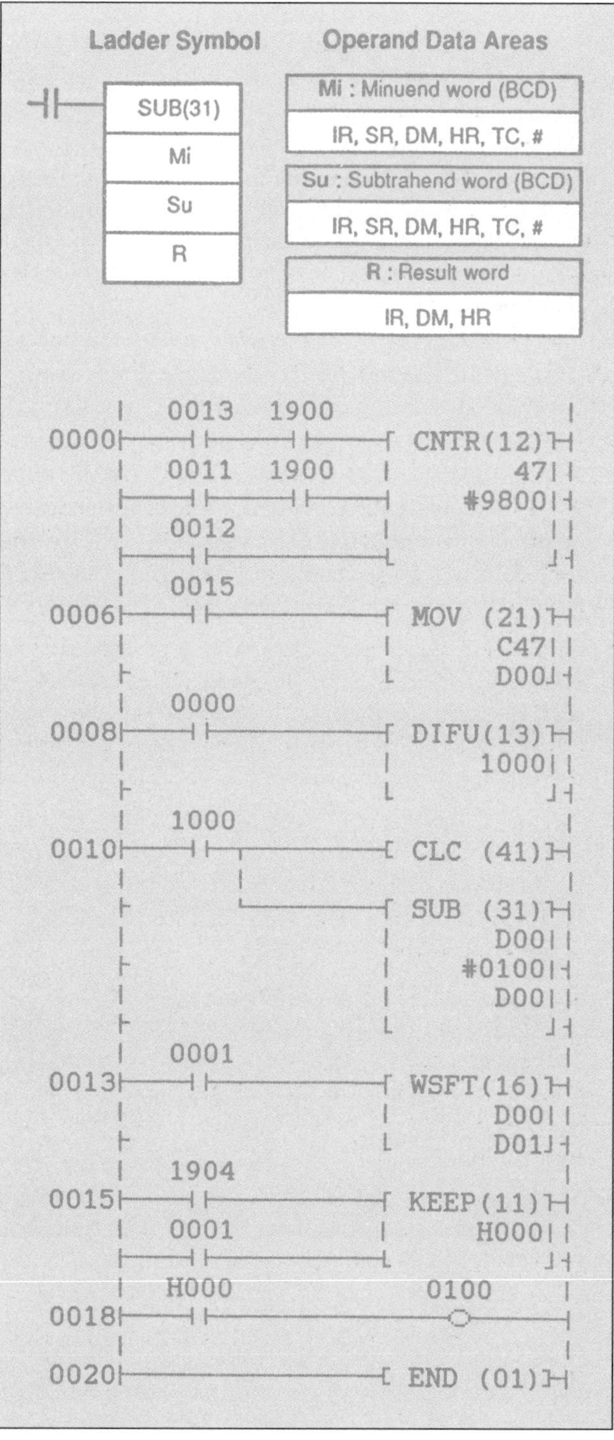

Figure 10-23. BCD subtraction function with application circuit.

PLC DOUBLE-PRECISION ADD AND SUBTRACT FUNCTIONS

Some brands and types of PLCs offer so-called double-precision, BCD, long addition and long subtraction functions. The double-precision add function (FUN 54 ADDL) uses two consecutive words to accommodate the augend data and two consecutive words to accommodate the addend data, and stores the obtained result in two consecutive result words.

Similarly, the BCD double-precision subtract function (FUN 55 SUBL) has two consecutive words for the minuend data and two consecutive words for the subtrahend data, and stores the obtained result in two consecutive result words. An example for a double-precision BCD add function is given in the ladder diagram belonging to control problem 1 at the end of this chapter (see figure 10-34, addresses 0033 and 0034). It must be noted that for these two double-precision arithmetic functions, one cannot use a constant in any of the augend, addend, minuend or subtrahend words. For operand data to be used with these two functions, see the Appendix for functions 54 and 55.

The programmer must be aware that any result greater than 9999 9999 creates an overflow, which turns "ON" the carry flag. Any existing carry at the time of function execution is added to the augend or minuend. If such a carry is not wanted, one must use the clear carry (FUN 41) before the function execution (see figures 10-20 and 10-22).

BCD MULTIPLY FUNCTION

The BCD multiply instruction (FUN 32) multiplies the data stored in the multiplicand word with the data stored in the multiplier word and places the result obtained into the result words. The result is stored in two consecutive words to accommodate any overflow generated by the multiplication (see figure 10-24). One must also make sure that the two result words are consecutive words, e.g. DM 06 and DM 07.

For the multiplication executed in figure 10-24, the result is stored in words (channels) HR 07 and HR 08. A circuit for practising such multiplications is given in figure 10-25.

Input 0000 in figure 10-25 causes the edge flag (differentiate-up FUN 13) to be actuated for one scan time. This edge flag 1000 is then used to actuate the ADD function for only one scan time. By this, the ADD function adds constant #0001 to the contents of data memory 05 (DM 05), which is always zero (DM 05 = 0000). This data memory will always remain zero because no number is being placed into it. The addition process therefore adds a "zero" value from the addend word to a "one" value in the augend word, and places the obtained result of this addition

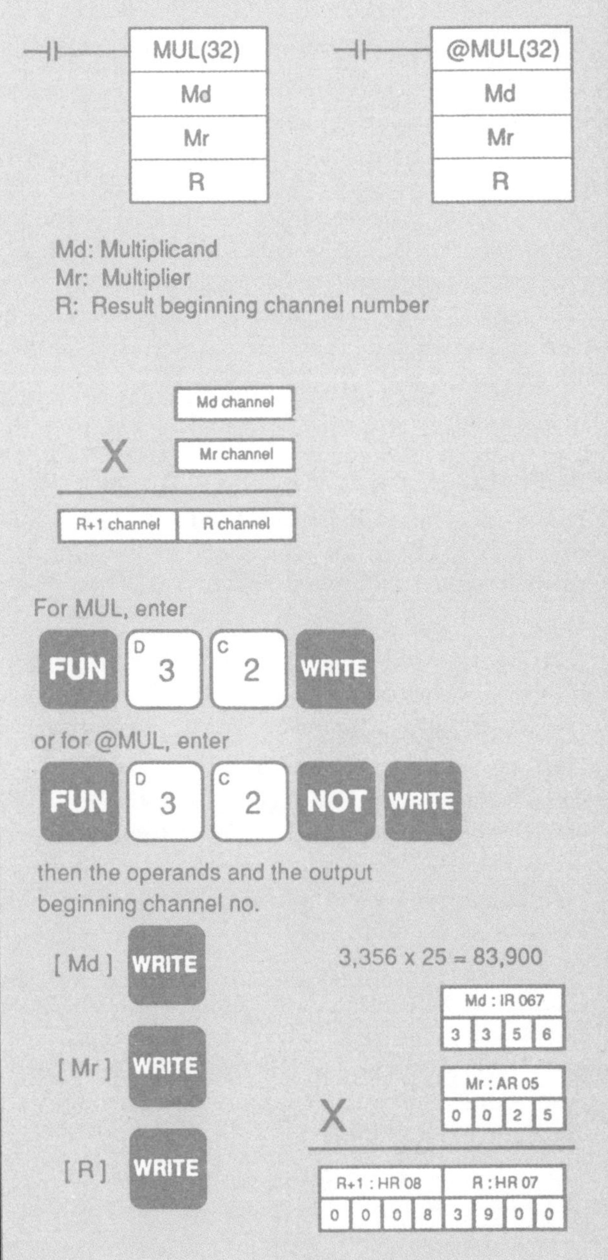

Figure 10-24. Programming procedures for BCD multiplication function.

into the result word, which for this exercise is DM 06.

Input 0015 causes a second edge flag (DIFU FUN 13) 1001 to be actuated for one scan time. This edge flag signal clears the carry (see addresses 0004 and 0005 in figure 10-25). Thereafter, by activating the multiply function with input 0015 via the edge flag 1001, constant #0002 is multiplied with data held in data memory 06 (DM 06) and the result obtained is

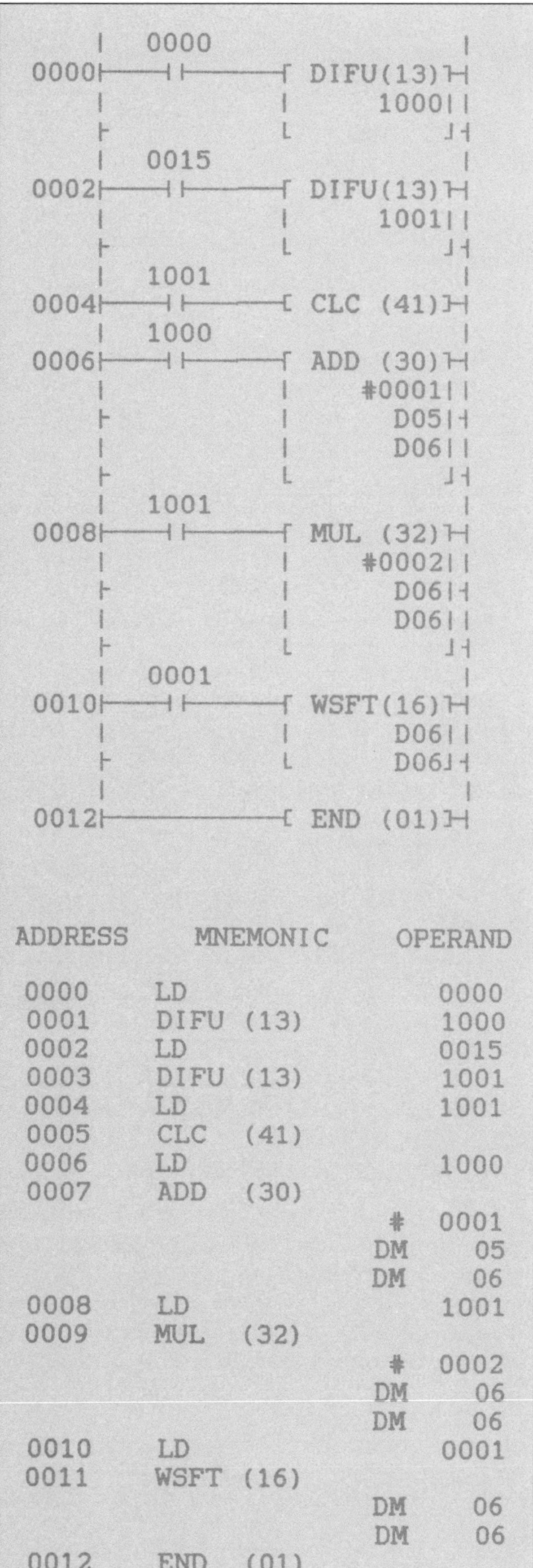

ADDRESS	MNEMONIC	OPERAND
0000	LD	0000
0001	DIFU (13)	1000
0002	LD	0015
0003	DIFU (13)	1001
0004	LD	1001
0005	CLC (41)	
0006	LD	1000
0007	ADD (30)	
		# 0001
		DM 05
		DM 06
0008	LD	1001
0009	MUL (32)	
		# 0002
		DM 06
		DM 06
0010	LD	0001
0011	WSFT (16)	
		DM 06
		DM 06
0012	END (01)	

placed back into DM 06. Thus, with every actuation of input 0015, the result held in DM 06 is being doubled (multiplied by #0002) until data memory 06 overflows (more than 9999) and an overflow is created. This overflow is automatically placed into DM 07. A word shift instruction (FUN 16) is added to this experimental circuit so that DM 06 can be cleared to 0000 for a new experiment to begin.

Operating the experimental circuit

After programming the circuit into a suitable PLC, one can now easily monitor the contents of DM 05, DM 06 and DM 07. This is accomplished in the monitor mode by pressing the keys CLR, DM, 5, MONTR, DM, 6, MONTR, DM, 7, MONTR. The display on the programming console should now show:

D07	D06	D05
0000	0000	0000

Now activate input 0000 briefly to place, by addition, a # 0001 into DM 06. (The result of #0001 + DM 05 is placed into DM 06. DM 06, therefore, becomes 0001.) This can now be observed by monitoring DM 06, which by now should show:

D07	D06	D05
0000	0001	0000

Now activate input 0015 slowly about 13 times until DM 06, through several successive multiplications, has reached an accumulated value of 8192. One further multiplication by #0002 must render a result of 16 384. Since DM 06 can hold results up to only 9999, a carry of 10 000 is created and placed into DM 07. The remainder will stay in DM 06. The programming console display should now show:

D07	D06	D05
0001	6384	0000

To clear all data from DM 06, the word shift function (FUN 16) needs to be activated. This shifts all data out of DM 06 and replaces it with four zeros. Now, activating input 0015 will also clear data held in DM 07 (1001 = CLC FUN 41). The experiment can now be repeated.

Figure 10-25. BCD multiplication function with experimental circuit.

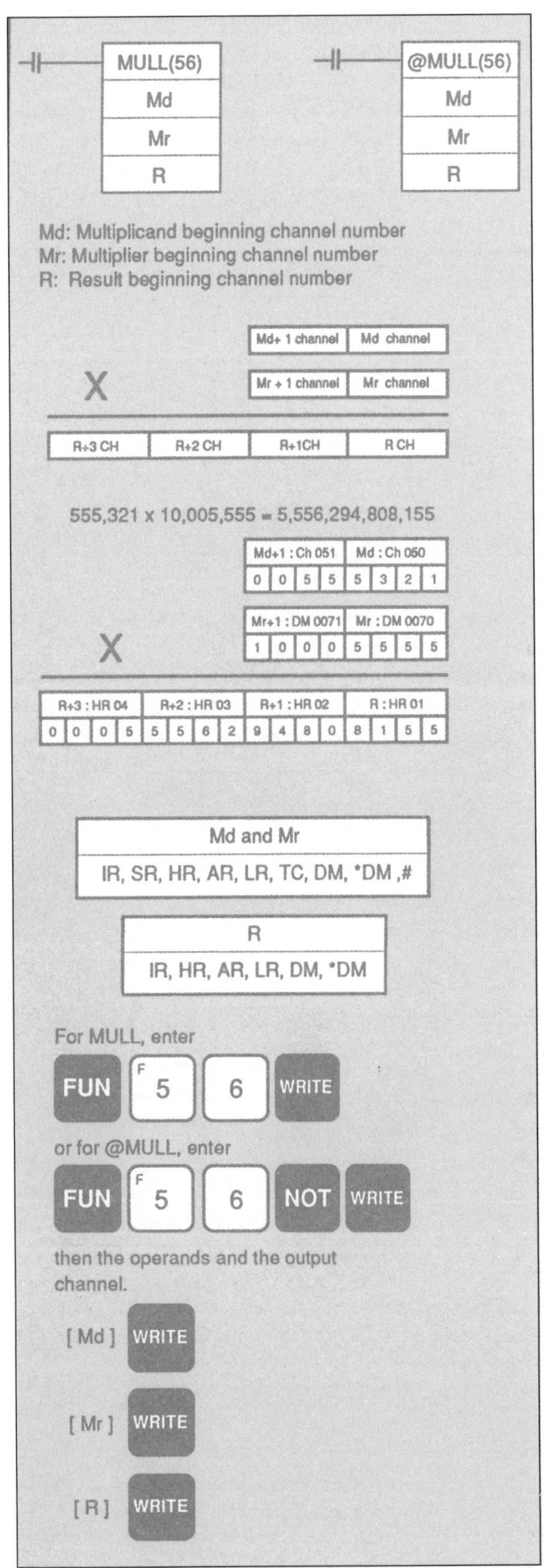

Figure 10-26. Double-precision BCD multiply (FUN 56).

```
      | 00002 25501                     |
00000|---| |-----| |-----------[ CNTR(12)]-|
      | 00001 25501           |      0471 |
      |---| |-----| |---------|    #9999 |-|
      | 00000                 |          | |
      |---| |-----------------L          J-|
      | 00014                            |
00006|---| |-------------------[ MULL(56)]-|
      |                       |     C0471 |-|
      |-                      |     D0000 | |
      |                       |     D0002 | |
      |-                      L           J-|
      | 00013                            |
00008|---| |-------------------[ WSFT(16)]-|
      |                       |     D0000 | |
      |-                      L     D0005 J-|
      |                                     |
```

ADDRESS	MNEMONIC	OPERAND
00000	LD	00002
00001	AND	25501
00002	LD	00001
00003	AND	25501
00004	LD	00000
00005	CNTR (12)	047
		# 9999
00006	LD	00014
00007	MULL (56)	
		CNT 047
		DM 0000
		DM 0002
00008	LD	00013
00009	WSFT (16)	
		DM 0000
		DM 0005
00010	END (01)	

Figure 10-27. Application circuit for double-precision BCD multiply.

BCD MULTIPLICATION WITH DOUBLE-PRECISION (LONG MULTIPLICATION)

For multiplication of larger numbers, it is often desirable to work with a PLC that offers double-precision arithmetic (see figure 10-26). With this advanced PLC function, the multiplicand is stored in two consecutively numbered words (channels). The multiplier also uses two consecutive words and the result obtained by this multiplication is stored in four consecutive words. Figure 10-26 shows an example of such a double-precision multiplication with 555 321 decimal value as multiplicand and 10 005 555 as multiplier. The result:

0005 5562 9480 8155

is stored in words HR 01, HR 02, HR 03 and HR 04

An experimental circuit for a double-precision multiplication calculation is given in figure 10-27. There, a reversible counter 47 is used to vary the multiplicand value. To enter data into the multiplier word, one may use the data change procedure explained under Manual Data Entry in this chapter (see figures 10-15 and 10-17). After completion of the experiment, all data held in DM 00 and DM 01 for the multiplier value and DM 02, DM 03, DM 04 and DM 05 for the result are cleared with the word shift instruction (FUN 16). For practice purposes, data for the multiplicand can easily be altered by incrementing or decrementing the counter to any value from 0001 to 9999. Remember that any multiplication with a zero value in the multiplicand or multiplier word gives a zero result.

NORMAL DIVIDE AND DOUBLE-PRECISION DIVIDE

The two arithmetic dividing instructions DIV (FUN 33) for normal divide and DIVL (FUN 57) for double-precision divide (long divide) are similar to the two multiplication functions explained before. Compare figure 10-24 with figure 10-28, and figure 10-26 with figure 10-29. Normal divide and normal multiply use only one word for multiplicand, multiplier, dividend and divisor, and two consecutive words to store the result. Double-precision multiply and double-precision divide, however, use two words for multiplicand, multiplier, dividend and divisor and four consecutive words for the result to be stored. The programmer, therefore, needs to take care when allocating these words that they are consecutive and not used for any other purpose, or overlapping. An experimental circuit for practising double-precision divisions is given in figure 10-30.

The divide function in figure 10-30 places the result obtained from dividing DM 06 and DM 07 into CNT 46 to DM 08, DM 09, DM 10 and DM 11. The quotient is stored in DM 08 and DM 09 and the remainder is stored in DM 10 and DM 11. The word shift function (FUN 16) is used to clear obtained results from words DM 08, DM 09, DM 10 and DM 11. To work this experimental circuit, move value 19 800 into DM 06 and DM 07 (0001 into DM 07 and 9800 into DM 06). Set CNT 46 to value 0320 (SV) and then execute the division with input 0014. The obtained result should be 61.875 decimal. The integer 0061 is stored in DM 08 and DM 09 contains 0000. The undivided remainder 0280 is stored in DM 10. DM 11 contains 0000. To obtain a correct decimal remainder, divide the undivided remainder in DM 10 with the value of the counter, 0320. This then renders the correct decimal remainder value of 0.875.

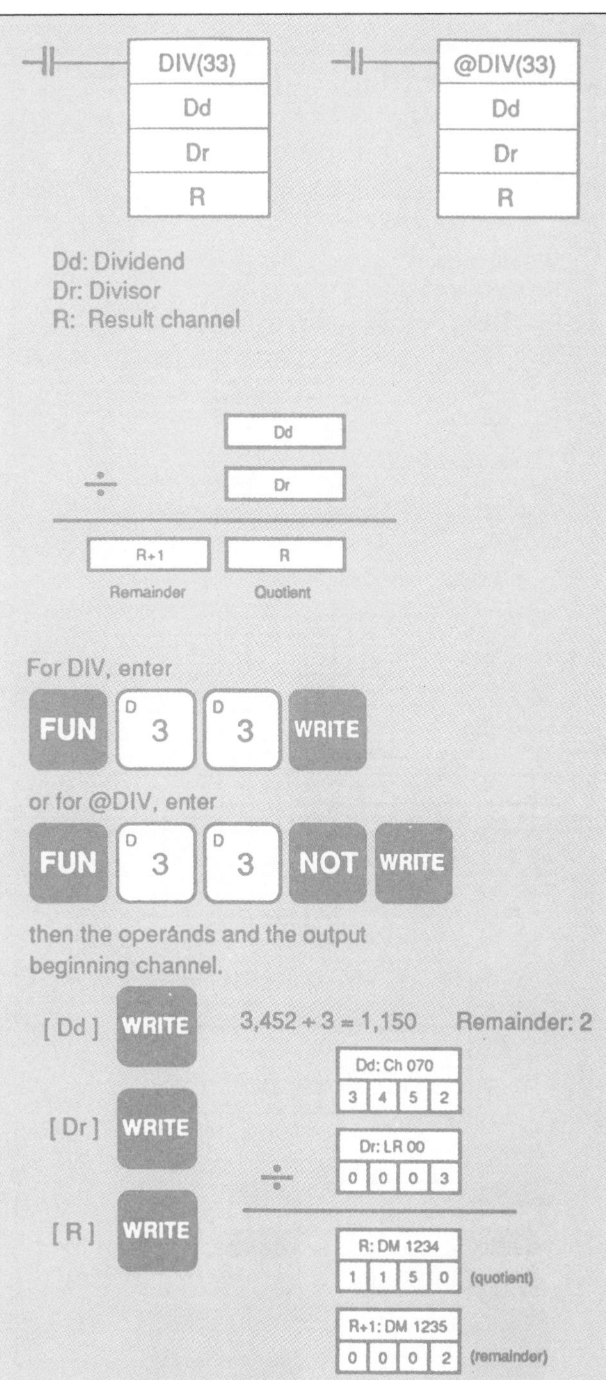

Figure 10-28. Normal divide function (FUN 33). During programming, A needs data for the dividend, B needs data for the divisor and C needs the word allocation for the result word.

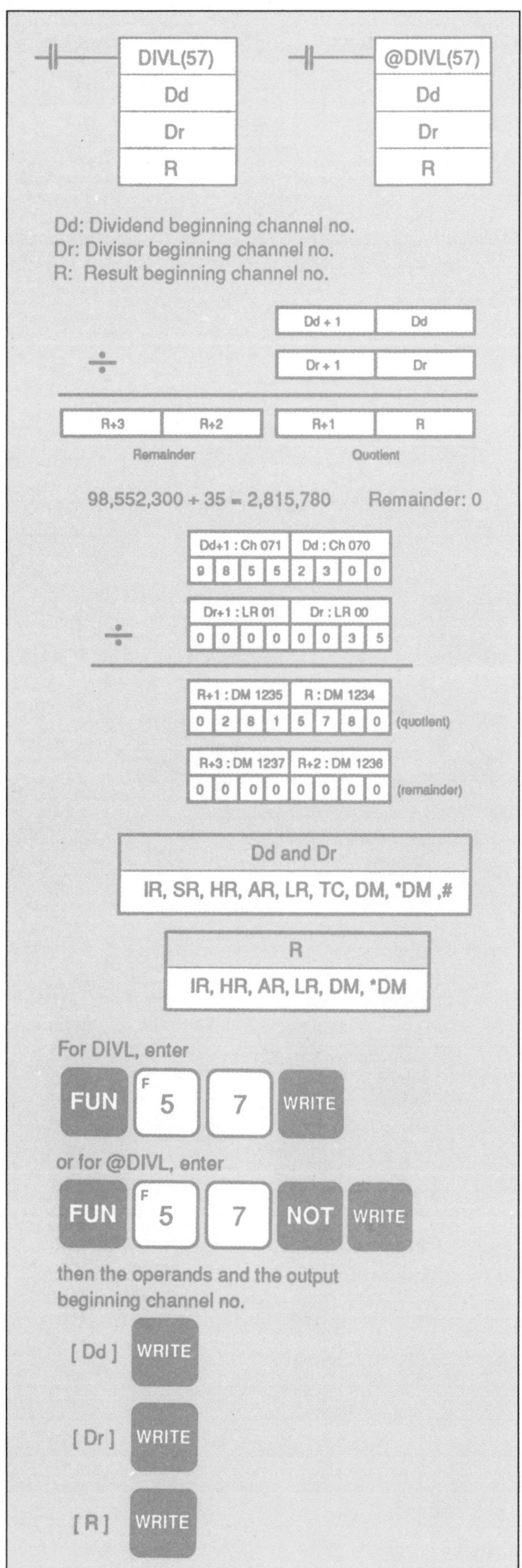

Figure 10-29. Double-precision divide function (FUN 57). Here, the dividend, the divisor and the result each use two words.

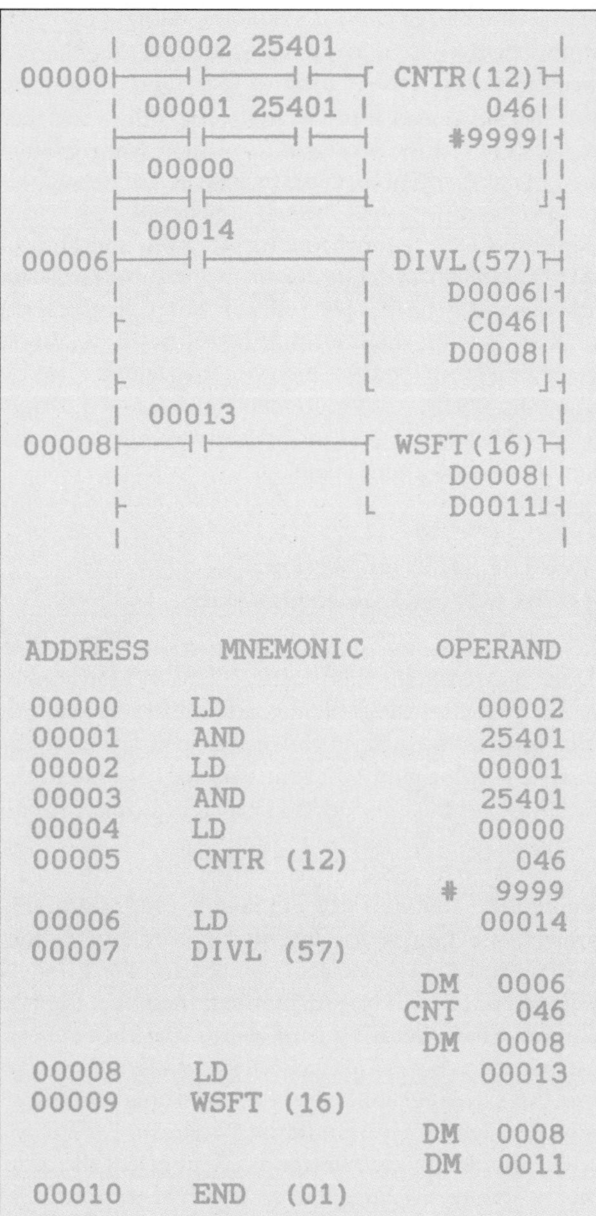

ADDRESS	MNEMONIC	OPERAND
00000	LD	00002
00001	AND	25401
00002	LD	00001
00003	AND	25401
00004	LD	00000
00005	CNTR (12)	046
		# 9999
00006	LD	00014
00007	DIVL (57)	
		DM 0006
		CNT 046
		DM 0008
00008	LD	00013
00009	WSFT (16)	
		DM 0008
		DM 0011
00010	END (01)	

Figure 10-30. Double-precision divide application circuit.

MAGNITUDE COMPARISON WITH A PLC

Magnitude comparison is used to compare digital values or data read from an input device, a timer or a counter with a reference value or data, and to differentiate the comparison result by expressing it with the concepts "greater than" (>), "equal to" (=), and "less than" (<). Indication of obtained results for Omron C20 and C28K

PLCs is given in flags 1905, 1906 and 1907. Flag 1905 is "ON" if the comparison shows that the second value is greater than the reference value. Flag 1906 is "ON" if the second value is equal to the reference value, and flag 1907 is "ON" if the second value is less than the reference value (see figure 10-31). Comparison is executed only if the enabling condition is "ON". In figure 10-31, this enabling contact is HR 915 (see address 0009).

When PLC magnitude comparison is applied to a boiling process, for example, if flag 1905 is "ON", the gas valve for the burners would be turned almost completely "OFF", or would stay on stand-by position to maintain a minimal or pilot flame. If flag 1906 is "ON", the gas flame would be increased to half maximum position to maintain simmer or low boiling temperature. If flag 1907 is "ON", the gas valve will be turned to full "ON" (maximum position), to obtain or maintain boiling temperature. A temperature probe or sensor in the liquid being boiled would send an analogue signal to an analogue to digital converter module on the PLC. This module converts the incoming analogue data to a digital number, which then is sent to an I/O word, or to the data memory, where it is compared with a reference constant, for example #180 (180°C). For operand data areas please check with figure 10-11.

ADVANCED PLC ARITHMETIC FUNCTIONS

The square root function (FUN 72) and the floating point divide function (FUN 79) belong to the more advanced PLC functions, which are usually found in larger PLCs, but recently also in modern small to medium PLCs.

Square root programming is relatively simple (see figure 10-33). The result of this computation is placed into one word only and the decimal part of a number is truncated. Square root taking is often found in computations for fluid flow, as applicable to hydraulics, pneumatics or waste water treatment. Figure 10-33 shows a calculation example, in which the square root function, when enabled by input bit 0000, takes the square root of decimal value 63 250 561 stored in source channels DM 00 and DM 01 and renders a truncated result of 7953, placed in channel 01.

The floating point divide (FUN 79) function is by nature not an easily understood PLC computation function. It permits data multiplication with decimal point values in the dividend and divisor word. If the result also renders a decimal point value, this may then also be evaluated as is shown in figure 10-32.

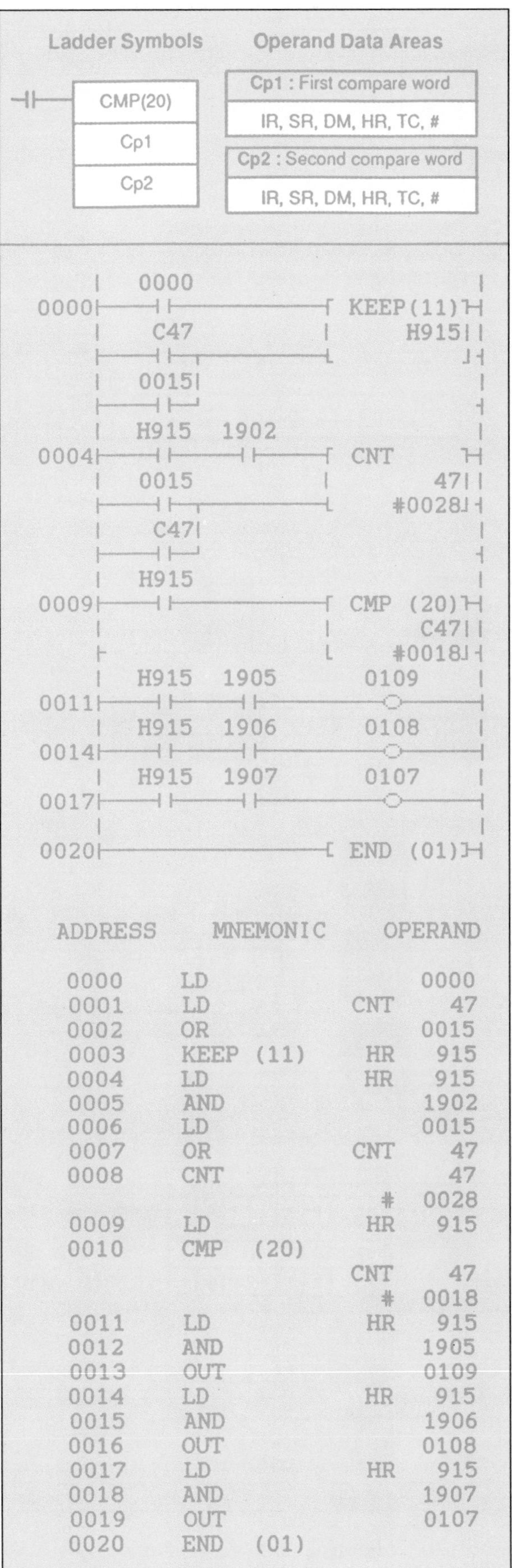

Figure 10-31. Magnitude compare function (FUN 20) with experimental circuit.

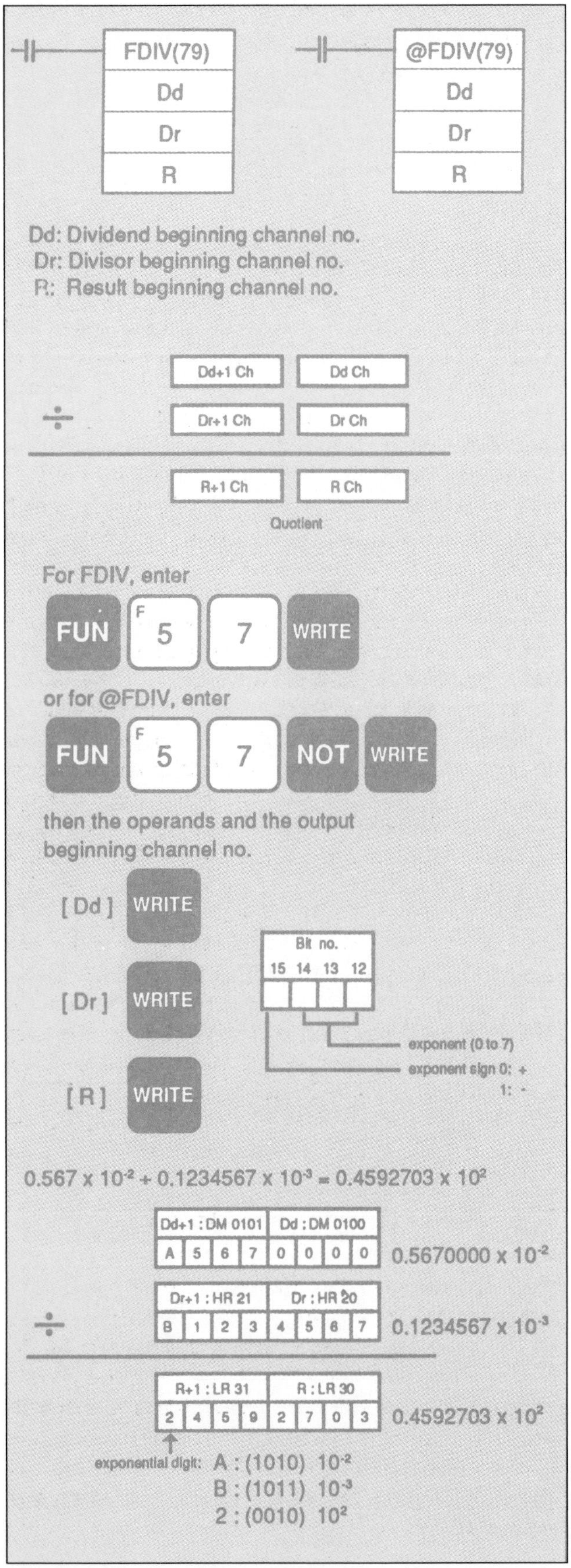

FDIV(79) — Dd, Dr, R

@FDIV(79) — Dd, Dr, R

Dd: Dividend beginning channel no.
Dr: Divisor beginning channel no.
R: Result beginning channel no.

Dd+1 Ch	Dd Ch
÷ Dr+1 Ch	Dr Ch
R+1 Ch	R Ch

Quotient

For FDIV, enter

FUN F 5 7 WRITE

or for @FDIV, enter

FUN F 5 7 NOT WRITE

then the operands and the output beginning channel no.

[Dd] WRITE

[Dr] WRITE

[R] WRITE

Bit no. 15 14 13 12

exponent (0 to 7)
exponent sign 0: +
1: -

$0.567 \times 10^{-2} \div 0.1234567 \times 10^{-3} = 0.4592703 \times 10^{2}$

Dd+1 : DM 0101				Dd : DM 0100				
A	5	6	7	0	0	0	0	0.5670000×10^{-2}

Dr+1 : HR 21				Dr : HR 20				
B	1	2	3	4	5	6	7	0.1234567×10^{-3}

R+1 : LR 31				R : LR 30				
2	4	5	9	2	7	0	3	0.4592703×10^{2}

exponential digit: A : (1010) 10^{-2}
B : (1011) 10^{-3}
2 : (0010) 10^{2}

Figure 10-32. Floating point divide function (FUN 79).

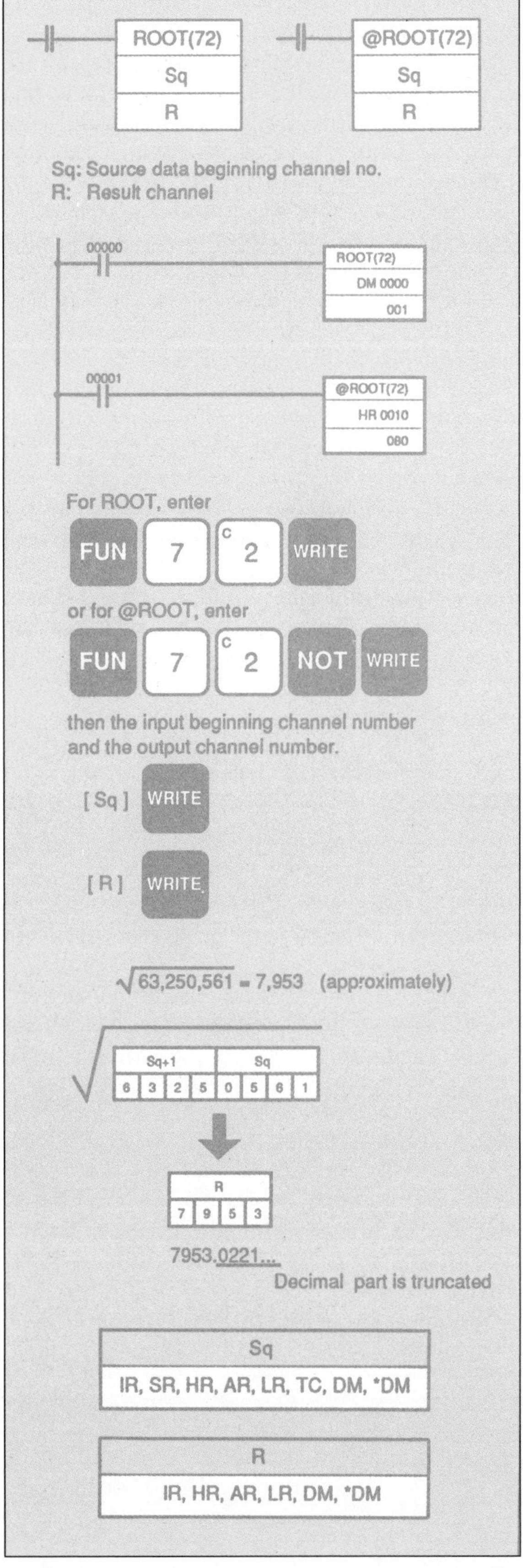

Figure 10-33. Square root function (FUN 72).

CONTROL PROBLEM 1

Two conveyors transport cardboard boxes. Each box contains eight plastic bottles filled with dishwashing liquid. Although the bottles can be filled with up to 900 mL (millilitres) of dishwashing liquid, depending on customer request filling quantities may vary from 600 mL to 900 mL. The cardboard box counting sensors are mounted on the exit side of the two conveyors. However, for the purpose of testing the PLC control problem circuit out of the factory, these two sensors are replaced by the two pulse generating bits 25 502 and 25 501. The pulse bits clock the two cardboard box counters CNT 031 and CNT 032 (see figure 10-34, addresses 0005 to 0012). In an Omron C200H PLC, bit 25 501 pulses with a 0.2 second clock and bit 25 502 pulses with a 1.0 second clock (see similar clocks for an Omron C28K, and Omron C20 PLC in figure 5-01). Bit 25 501, therefore, simulates a box moving past the sensor every 0.2 second and bit 25 502 pretends to count a box every second. In reality, on the machine, these sensor pulses would appear much further apart. Hence, counter 031 counts boxes on conveyor 1 and counter 032 records boxes on conveyor 2. Both counters are so-called down-counters, which decrement from SV 9999 to zero (0000).

Most PLCs have up-counters, down-counters and reversible counters. For the given hypothetical control problem, however, designed to demonstrate arithmetic functions, it is assumed that the PLC has only two reversible counters that need to be used for the simulation of liquid filling percentages. One therefore has to use down-counters for counting the cardboard boxes.

To keep accounts of daily production, the manufacturers wish to obtain accurate filling and processing data at the end of each production shift. The PLC's data accounting process may be made visible with a seven-segment display or through the PLC's programming console. For example, the PV of both counters could be monitored with the programming console keys CLR, CLR, CNT, 3, 1, MONTR, CNT, 3, 2, MONTR. The conveyors are started with input 0001, and stopped with input 0000. Machine "Running" is displayed with lamp 0500.

For the purpose of arithmetic function demonstration only, it is assumed that both counters are down-counters. Their real count value (PV real) is, therefore, obtained by subtracting PV from 9999 (SV). These subtractions (one for each counter), are made with FUN 31 at addresses 0029 and 0031, and the real cardboard box count figures from the two counters are stored in DM 02 and DM 04. Total production of both conveyors is added with a double-precision FUN 54 add function, and its result is stored in DM 06 and DM 07. Its maximum addition capacity could, therefore, be:

$$\begin{array}{r} 9999 \\ +\ \underline{9999} \\ =\ \underline{\underline{19998}} \end{array}$$

In this case, DM 06 would contain 9998 and DM 07 would contain 0001. One may therefore say that DM 06 and DM 07 store the total cardboard box production of one shift. To obtain the total of containers packed and filled and placed into cardboard boxes, one now needs to multiply the result in DM 06 and DM 07 by decimal value 8 (eight containers per box). The double-precision function cannot be programmed with a constant (#). For this reason, value 8 is entered into DM 00 during PLC programming and remains unchanged (for as long as each box contains eight containers). When FUN 54 is executed, the total cardboard box quantity held in the two multiplicand words DM 06 and DM 07 is multiplied by the value of the two multiplier words DM 00 and DM 01. The result obtained is placed into DM 08, DM 09, DM 10 and DM 11.

To obtain an accurate figure of the total amount of dishwashing liquid processed and filled per shift, the PLC is now programmed to multiply the total number of bottles filled and packed by the percentage of liquid filled into each bottle. If, for the present customer production run, each bottle is filled with say 730 mL (0.73 litre), the PLC is then instructed to multiply data held in DM 08 and DM 09 with 730 (multiplier) and divide the result obtained by 1000. The result will then be the total amount of dishwashing liquid filled in litres per shift.

To obtain 73% of filling capacity, one could use the floating point divide function FDIV (FUN 79) to reach the same result, but not all PLCs have such a function. With the floating point divide function, MULL FUN 56 on address 37 and DIVL FUN 57 on address 39 could have been amalgamated into one function.

Experimental procedures for control problem 1 (figure 10-34)

Enter manually a hypothetical count from counter 031 into DM 02 (e.g. 9000). Enter manually a hypothetical count from counter 032 into DM 04 (e.g. 8000). To enter these two counter figures, use procedures shown in figures 10-15 and 10-16. Enter 0008 also manually into DM 00 to multiply total cardboard boxes counted by eight containers per box. Now, set counter 033 to a PV of 730 (730 mL per bottle for the present customer), and counter 034 to 1000 (1000 mL per litre).

$$\frac{730\ \text{mL}}{1000} = 0.73\ \text{L}$$

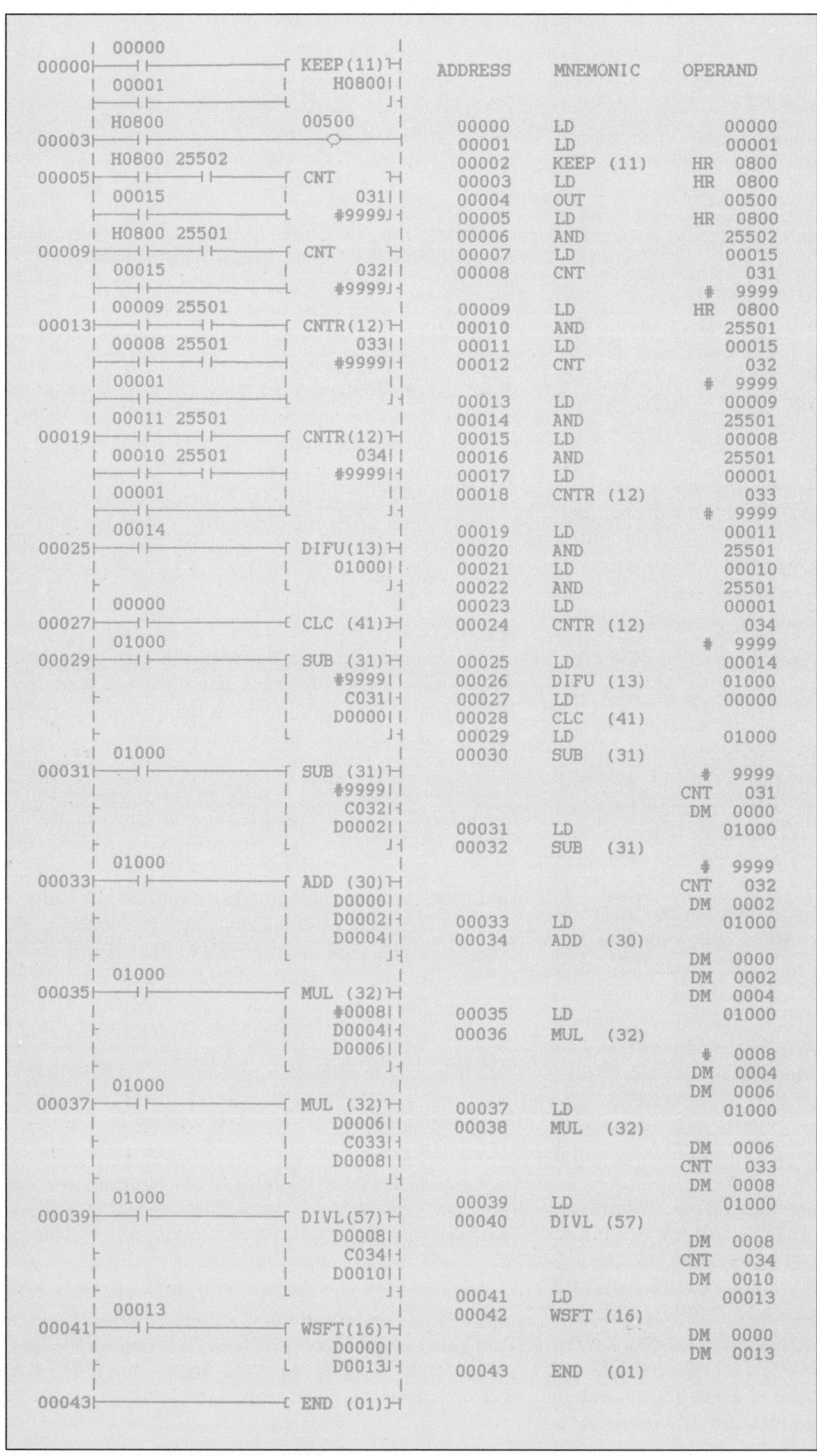

ADDRESS	MNEMONIC	OPERAND
00000	LD	00000
00001	LD	00001
00002	KEEP (11)	HR 0800
00003	LD	HR 0800
00004	OUT	00500
00005	LD	HR 0800
00006	AND	25502
00007	LD	00015
00008	CNT	031
		# 9999
00009	LD	HR 0800
00010	AND	25501
00011	LD	00015
00012	CNT	032
		# 9999
00013	LD	00009
00014	AND	25501
00015	LD	00008
00016	AND	25501
00017	LD	00001
00018	CNTR (12)	033
		# 9999
00019	LD	00011
00020	AND	25501
00021	LD	00010
00022	AND	25501
00023	LD	00001
00024	CNTR (12)	034
		# 9999
00025	LD	00014
00026	DIFU (13)	01000
00027	LD	00000
00028	CLC (41)	
00029	LD	01000
00030	SUB (31)	
		# 9999
		CNT 031
		DM 0000
00031	LD	01000
00032	SUB (31)	
		# 9999
		CNT 032
		DM 0002
00033	LD	01000
00034	ADD (30)	
		DM 0000
		DM 0002
		DM 0004
00035	LD	01000
00036	MUL (32)	
		# 0008
		DM 0004
		DM 0006
00037	LD	01000
00038	MUL (32)	
		DM 0006
		CNT 033
		DM 0008
00039	LD	01000
00040	DIVL (57)	
		DM 0008
		CNT 034
		DM 0010
00041	LD	00013
00042	WSFT (16)	
		DM 0000
		DM 0013
00043	END (01)	

Figure 10-34. Control problem 1, ladder diagram and mnemonic list.

The PLC is now ready to perform the arithmetic computations it is programmed for. To start such computations, activate input 0000. This in turn activates DIFU 1000 (edge pulse flag 1000).

	9000		ADDL	(FUN 54)
+	8000			
=	17000	⇒	0001	7000
			DM 07	DM 06

17000 x 0008		MULL	(FUN 56)
= 136000	⇒	0013	6000
		DM 09	DM 08

136000 x 0730		MULL	(FUN 56)
=99280000	⇒	9928	0000
		DM 13	DM 12

99280000		DIVL	(FUN 57)
÷ 1000	⇒	9928	
		DM 17	DM 16

One may now monitor all relevant DM words with the keys DM, 0, MONTR. Pressing the DOWN key at the right-hand bottom corner of the programming console causes the console's display window to indicate the data held in every monitored DM word.

DM 00 ⇒ 0008	DM 10 ⇒ 0000
DM 01 ⇒ 0000	DM 11 ⇒ 0000
DM 02 ⇒ 9000	DM 12 ⇒ 0000
DM 03 ⇒ 0000	DM 13 ⇒ 9928
DM 04 ⇒ 8000	DM 14 ⇒ 0000
DM 05 ⇒ 0000	DM 15 ⇒ 0000
DM 06 ⇒ 7000	DM 16 ⇒ 9928
DM 07 ⇒ 0001	DM 17 ⇒ 0000
DM 08 ⇒ 6000	DM 18 ⇒ 0000
DM 09 ⇒ 0013	DM 19 ⇒ 0000

If the total numbers of bottles had to be displayed on a seven-segment data display panel, one could use the seven-segment decoder function (SDEC FUN 78) to display the relevant information in DM 08 and DM 09. Similarly, on another display, one could show data held in DM 16 and DM 17 to display the total amount in litres bottled per shift. PLCs are excellently equipped with such decoding functions to retrieve information from their data memory and display it on monitors or seven-segment display screens. For decoding functions, see the Appendix for FUN 78 and FUN 86.

CONTROL PROBLEM 2

A container filling and weighing machine is controlled by a PLC. Actuator Ⓐ stops and releases containers at the entrance of the weighing station (figure 10-35). Actuator Ⓑ "sweeps" the weighed containers from the weighing station onto the exit conveyor. Actuator Ⓒ pushes underweight containers onto exit conveyor X, where they are manually topped-up to the correct weight of 1000 grams. Actuator Ⓓ pushes overweight containers onto exit conveyor Y, where they are manually emptied and their content is pumped back to the bulk liquid tank. To simulate the actual weighing process, counter 47 is used to mimic weighing data. Its PV may be incremented or decremented with inputs 0013 and 0011 respectively. PV 1011 represents the least overweight value and PV 0990 represents the greatest underweight value (1011 grams and 990 grams respectively). Any weighing value (in grams) from 0990 to 1010 is regarded as acceptable. The tolerance range is therefore 10 grams. Bit 1907 turns "ON" if the comparison function of the PLC (FUN 20) indicates "greater than" #1010 (upper tolerance). Bit 1905 turns "ON" if the comparison function (FUN 20) indicates "less than" #0990 (lower tolerance). Bits 1905 and 1907 are not "ON" if the weighing process shows a weighing result between 0990 minimum and 1010 maximum. Bit 1906 will turn "ON" if the weighing result is exactly 1000. For this control problem, however, this bit is not used, as the evaluation is only of exceeded maximum or exceeded minimum. Any container filled with liquid within these two boundaries is regarded as correctly filled and is permitted to pass through exit 2. It is then registered and counted by sensor 009.

An underfilled (underweight) container is pushed-out after 10 seconds' travel time on the exit conveyor. An overfilled (overweight) container is pushed out after 8 seconds' travel time. An underweight container, for example, if detected by the weighing station and the PLC's comparison function, activates bit 1907. This in turn sets flip-flop HR 03, which enables timer 02. Timer 02 turns "ON" after 10 seconds and activates actuator Ⓒ to extend and force the underfilled container onto exit conveyer X. Actuator Ⓒ retracts automatically as soon as it has reached limit switch 006. A similar process would cause actuator Ⓓ to force an overfilled container onto exit conveyor Y.

To synchronise this machine and cause it to run automatically and reliably, actuator Ⓑ will sweep only when input 000 (container present), input 003 (correct actuator Ⓑ starting position) and input 1001 (exit conveyor cleared and comparison function reset) are signalled. Signal 1001 is present only when either an overweight container has activated timer 2 or a correct weight container has been registered by sensor 009.

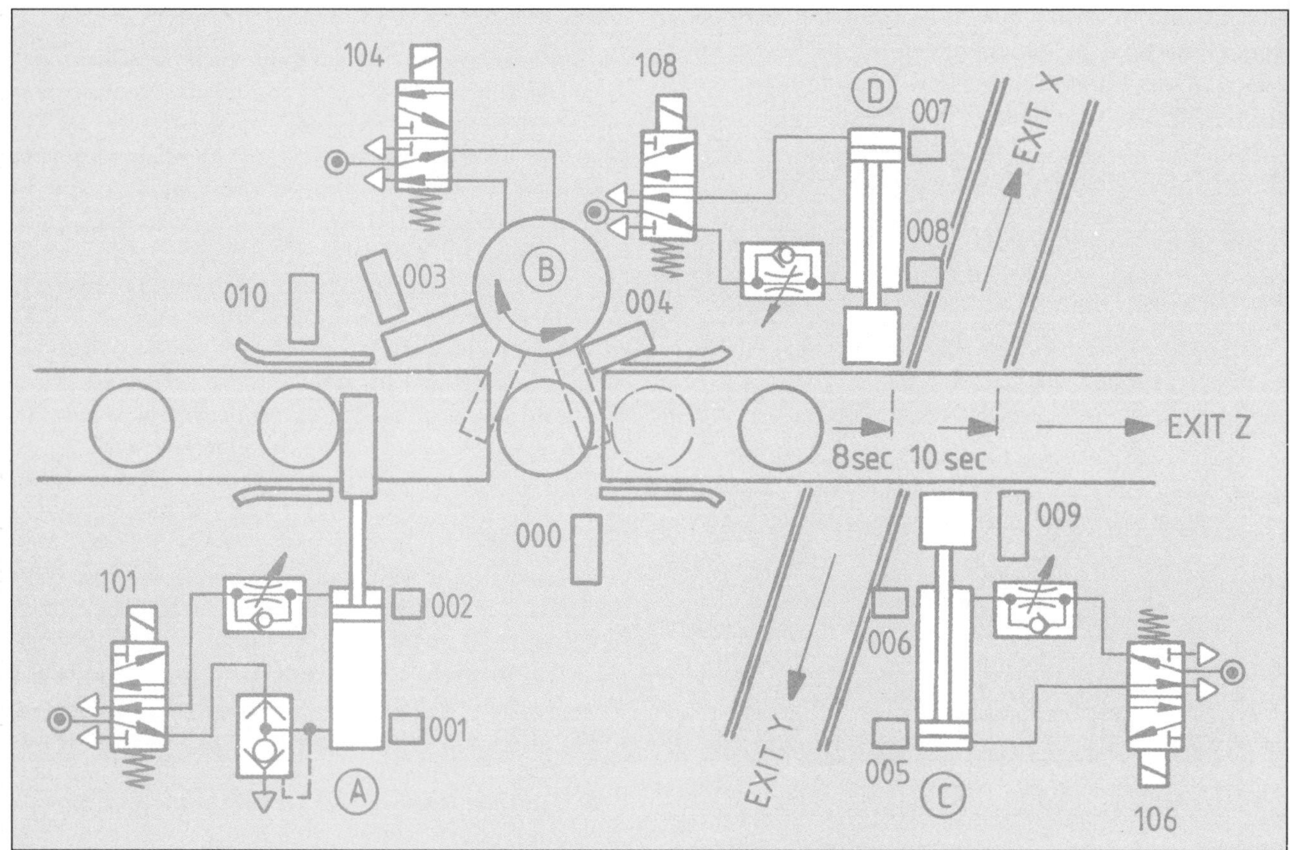

Figure 10-35. Control problem 2, machine layout for container weighing and separating station.

Actuator Ⓐ will release a container into the weighing station only when the weighing station is empty ($\overline{0000}$), actuator Ⓑ gives free passage (003), a container is at the barrier (010) and the machine is started (1000).

Lamp output 109 is "ON" if HR 01 or HR 03 signals an incorrectly filled container being present between sensors 000 (weighing station) and 009 (exit for correctly filled containers). Lamp output 110 is "ON" if a container is weighed and swept off the weighing station, and the comparison function (FUN 20) has rendered a correct weighing result ($\overline{\text{HR 003}} \bullet \overline{\text{HR 001}}$). De Morgan's theorem is used here, which says:

$$\overline{\text{HR 003}} \bullet \overline{\text{HR 001}} = \overline{\text{HR 003}} + \overline{\text{HR 001}}.$$

Internal relay 1001 is "ON" if no container is in the machine between sensor 000 and sensor 009 (weighing station and remotest exit point).

Flip-flop HR 001 and HR 003 can be reset only if internal relay 1001 is "ON".

The weighing process may be started only if actuator Ⓑ starts the sweeping motion and flip-flops HR 001 and HR 003 are reset.

Internal relay 1002 is "ON" if weighing is in progress, and the weight is found to be correct (HR 003 and HR 001 not "ON").

HR 002 is "ON" only if the weighing and comparing process has started and the container is found to be of correct weight (within the tolerance).

Lamp output 110 is "ON" if a correct weight container is in the machine.

Correct weight containers, having been weighed and processed, are counted by counter 03. Containers with incorrect weight are counted by counter 04. Both counters are enabled when the machine is started (KEEP 1000).

All four conveyor motors are started with input 015 activating KEEP 1000, which thereupon activates output 101 (address 0052). When machine "STOP" is signalled by input 014, all conveyors keep running for 30 seconds to clear out containers in the weighing process, but actuator Ⓐ is then no longer permitted to retract. Timer 00 stops the conveyor motors after an elapse of 30 seconds. The two totalising counters CNT 03 and CNT 04 are automatically reset when the machine is stopped.

Counter 47 is used only to simulate weighing results. Under real conditions, a load cell supplying an analogue signal would be used to enter data for the compare function.

Figure 10-36. Control problem 2, ladder diagram and mnemonic list.

```
     | 0015   0003                          |
0000 |--| |----| |------------------------[ KEEP(11) ]
     | 0014                                   1000
     |--| |-------------------------------[         ]
     | 0000   1002                          |
0004 |--| |----| |--+---------------------[ CMP (20) ]
     |              |                         #1010
     |              |                           C47
     |              | 1907
     |              +--| |----------------[ KEEP(11) ]
     | 1001                                   H003
     |--| |-------------------------------[         ]
     | 0000   1002                          |
0010 |--| |----| |--+---------------------[ CMP (20) ]
     |              |                         #0990
     |              |                           C47
     |              | 1905
     |              +--| |----------------[ KEEP(11) ]
     | 1001                                   H001
     |--| |-------------------------------[         ]
     | 0000   1002   H001   H003            |
0016 |--| |----| |----|/|----|/|----------[ KEEP(11) ]
     | 1001                                   H002
     |--| |-------------------------------[         ]
     | T01                              1001 |
0022 |--| |--+------------------------------( )----|
     | T02   |
     |--| |--+
     | 0009  |
     |--| |--+
     | 0104   H003   H001               1002 |
0026 |--| |----|/|----|/|-------------------( )----|
     | 1000
0030 |--|/|-------------------------------[ TIM     ]
     |                                          00
     |                                       #0300
     | H001
0032 |--| |-------------------------------[ TIM     ]
     |                                          01
     |                                       #0080
     | H003
0034 |--| |-------------------------------[ TIM     ]
     |                                          02
     |                                       #0100
     | 0009   1000
0036 |--| |----| |------------------------[ CNTR(12) ]
     | 1812                                     03
     |--| |-------------------------------[    #9999 ]
     | 1000
     |--|/|-------------------------------[         ]
     | 0109   1000
0041 |--| |----| |------------------------[ CNTR(12) ]
     | 1812                                     04
     |--| |-------------------------------[    #9999 ]
     | 1000
     |--|/|-------------------------------[         ]
     | 1000
0046 |--| |-------------------------------[ KEEP(11) ]
     | T00                                    0100
     |--| |-------------------------------[         ]
     | 0000   0003   0010   1000
0049 |--|/|----| |----| |----| |----------[ KEEP(11) ]
     | 0001                                   0101
     |--| |-------------------------------[         ]
     | 0000   0003   1001
0055 |--| |----| |----| |-----------------[ KEEP(11) ]
     | 0004                                   0104
     |--| |-------------------------------[         ]
     | T02
0060 |--| |-------------------------------[ KEEP(11) ]
     | 0006                                   0106
     |--| |-------------------------------[         ]
     | T01
0063 |--| |-------------------------------[ KEEP(11) ]
     | 0008                                   0108
     |--| |-------------------------------[         ]
     | H001                             0109 |
0066 |--| |--+------------------------------( )----|
     | H003  |
     |--| |--+
     | H002                             0110 |
0069 |--| |---------------------------------( )----|
     | 0013   1900
0071 |--| |----| |------------------------[ CNTR(12) ]
     | 0011   1900                              47
     |--| |----| |------------------------[    #1100 ]
     | 0012
     |--| |-------------------------------[         ]
```

ADDRESS	MNEMONIC		OPERAND
0000	LD		0015
0001	AND		0003
0002	LD		0014
0003	KEEP (11)		1000
0004	LD		0000
0005	AND		1002
0006	CMP (20)		
		#	1010
		CNT	47
0007	AND		1907
0008	LD		1001
0009	KEEP (11)	HR	003
0010	LD		0000
0011	AND		1002
0012	CMP (20)		
		#	0990
		CNT	47
0013	AND		1905
0014	LD		1001
0015	KEEP (11)	HR	001
0016	LD		0000
0017	AND		1002
0018	AND NOT	HR	001
0019	AND NOT	HR	003
0020	LD		1001
0021	KEEP (11)	HR	002
0022	LD	TIM	01
0023	OR	TIM	02
0024	OR		0009
0025	OUT		1001
0026	LD		0104
0027	AND NOT	HR	003
0028	AND NOT	HR	001
0029	OUT		1002
0030	LD NOT		1000
0031	TIM		00
		#	0300
0032	LD	HR	001
0033	TIM		01
		#	0080
0034	LD	HR	003
0035	TIM		02
		#	0100
0036	LD		0009
0037	AND		1000
0038	LD		1812
0039	LD NOT		1000
0040	CNTR (12)		03
		#	9999
0041	LD		0109
0042	AND		1000
0043	LD		1812
0044	LD NOT		1000
0045	CNTR (12)		04
		#	9999
0046	LD		1000
0047	LD	TIM	00
0048	KEEP (11)		0100
0049	LD NOT		0000
0050	AND		0003
0051	AND		0010
0052	AND		1000
0053	LD		0001
0054	KEEP (11)		0101
0055	LD		0000
0056	AND		0003
0057	AND		1001
0058	LD		0004
0059	KEEP (11)		0104
0060	LD	TIM	02
0061	LD		0006
0062	KEEP (11)		0106
0063	LD	TIM	01
0064	LD		0008
0065	KEEP (11)		0108
0066	LD	HR	001
0067	OR	HR	003
0068	OUT		0109
0069	LD	HR	002
0070	OUT		0110
0071	LD		0013
0072	AND		1900
0073	LD		0011
0074	AND		1900
0075	LD		0012
0076	CNTR (12)		47
		#	1100
0077	END (01)		

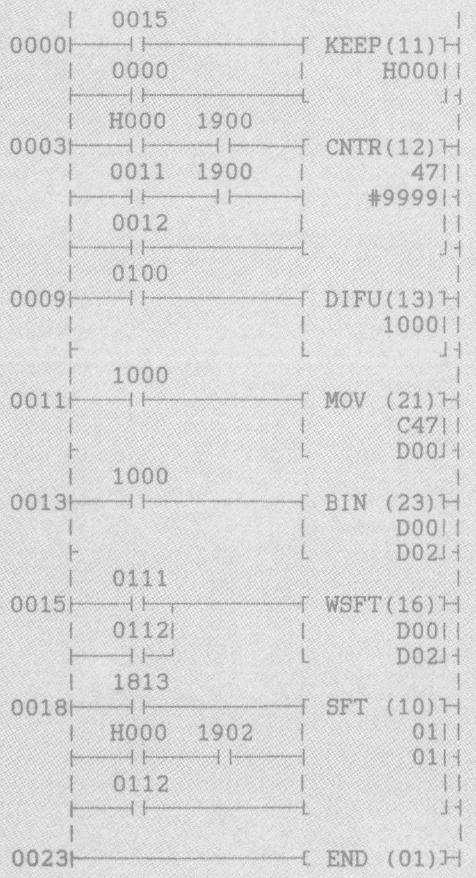

```
     | 0015                          |
0000 |---| |---------------------[ KEEP(11) ]
     | 0000                          | H000 |
     |---| |-------------------------|      |
     | H000    1900                  |
0003 |---| |----| |--------------[ CNTR(12) ]
     | 0011    1900                  |   47 |
     |---| |----| |------------------| #9999|
     | 0012                          |      |
     |---| |-------------------------|      |
     | 0100                          |
0009 |---| |---------------------[ DIFU(13) ]
     |                               | 1000 |
     |                               |      |
     | 1000                          |
0011 |---| |---------------------[ MOV  (21) ]
     |                               |  C47 |
     |                               |  D00 |
     | 1000                          |
0013 |---| |---------------------[ BIN  (23) ]
     |                               |  D00 |
     |                               |  D02 |
     | 0111                          |
0015 |---| |--+------------------[ WSFT(16) ]
     | 0112   |                      |  D00 |
     |---| |--+                      |  D02 |
     | 1813                          |
0018 |---| |---------------------[ SFT  (10) ]
     | H000    1902                  |   01 |
     |---| |----| |------------------|   01 |
     | 0112                          |      |
     |---| |-------------------------|      |
     |                               |
0023 |---------------------------[ END  (01) ]
```

ADDRESS	MNEMONIC	OPERAND
0000	LD	0015
0001	LD	0000
0002	KEEP (11)	HR 000
0003	LD	HR 000
0004	AND	1900
0005	LD	0011
0006	AND	1900
0007	LD	0012
0008	CNTR (12)	47
		# 9999
0009	LD	0100
0010	DIFU (13)	1000
0011	LD	1000
0012	MOV (21)	
		CNT 47
		DM 00
0013	LD	1000
0014	BIN (23)	
		DM 00
		DM 02
0015	LD	0111
0016	OR	0112
0017	WSFT (16)	
		DM 00
		DM 02
0018	LD	1813
0019	LD	HR 000
0020	AND	1902
0021	LD	0112
0022	SFT (10)	01
		01
0023	END (01)	

Figure 10-37. Control problem 3, ladder diagram and mnemonic list.

CONTROL PROBLEM 3

A continuous, automatic, decimal-to-binary conversion experimental circuit is given in figure 10-37. Input 0015 starts the conversion process and input 0000 stops it. Reversible counter 47 is used to produce a continuously changing PV for as long as the counter is enabled. Counter input HR 00 increments the counter with pulse bit 1900 and input 011 decrements the counter with pulse bit 1900. Input 012 resets the counter. Shift register 01 is clocked by HR 00, as enabling signal, and by pulse bit 1902, which creates clock pulses every second. The shift register can be monitored while in operation, since its outputs from 0100 to 0111 can be monitored by the PLC's output LEDs. The shift register resets itself with output 0112 (self-reset). Every time the shift register passes through output 0100, it activates differentiated internal output 1000 for one scan pulse on the leading edge of shift register signal 0100. This edge pulse activates the move process function (FUN 21), which moves the present counter PV into DM 00. The same edge pulse converts the moved decimal data in DM 00 to binary (hexadecimal) data in DM 02.

To monitor and display the process, one can operate the programming console keys CLR, DM, 0, MONTR, DM, 2, MONTR, CNT, 4, 7, MONTR. The display on the programming console will now show:

C47	D02	D00
6766	1A5F	6751

if the process is halted shortly after the counter has passed PV 6751 and before the shift register resets the display to the next conversion image. Hence, decimal 6751 is equivalent to hexadecimal (binary) 1A5F. The control process may of course be halted at any time with input 000.

CONTROL PROBLEM 4

A PLC long divide function (FUN 57) must divide a dividend value of 178 492 decimal by the current PV of counter 46 as divisor (see figure 10-38). The result is placed into DM 08, DM 09, DM 10 and DM 11. The value 8492 is placed into DM 06, and 0017 is placed into DM 07. Assuming that the counter at the time of arithmetic function execution has a PV of 0038 (when input 0014 is "ON"), the mathematical result would then render 4697.1579 decimal. Hence, the quotient 0000 4697 is placed into DM 08 and DM 09, and the undivided remainder (overflow of 0006) is placed into DM 10 and DM 11. This remainder now has to be divided again by counter 46 PV value, to obtain the correct remainder of 0.1578947 decimal. To accomplish this, DM 10, containing the undivided overflow of 0006, is moved to DM 13 (see address 0008 in figure 10-38). DM 12 and DM 13 are now divided again by counter 46 PV and the result is placed into DM 14, DM 15, DM 16 and DM 17. DM 14 and DM 13 hold the quotient, which is 1579, and an overflow as remainder is again placed into DM 16 and DM 17. To gain even greater accuracy for the remainder, the remaining remainder of this second dividing process could be divided again. To obtain a seven-segment readout, one could connect DM 08 and DM 14 as follows:

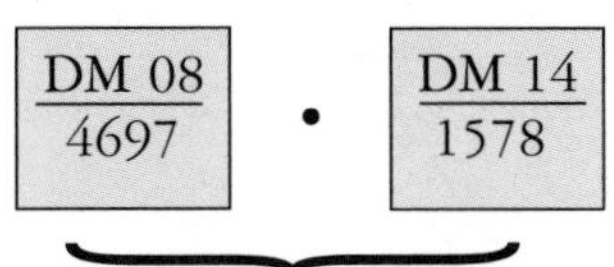

seven-segment display

In order to monitor the obtained result from the division, one must press the programming console keys:

CLR, CLR, DM, 8, MONTR.

To obtain a readout of the data in each channel involved with this computation, one now simply presses the down-arrow key and thus checks each data memory channel in successive order. The monitored results should read as follows:

DM 08 ⇒ 4697	DM 13 ⇒ 0006
DM 09 ⇒ 0000	DM 14 ⇒ 1578
DM 10 ⇒ 0006	DM 15 ⇒ 0000
DM 11 ⇒ 0000	DM 16 ⇒ 0036
DM 12 ⇒ 0000	DM 17 ⇒ 0000

The rounded-off result of this division should be 4697. The not rounded-off computation would render:

$$\frac{178492}{000038} = 4967.1578$$

ARITHMETIC FUNCTION PROGRAMMING SUMMARY

The programming of arithmetic functions on PLCs needs some basic knowledge of numbering systems such as decimal, binary, hexadecimal and BCD. The BCD system is the most widely used PLC numbering system. Modern PLC hardware permits arithmetics for add, subtract, multiply, divide, square root and floating point divide. Data for such functions are obtained from timers, counters, thumbwheel switches and factory intelligent terminals (FITs). Some PLC brands also provide binary to BCD and BCD to binary conversion functions. Arithmetic function processing is an imperative tool for today's data-seeking and totally automated society.

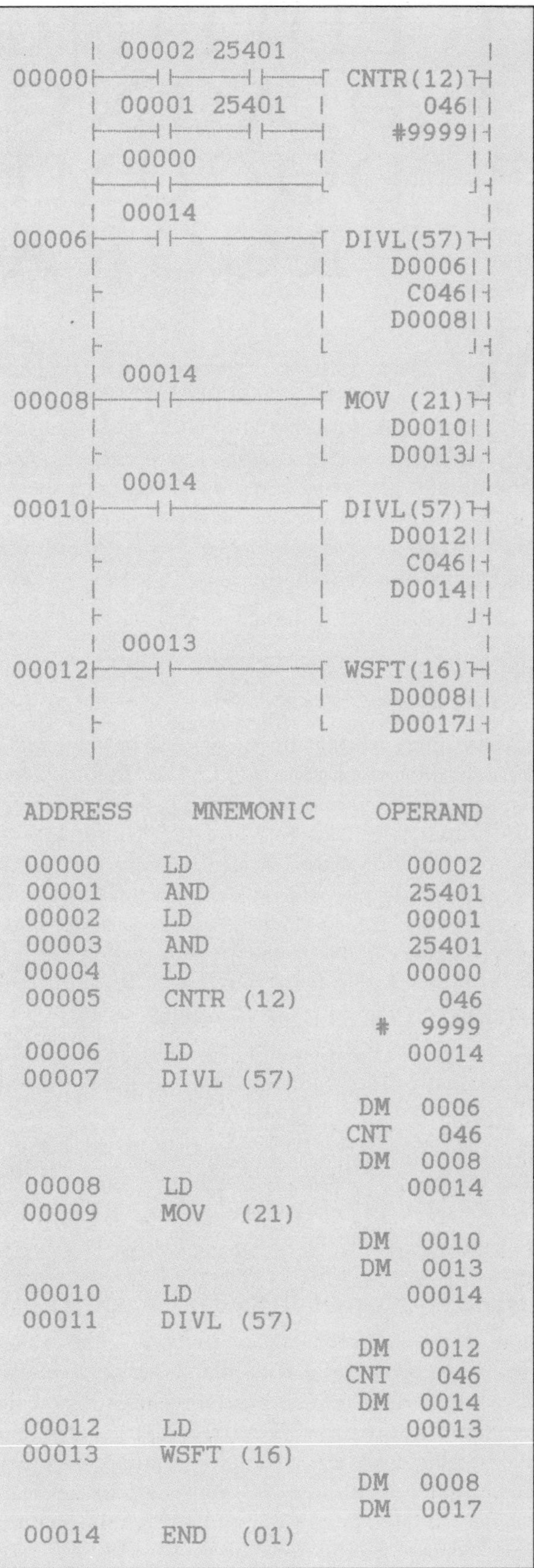

Figure 10-38. Control problem 4, ladder diagram and mnemonic list.

11 EDITING, DIAGNOSING AND DOCUMENTING PLC PROGRAMS

When programming a PLC, one invariably makes mistakes, especially so if the person programming the PLC is inexperienced. Such programming mistakes can occur when designing the program or when entering it into the PLC's user memory, and often don't become obvious until the program is being tested and the machine controlled by the PLC is installed and commissioned. Eradicating mistakes (debugging) usually involves:

- changing an existing program instruction
- deleting an existing program instruction
- inserting a program instruction.

These three editing functions must be mastered by anyone wishing to program a PLC. They need not, however, be difficult if approached methodically and logically. It must be remembered that the PLC instruction is stored in a location within the user memory and that that location has a discrete address, hence the information can be searched for and found whenever necessary (see figures 1-11, 9-20, 11-01, 11-02 and 11-03).

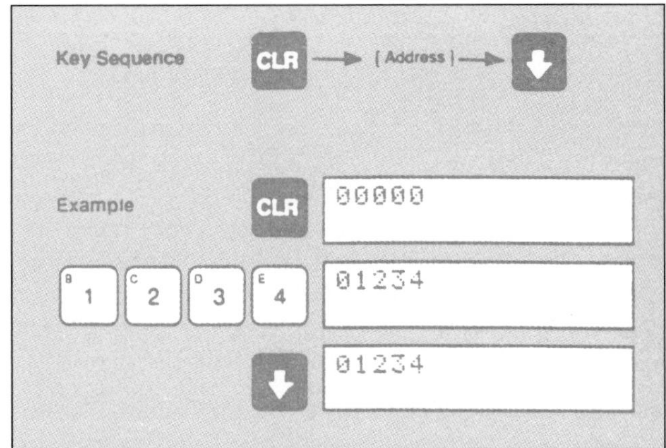

Figure 11-01. Searching for and displaying a programming address with the PLC's in-built address search facility.

ADDRESS LOCATION SEARCH

The PLC's in-built address-finding or -searching feature can be used to find a programming instruction for the purpose of reading it, changing or deleting it, or inserting an additional instruction next to it. To do this, press the CLR key, nominate the address (e.g. 0087), then press the down-arrow key once. Leading zeros can be left out (e.g. for 0087 key in 87).

ADDRESS-BY-ADDRESS PROGRAM READ

To read the content in the program (in mnemonic form), one needs to specify the target address from which one wishes to read program contents, and then press the "down-arrow" key (see figures 11-01 and 11-02). This down-arrow key needs only to be pressed once initially, to display and read the contents of the specified address. Once the specified address content is displayed, the up- and down-arrow keys serve as decremental or incremental move keys, to move the address counter up or down to consecutive address locations (see figure 11-02). This means that the "up-arrow" key will

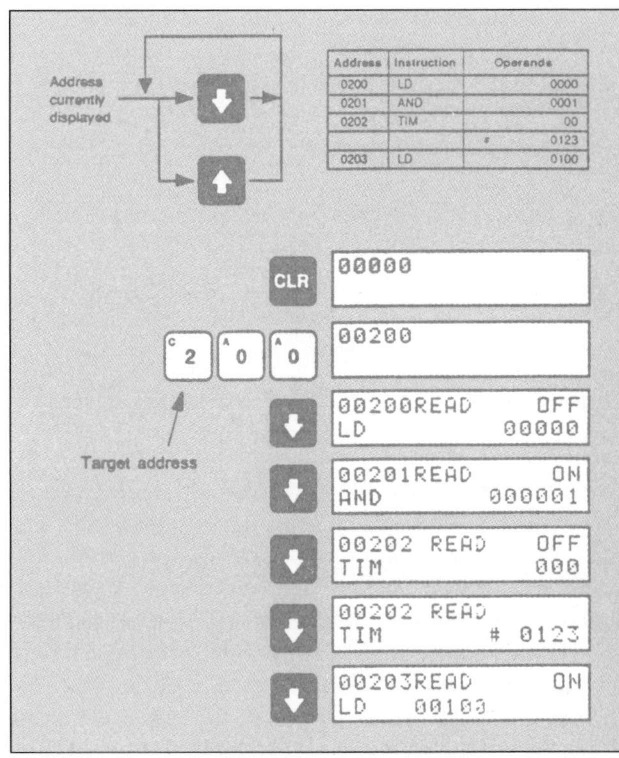

Figure 11-02. Stepping through the mnemonic list and displaying each instruction through the programming console display window address by address.

display the contents of the current address less 1 (n – 1), and the down-arrow key will display the contents of the current address plus 1 (n+1). A detailed key-in example for this procedure is given in figure 11-02.

The step-by -step reading of a segment of program mnemonic code, such as shown in figure 11-02, through the Programming console display window, is given in figure 11-02. Each individual window display shows the programming address, the mnemonic instruction and the data (operand), and for monitoring or diagnostic purposes it shows also whether the bit is "ON" or "OFF" (logic 1 or logic 0).

SPECIFIC MNEMONIC INSTRUCTION SEARCH

To do a step-by-step search of the user memory for a specific programming instruction such as:

OR HR 201, LD 000, AND TIM 03

as is shown in figure 11-02, is certainly possible, but can be very tedious and slow. A much simpler approach is to use the PLC's search facility, as is shown in figure 11-03. This searching process can be performed in:

RUN, MONITOR or PROGRAM

mode. For example, to search for:

LD 0000,

press the LD key followed by the operand (see figure 11-03). Having completed this task, press the SRCH key (search key). If the SRCH key is pressed repeatedly, all addresses containing the specified instruction (LD 0000) are successively found and displayed until either the END instruction (FUN 01) or the last used program instruction is encountered (figure 11-03).

To search for the set value (SV) of a timer or counter, first search for the timer or counter flag number (e.g. TIM 09, CNT 24), and then use the down-arrow key to find the set value of this timer or counter (see lower half of figure 11-03: multiword instruction search for timer 01). Pressing any key other than the SRCH key terminates the search.

OPERAND SEARCH (BIT SEARCH)

The operand search or bit search is another fast way to locate a programming instruction within the user memory of the PLC. As the instruction being searched for is not a mnemonic but an operand or bit, one needs to press the contact key CONT before nominating the operand (see figure 11-04). As the contact sign is on the upper half of a dual purpose key, the SHIFT key must be pressed before the contact key (see also figure 1-13). (The shift key and the contact key need not be pressed at the same time but the shift key must always be pressed first.) Thereafter, key in the bit number or operand.

To search for a bit in the LR, HR, AR, TIMER or COUNTER area, press the acronym key first and then nominate the bit number (see left-hand half of figure 11-04).

The three essential editing functions mentioned at the start of this chapter may now be explained. Note that, unlike searching, all three must be executed in the PROGRAM mode. They cannot be accomplished in the RUN or MONITOR mode.

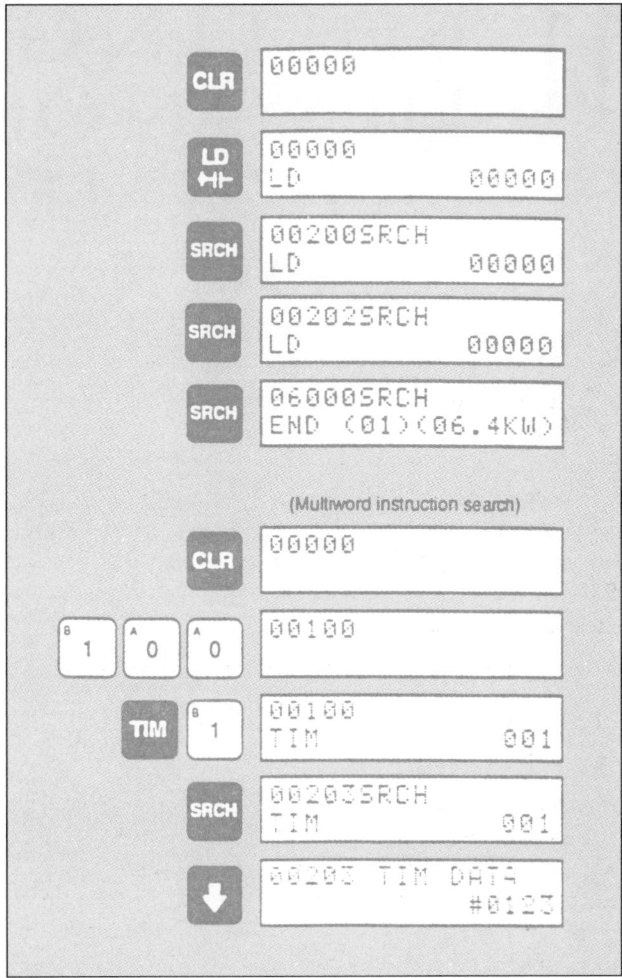

Figure 11-03. Specific mnemonic instruction search and multiword instruction search.

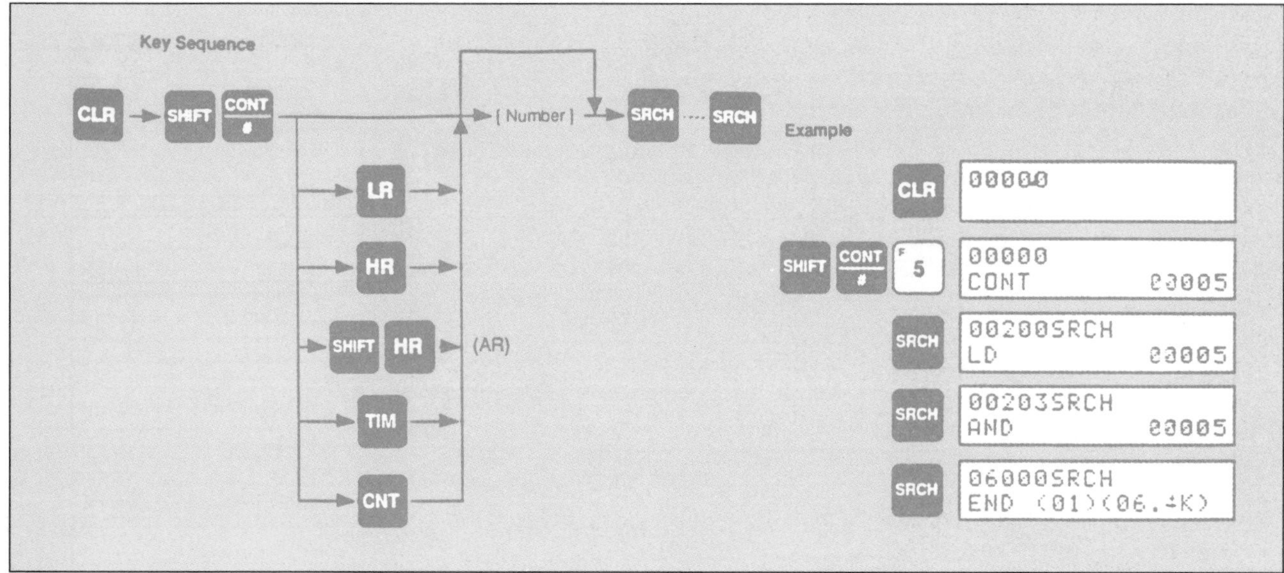

Figure 11-04. Operand or bit search.

CHANGING AN EXISTING PROGRAMMING INSTRUCTION

To change an existing instruction, search for it as explained above under "Specific Mnemonic Instruction Search". If, for example, instruction:

LD 000

in address 0202 in figure 11-03 (upper half) had to be changed to LD 004, one would follow the first four steps shown in figure 11-03 (upper half). When the display on the programming console shows address 0202 with LD 0000, one simply replaces this previously entered instruction by keying in LD and 4 and then pressing the WRITE key (see also figure 1-05). The old instruction will then automatically be replaced by the new instruction.

DELETING A PROGRAM INSTRUCTION

The deleting process removes a program instruction altogether and closes the created gap in the mnemonic list by moving all instructions below the gap up by one increment.

Before deleting an instruction, carefully read the program around the instruction earmarked for deletion to get a clear picture of this section of program. Better still, inspect the ladder diagram section where the contact to be deleted is located. If a ladder diagram is not available, reconstruct one from the "address-by-address program read" (figure 11-02). If deletion is justified, move to the instruction to be deleted (with the up- or down-arrow keys) and simply press the delete key (DEL) followed by the up-arrow key (figure 11-05). One should always finish the deleting process by pressing the up-arrow key to check that the instruction has been removed.

If the deleted instruction is tied to a block formation and block connection instruction such as "OR LD" or "AND LD" (see figures 2-11 to 2-18), one sometimes also needs to delete the block connection instruction and change the block formation instruction.

Example 1: If in figure 2-19 the input contact OR 0003 in address 0011 had to be deleted, one would also need to delete the "AND LD" instruction in address 0012 and to change the instruction LD 0001 to AND 0001.

Example 2: If in figure 2-19 the timer 01 contact had to be deleted, one would also have to remove the "OR LD" instruction, and change the instruction LD NOT 0000 to OR NOT 0000 (address 0005).

INSERTING A PROGRAM INSTRUCTION

The insertion process inserts an additional program instruction at a specified point in the user program. The simplest approach to inserting an additional program instruction is to follow the ladder diagram or mnemonic list downwards (from lower address numbers to higher address numbers) to the precise point where the insertion must happen, then to move one address increment further down (figure 11-06). At this point, write into the program the instruction to be inserted (e.g. AND 105) and then press the insert key (INS), followed by the down-arrow key (see upper half of figure 11-06).

In figure 11-06, the insert key wedges the additional programming instruction between addresses 0005 and

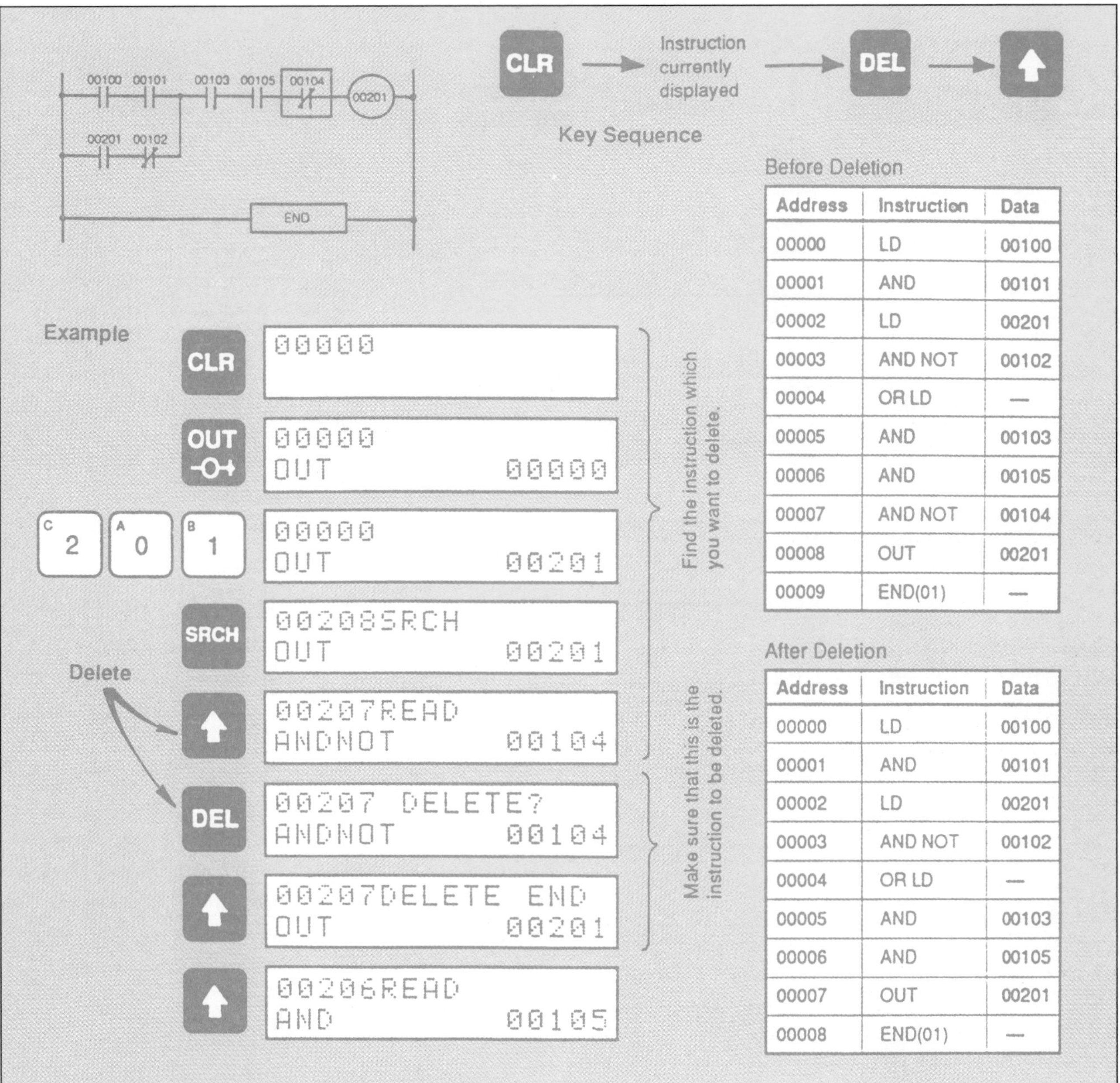

Before Deletion

Address	Instruction	Data
00000	LD	00100
00001	AND	00101
00002	LD	00201
00003	AND NOT	00102
00004	OR LD	—
00005	AND	00103
00006	AND	00105
00007	AND NOT	00104
00008	OUT	00201
00009	END(01)	—

After Deletion

Address	Instruction	Data
00000	LD	00100
00001	AND	00101
00002	LD	00201
00003	AND NOT	00102
00004	OR LD	—
00005	AND	00103
00006	AND	00105
00007	OUT	00201
00008	END(01)	—

Figure 11-05. Deleting a programming instruction.

0006, and the down-arrow key moves all programmed addresses below address 0005 one increment down to make room for the inserted instruction.

As with deletion, one should always first have a clear picture of the ladder diagram section where the insert has to take place. Without such a ladder diagram one could easily make mistakes.

Example: In figure 2-16 one needs to insert an additional contact to the right of contact $\overline{0500}$. This additional contact is HR 007. One would therefore move to address 0004 and insert the additional contact at that point by pressing the keys:

AND, HR, 7, INS, DOWN-ARROW.

As this lower rung containing this "OR"-connected "AND" function now contains more than one contact, one also needs to insert the "OR LOAD" instruction (see figures 2-11 to 2-13) by pressing the keys:

OR, LD, INS, DOWN-ARROW.

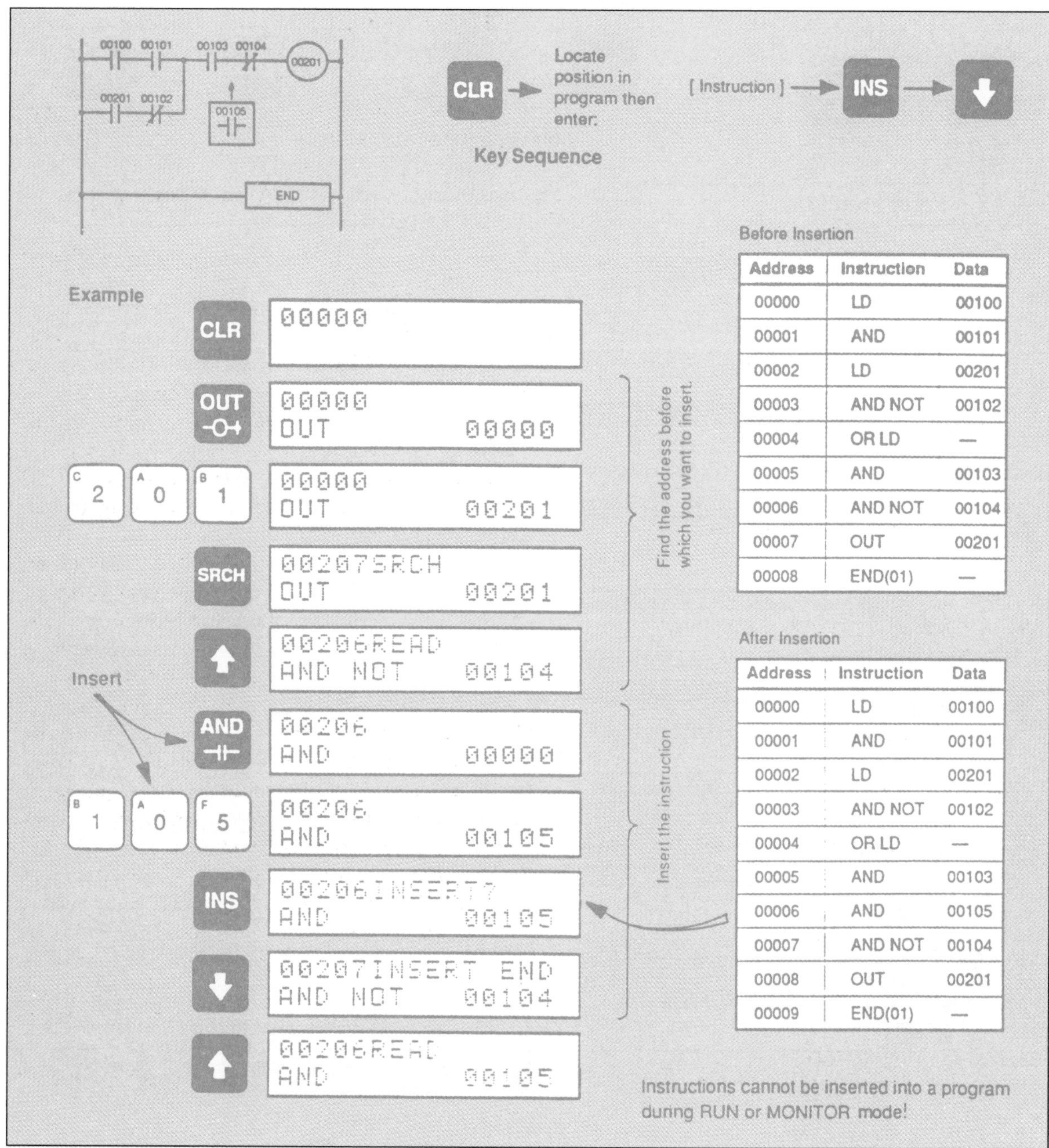

Before Insertion

Address	Instruction	Data
00000	LD	00100
00001	AND	00101
00002	LD	00201
00003	AND NOT	00102
00004	OR LD	—
00005	AND	00103
00006	AND NOT	00104
00007	OUT	00201
00008	END(01)	—

After Insertion

Address	Instruction	Data
00000	LD	00100
00001	AND	00101
00002	LD	00201
00003	AND NOT	00102
00004	OR LD	—
00005	AND	00103
00006	AND	00105
00007	AND NOT	00104
00008	OUT	00201
00009	END(01)	—

Figure 11-06. Inserting an additional programming instruction.

Then one needs to move up to address 0003 and change the instruction OR NOT 0500 to LD NOT 0500.

PLC START-UP PASSWORD

To gain access to the PLC's programming function, one first needs to enter the programming password, usually unknown to anyone not familiar or not involved with PLC programming. This prevents unauthorised access to the PLC's program (see figures 11-07 and 1-05).

The PLC programmer is prompted for a password when the power is switched on to the PLC or after the programming console has been connected to the PLC. To enter the password, the PLC may be in PROGRAM, MONITOR or RUN mode. Thereafter, one can change it with the black mode switch on the programming console to RUN or MONITOR mode (see figure 1-05). To enter the password, key-in:

CLR, MONTR, CLR.

ERASING AN OLD PROGRAM

Before programming a completely new program (if the old program is no longer required), any old or remnant

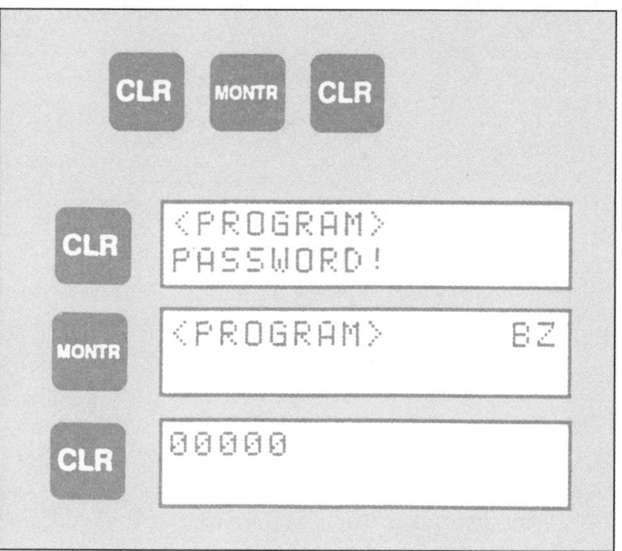

Figure 11-07. Programming access password.

program sections must be erased. To erase, switch the PLC to PROGRAM mode (figure 1-05), and then press the following six keys in correct order:

SLR, SET, NOT, RESET, MONTR, CLR.

Although the SET and RESET keys are dual purpose keys, for erasing, the function key (FUN) does not need to be pressed first (see figure 11-08). Partial memory clearing is also possible (see PLC operation manual).

FORCING AN OUTPUT RELAY, A TIMER FLAG OR A COUNTER FLAG

Most modern PLCs are equipped with a "force function". This force function is essentially a manual override function. It gives the PLC programmer or the technician working on a PLC-controlled machine the opportunity to "force" the status of:

IR, SR, HR, AR or LR relays.

All of these relays are by nature internal or external output relays. One can also force TIMER and COUNTER flags (see figures 1-12, 5-02, 6-01 and 11-09). Note, however, that external inputs cannot be forced, and none of the forced relays or timer and counter flags can be forced for longer than one scan time. Normal forcing is possible only when the PLC is in MONITOR mode.

There is also a force instruction that forces output relays permanently "ON". For this, the PLC is to be in the PROGRAM mode. This force function is usually used to check the correct operation of solenoid valves and motors. As soon as the PLC is switched to TURN or MONITOR mode the forced outputs are automatically forced to "OFF".

For educational purposes, teaching boxes are available with a commercial PLC mounted, and a power unit supplying inputs and outputs of the PLC inside the box. These teaching boxes usually have simulation input switches to enable the student to simulate inputs and test the program on the student's desk.

Forcing a relay bit is an essential and extremely useful function for fault-finding, system debugging and testing a PLC program before the PLC is connected to the machine (see also the section "Testing the PLC Program with the 'Dry-Run Test'" in Chapter 8). As explained in Chapter 8, a step-counter always needs its last flip-flop to be forced "ON". Should that last flip-flop (HR) not be forced "ON", the step-counter could then not be started as the preparation signal in the logic set line for HR 001 would be missing (figure 8-07).

Forcing an HR or latched output "ON" may therefore have both desirable and undesirable consequences. All contacts driven by such a forced relay change logic state. This means that all N/C (⊣/⊢) contacts depending on this relay will turn "OFF", and all N/O contacts (⊣⊢) will turn "ON". Forcing must therefore be used with **utmost caution**, especially if the PLC is already connected to the machine and forcing could cause actuators to start moving. This could lead to equipment damage or, worse still, injury to an operator or to a person standing close to the machine!

To illustrate the advantages and disadvantages of relay bit forcing, three practical examples are given.

Example 1: In the control for the clamp and drill machine of figure 8-07 it is assumed that the drill is

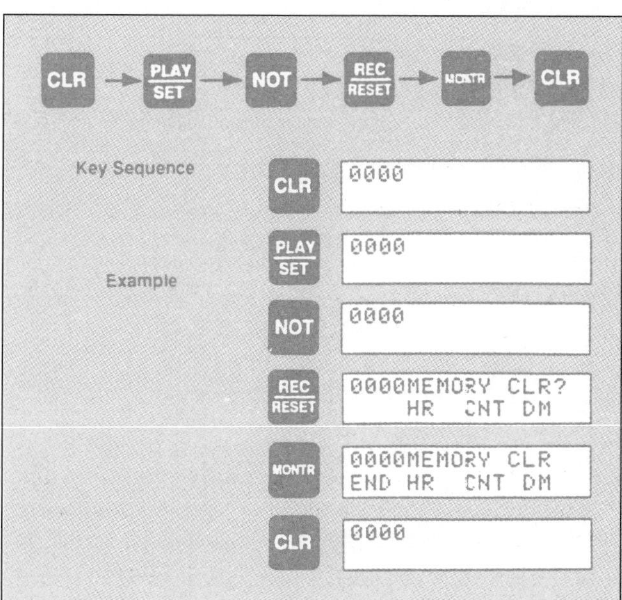

Figure 11-08. Erasing program memory.

damaged (broken or blunt) and advancing of actuator Ⓑ to limit switch 0004 is therefore very slow or impossible. The machine operator or maintenance technician could therefore, in MONITOR mode, force step-module 3 (flip-flop HR 003) by pressing the keys:

CLR, SHIFT, CONT, HR, 3, MONTR, SET.

This would force HR 003 "ON" and the sequential process would continue to the end of its prescribed sequence. The drill bit could then be replaced and production could resume.

Example 2: It is assumed that HR 000 in figure 4-10 is forced "ON". This would start the machine immediately without the guard perhaps being closed and could cause disastrous consequences for machine, product, operator or persons standing close by.

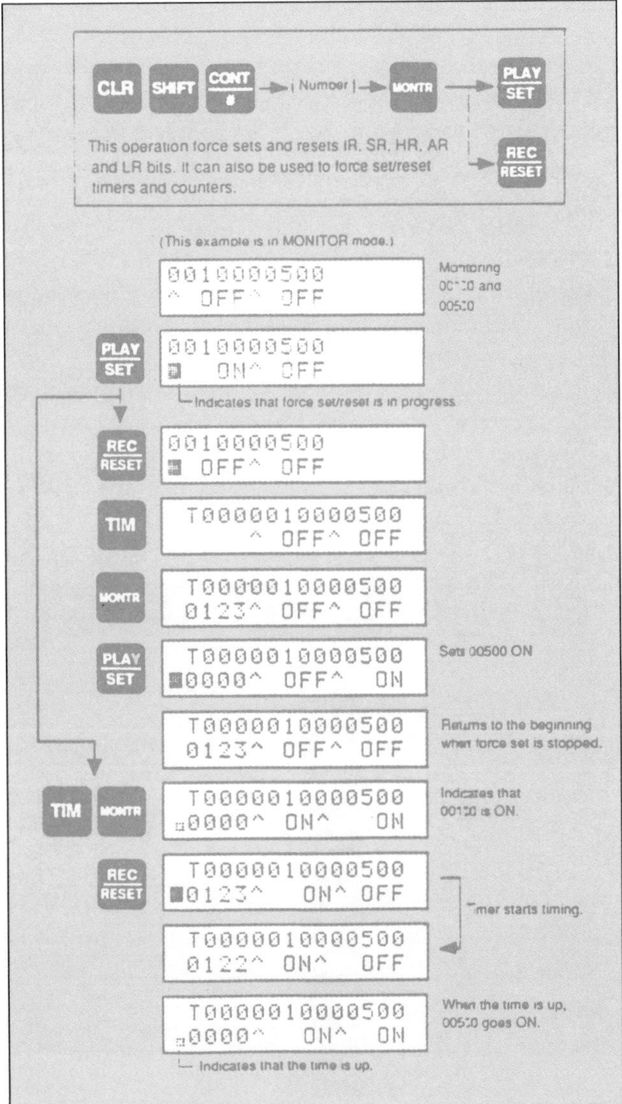

Figure 11-09. Forcing a relay, timer or counter for one scan time.

Example 3: Similarly, if HR 901 in figure 9-12, section 1, were forced "ON", the machining process would start without a proper and wilful start command being given. Again, such forcing action could have inadvertent consequences.

CHANGING TIMER OR COUNTER SET VALUES (SV CHANGE)

Set values (SVs) of timers and counters may be determined and set when the program is written into the user memory during the initial programming process (see Chapters 5 and 6). They may, however, be altered in MONITOR mode while the machine is being controlled by the PLC and is running. This "on-line" function is essential for time-dependent machine sequencing and processing functions, such as container filling or glue melting. As for counters, the speeding-up or slowing-down of conveyor belts whose speed depends on motor r.p.m. needs counter SV changes while the machine is being tested and commissioned. Note that on-line changing of timer and counter SVs can be accomplished only in MONITOR mode. The procedural steps for this are given in figure 11-10.

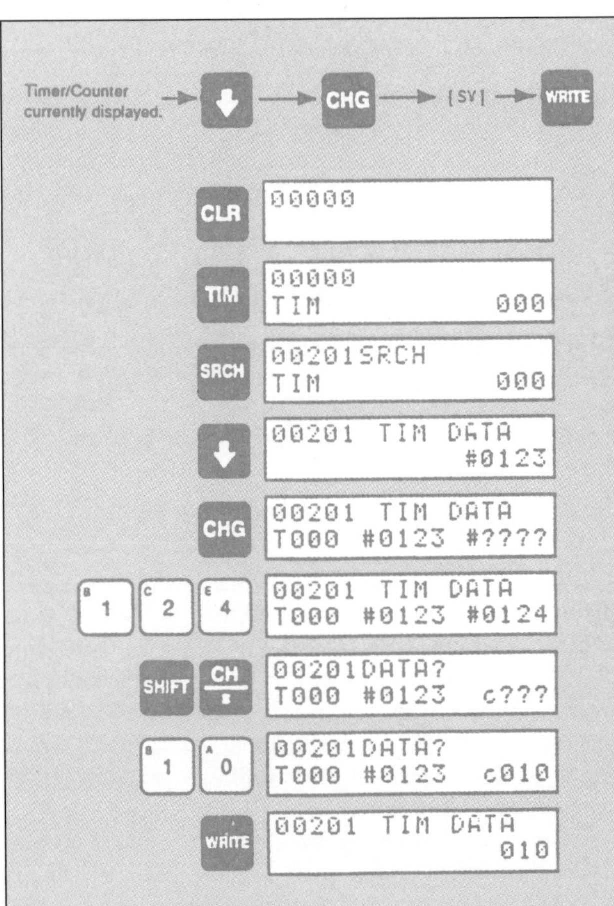

Figure 11-10. Changing timer or counter set value (SV) in on-line operation.

Figure 11-11. Monitoring procedures to monitor (look at) a bit, timer, counter or data memory channel.

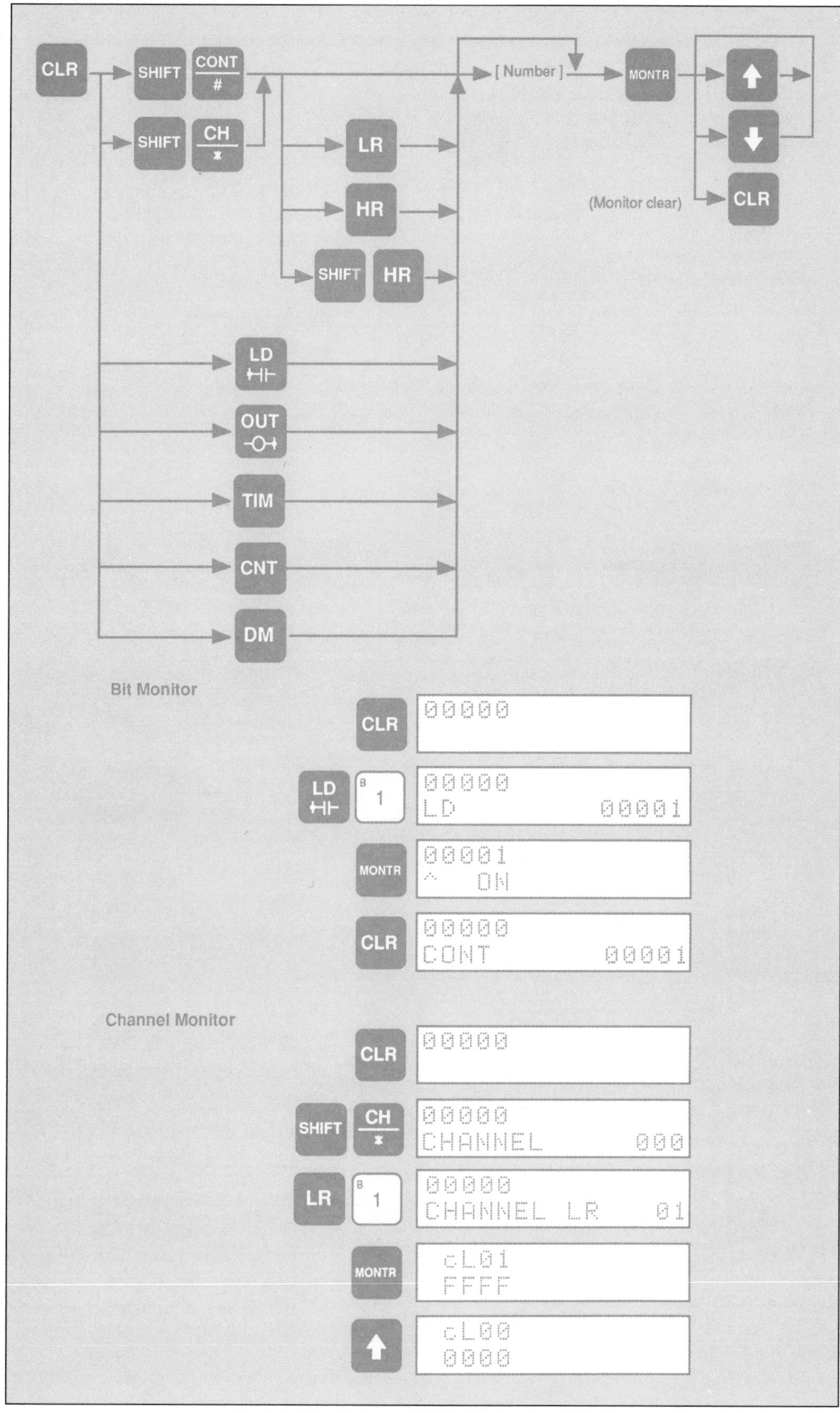

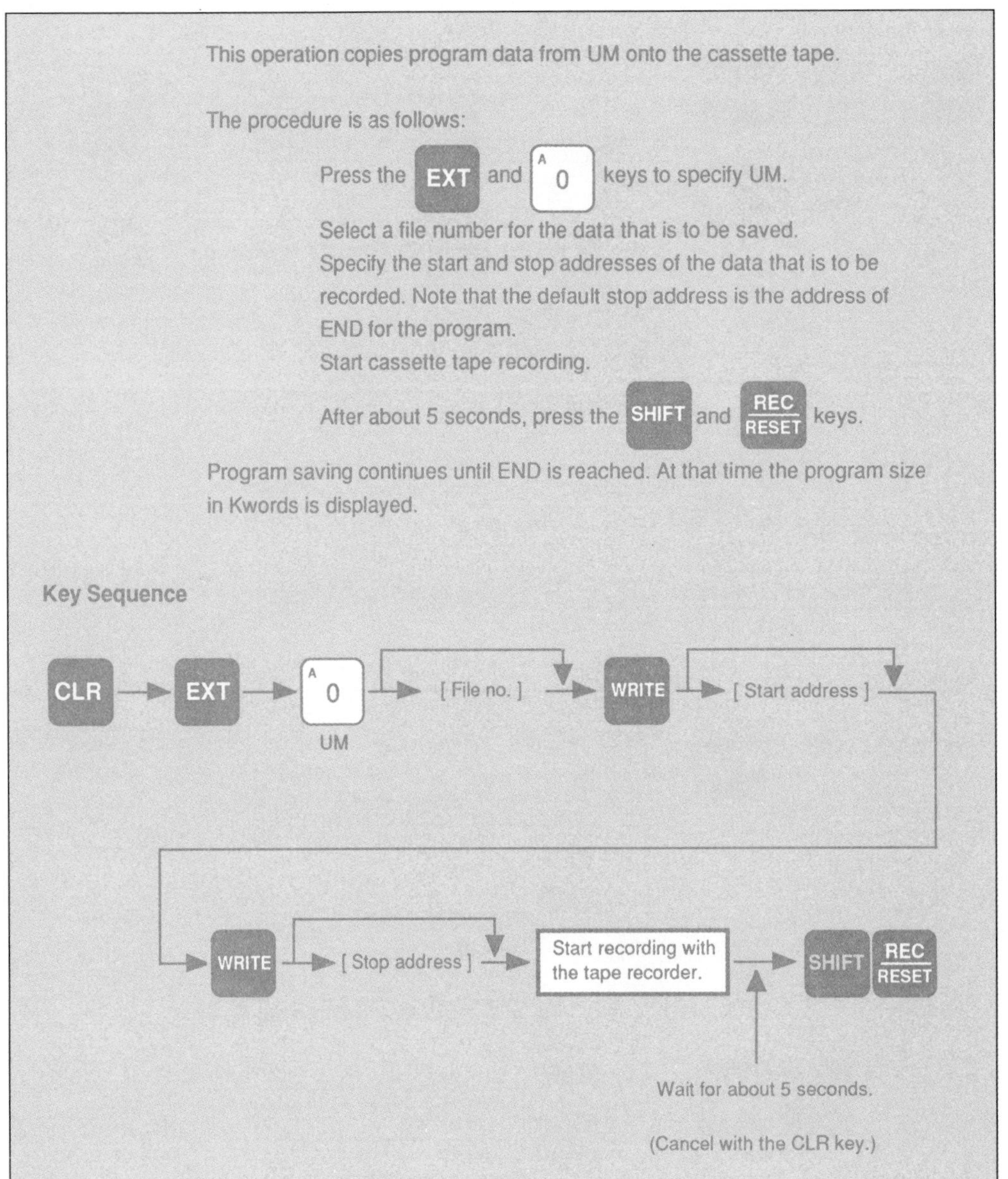

Figure 11-12. Procedural steps to save (record) a PLC program onto a cassette tape. The tape may then be used to restore an accidentally lost or erased program or to maintain a record of the original program.

MONITORING OPERATIONS

The monitoring operation allows the programmer or system diagnostics engineer to monitor or carry out surveillance on processes going on within the PLC's complex operations. Monitoring could therefore be replaced by the words "look at" or "check what it is like". But as most PLCs use "monitor", this book also uses this expression. The monitoring operation permits one to look at individual bits, entire channels (words), present value (PV) or set value (SV) of timers and counters. Monitoring also supplements the function provided by the "Program read", "Instruction read" and "Bit search" operations described in this chapter. In all such cases, monitoring always requires the programmer to specify first the bit, channel, timer or counter operand to be monitored (looked at), and then one must press the "Monitor" key (MONTR, see figure 11-11). The monitoring operation described in this chapter may be performed in the "RUN", "MONITOR" or "PROGRAM" mode and can be cancelled by pressing the clear key (CLR).

DIAGNOSING THE TRUENESS OF A LOGIC LINE

Where an output of a logic line seems not to be "ON" (logic 1) when it should be, one may check for missing logic 1 states in one of the contacts along the network of this logic line. (To explain the checking procedure we use the ladder diagram presented in figure 11-6.) To search for such "OFF" contacts:

1. Search for the beginning or end of the logic line (the beginning in the example would be address 0000, the end would be address 0007).

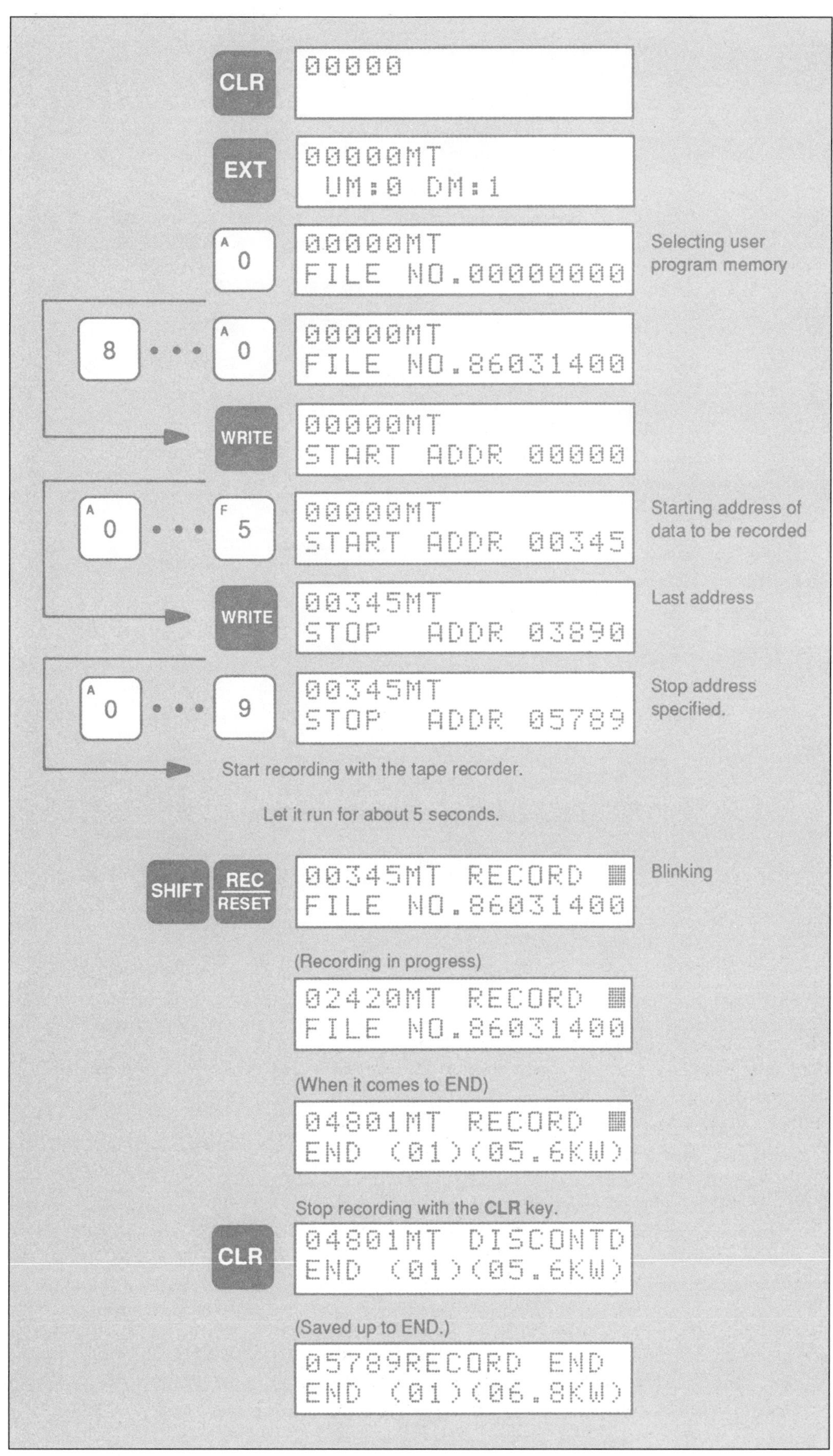

Figure 11-13. Example of recording a PLC program onto cassette tape.

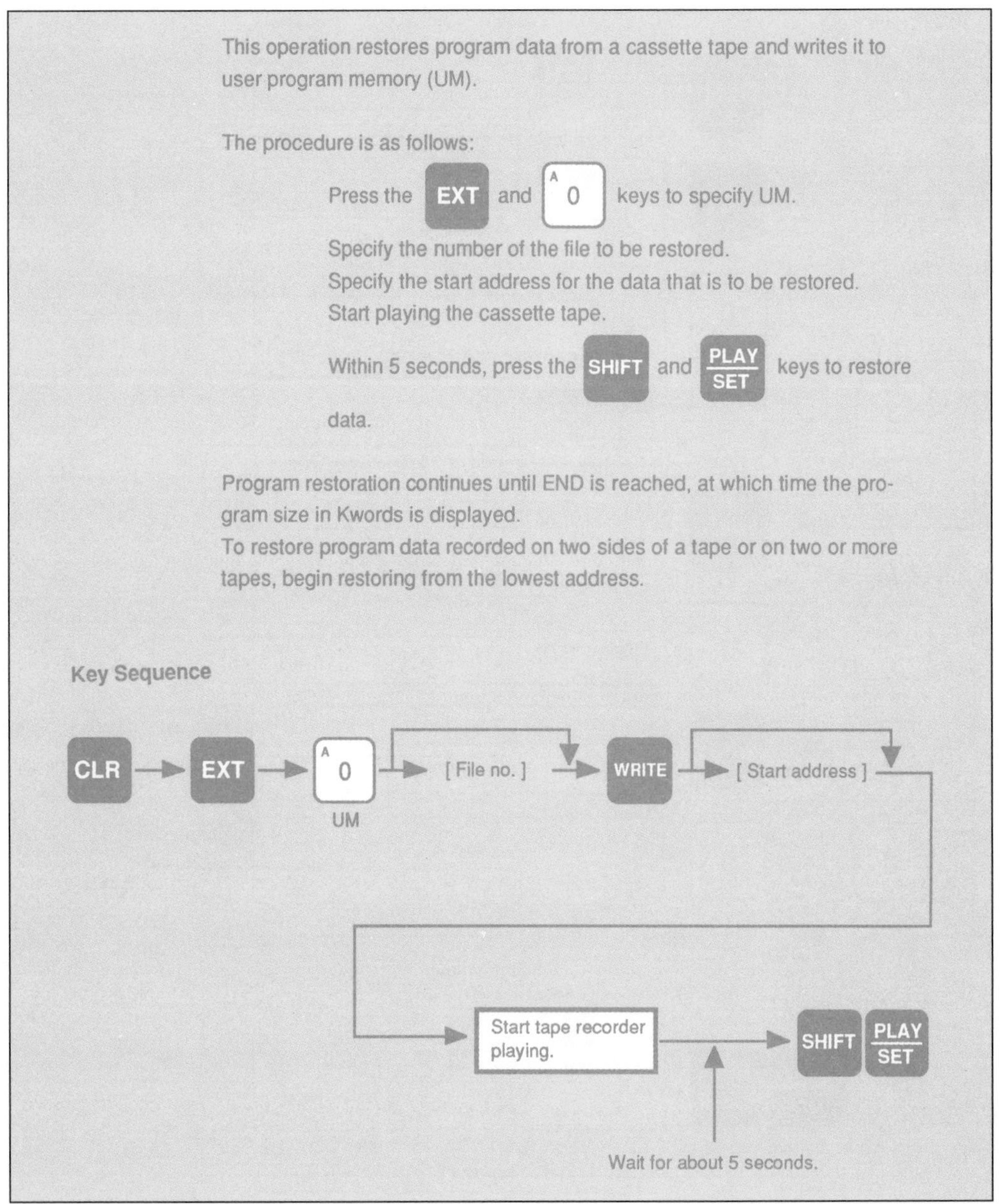

Figure 11-14. Procedural steps to load (transfer) a PLC program from cassette tape into the user memory of a PLC. The PLC user memory must first be cleared (erased) as shown in figure 11-08.

2. Proceed along the logic line by checking every contact to see if it is "ON" or "OFF". The contacts or conditions that are not "ON" are obviously the inhibiting links that prevent the output (0201) from turning to logic 1 ("ON"). To step or move along the logic line, simply press the up- or down-arrow key slowly until the condition that is not "ON" is found.
3. Once this missing condition (or conditions) is found, then check for the reason why it is not "ON". One reason could be that the limit switch or sensor is not actuated, or the signal wire between PLC and switch is broken, or an internal bit acting as a contributing contact is not "ON".
4. If the missing "ON" condition is an internal output, one may use the force instruction (forcing an output "ON") to turn the inhibiting contact to logic 1 (forcing it "ON"). When this is achieved, the output should also turn "ON".

DIAGNOSING THE INADVERTENT HALTING OF A STEP-COUNTER CHAIN

When a sequential process being controlled by a step-counter grinds to an unexpected halt within its sequence, use the following simple procedure to find the cause of the stoppage:

1. As the step-counter is of modular structure (see figures 8-06 and 8-15), only one step within the modular chain is "ON" at a given time in the sequence. Should the advancing process come to a halt, simply interrogate the PLC to find out where this stoppage occurred by pressing the following keys in the displayed order:

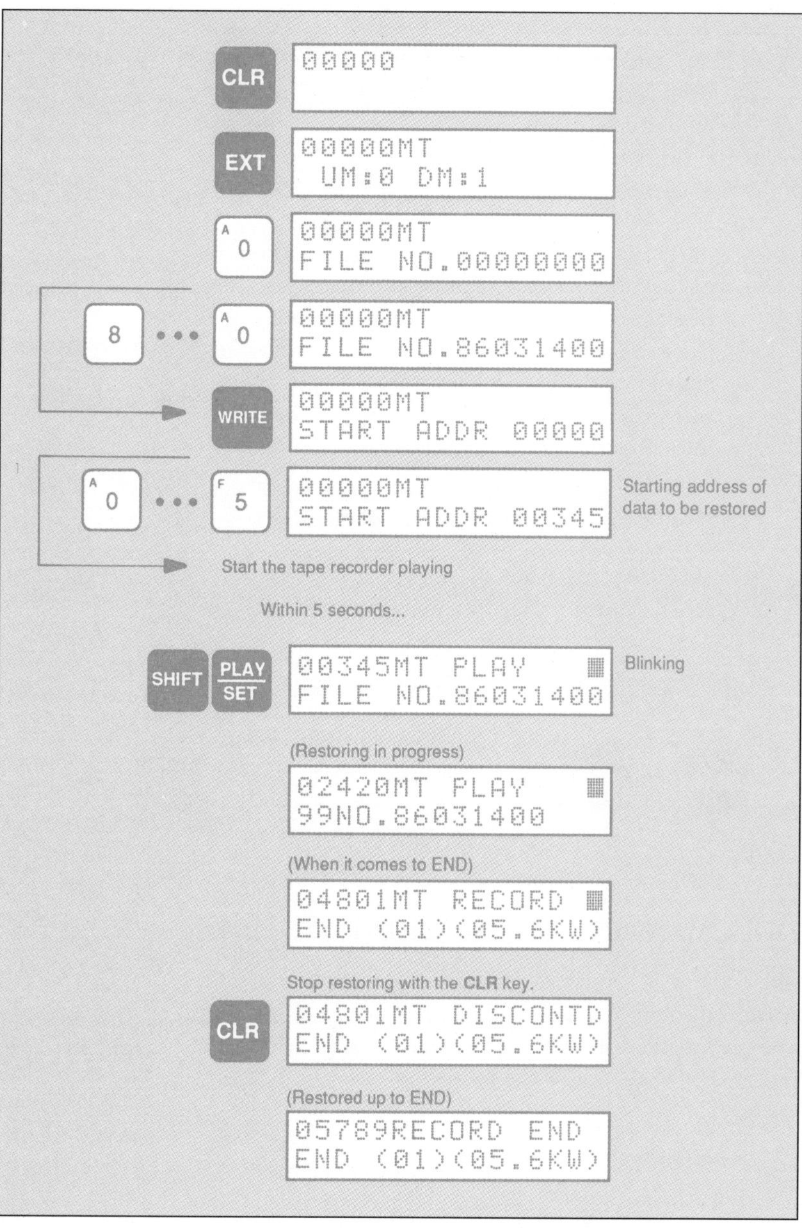

Figure 11-15. Example of loading (transferring) a PLC program from tape into the user memory of a PLC.

CLR, SHIFT, CONT, HR, 1, MONTR.

This procedure is also illustrated in figures 11-04 and 11-09 (top). As soon as the MONTR key is pressed, the programming console displays the state of HR 001, which in this case is most likely "OFF" as the step-counter has probably gone beyond step 1.

2. Now press the down-arrow key until the HR that is "ON" at this time is found (the HR is the step-counter flip-flop). If, for example, HR 005 is "ON", this would then mean that the step-counter has stopped at sequence step 5 (see figures 8-14 and 8-15).
3. Logical reasoning would therefore suggest that an essential contact condition in the set rung for HR 006 flip-flop is missing. From the step-action diagram for this control (figure 8-14) and the matching ladder diagram (figure 8-15), one can easily see that the essential contacts for HR006 flip-flop to be set are:

   LD HR 005, AND 0004, AND 0505.

4. There are now several options for finding the missing logic 1 contact (or contacts). The easiest approach is to inspect the diagnostic panel of the PLC. Every modern PLC has such a diagnostic panel with LED display of incoming and outgoing signals. The diagnostic panel at this stage of the sequence should show illuminated LEDs for the inputs of 0004 and 0002 and output 505 rendering output command 504 (see figure 8-14, time-line 6). Of these three confirmation signals, however, only external input 0004 and output 505 are essential for the switching of the missing step (step 6). Output 505 serves here as an essential input and confirms the running of air motor © (see

electropneumatic circuit in figure 8-14). Whichever of these two confirmation signals is missing causes the stoppage of the PLC step-counter and the machine controlled by the PLC.

5. If the missing essential confirmation signal is input 0004, one then has to find out why. Reasons could be:
   - the actuator has not responded to the 503 output signal in step 5 because the machine structure prevents actuator Ⓑ from retracting
   - failure of solenoid 503 owing to an electrical fault on the solenoid
   - valve failure owing to valve spool not moving into alternate position because it is jammed or has seized up
   - output signal wire to solenoid 505 broken between PLC and solenoid
   - output relay within the PLC is defective. replace output relay (see manufacturer's maintenance instructions)
   - switch or sensor rendering signal 0004 is defective or its wire to the PLC is broken.
6. If the missing essential confirmation signal is output 505 serving as an input to the PLC's processor (CPU), the problem would be less obvious and certainly not easy to find, as this would constitute a PLC internal fault.

DOCUMENTING THE MACHINE SYSTEM

Once the PLC-controlled machine is operating, fine-tuned and commissioned, the final PLC user program and user manual need to be documented. The documentation should consist of the following essential parts:

- Original machine specification to which the machine was designed and built.
- Relevant machine construction drawings, calculations for important structural parts, sketches and so on.
- Step-action diagrams depicting the sequencing actions fo the machine.
- Hydraulic or pneumatic control circuits depicting the machine actuators and valves in rest position and with pressure "OFF".
- Manuals for PLC and fluid power equipment in the machine.
- Ladder diagram print-out of the PLC user program with relevant outputs and inputs labelled.
- Mnemonic list and cross-reference list of contacts, outputs, timers, counters and functions.
- Precise PLC input/output assignment list including relevant internal outputs such as KEEP relays, timers and counters.
- Precise list of all KEEP relays that need to be forced ("ON" or "OFF") for machine start-up.
- Machine routine maintenance schedule.
- Machine start-up and shut-down procedures.

As no machine follows a regular design pattern, this book does not provide fine details for the design of documentation (a packaging machine is different from a harvesting machine or an automotive manufacturing machine).

Relevant to this book, however, is the printing of ladder diagrams and menmonic lists and saving a user program onto cassette tape. Printing a program is also useful during the design stage of the ladder diagram to obtain an overall "view" of the PLC or for making corrections when the circuit or circuit sections are being tested.

SAVING (RECORDING) OR LOADING (TRANSFERRING) A PLC USER PROGRAM WITH CASSETTE TAPE

PLC programs (from user program memory) or DM data may be backed up on a standard, commercially available, cassette tape recorder. Any kind of magnetic tape of adequate length will suffice. (Note: To save an 8 kiloword program, the tape must be 15 minutes long.) Always allow five seconds of blank tape leader before the taped data begin. Store only one program on a single side of the tape. Use patch cords to connect the cassette recorder earphone (or line-out) jack to the Programming Console EAR jack

ERROR MESSAGES

The following error messages may appear during cassette tape operations:

Message	Meaning and appropriate response
0000 ERR ******* FILE NO. ********	File number on cassette and designated file number are not the same. Repeat the operation using the correct file number.
**** MT VER ERR	Cassette tape contents differ from that in the PLC. Check content or tape, PLC or both.
**** MT ERR	`Cassette tape is faulty. Replace it with another.

and the cassette recorder microphone (or line-in jack) to the Programming Console MIC jack. Set the cassette recorder volume and tone controls to maximum levels.
Note: For all operations—saving, loading and verifying:

- the PLC must be in the PROGRAM mode
- while the operation is in progress, the cursor blinks and the block count is incremented on the display
- operation may be halted at any time by pressing the CLR key. (see figs. 11-12 to 11-15)

CONTROL PROBLEM 1

A PLC-controlled machine sequence needs to be altered. The original sequence is given in figure 11-16, which depicts the control application with its six sequential steps.

The altered circuit should cause sequence step 5 to be delayed for 4 seconds, starting from the completion of sequence step 4 (actuator Ⓑ being fully retracted). The timer must be placed into the output signal of the PLC electronic step-counter (see ladder diagram circuits in figures 11-15 and 11-16). Additionally, the circuit must be equipped with an extended cycle selection module, as shown in figures 9-05, 9-21 and 9-22, and a complex sequencing interrupt module, as shown in figures 9-01, 9-03, 9-21 and 9-22. Signalling cycle interrupt causes the machine to stop at the end of the step at which stopping is signalled. The machine sequence must continue from wherever it has been signalled to stop once a reset signal and a start signal are given (sequencing resumption needs both of these signals).

Procedure

First, insert the extended cycle selection module and the complex sequencing interrupt module. This is started at address 0000. Insertions are always completed with the two keystrokes INS and down-arrow (⇓). When completed, continue inserting from steps 2 to 6 an HR 114 contact as is shown in figure 11-16. This contact inhibits the setting of any of the HRs (flip-flops) when stopping is signalled. Thereafter, change contact 0000 in the set rung to HR 001 to contact 1000 (AND 1000). Relay 1000 contact represents the extended cycle selection module (see figure 9-05). When this task is completed, insert the memory-retentive timer, which causes the 4-second delay and maintains elapsed time during power failure or during machine interrupt being signalled by the cycle interrupt module. Contact 1902 in the enabling line leading to stopping lamp 511 causes the stopping lamp to blink at one pulse per second (see also figure 5-01 for pulse bits).

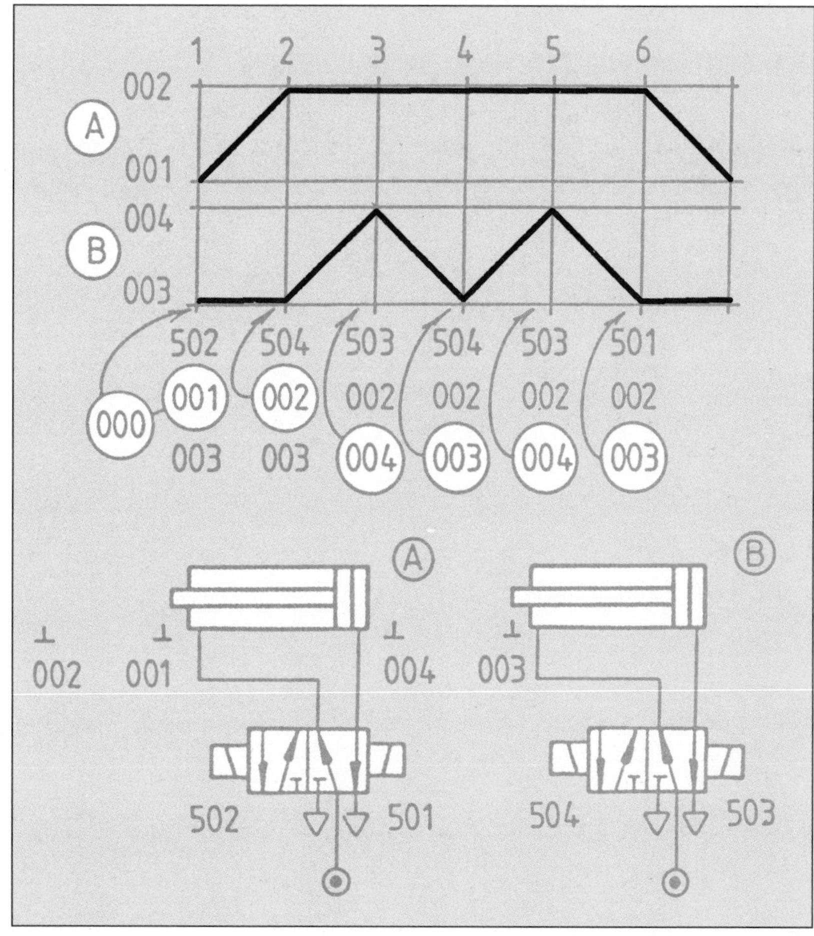

Figure 11-16. Step-action diagram for starting conditions (original sequence) of control problem 1.

```
           LADDER DIAGRAM          PAGE-001
     | H006   0001   0000                    |
0000 |--| |----| |----| |---[ KEEP(11)]-|
     | H002                        H001  |
     |--| |-----------------------       |
     | H001   0002                       |
0005 |--| |----| |------------[ KEEP(11)]-|
     | H003                        H002  |
     |--| |-----------------------       |
     | H002   0004                       |
0009 |--| |----| |------------[ KEEP(11)]-|
     | H004                        H003  |
     |--| |-----------------------       |
     | H003   0003                       |
0013 |--| |----| |------------[ KEEP(11)]-|
     | H005                        H004  |
     |--| |-----------------------       |
     | H004   0004                       |
0017 |--| |----| |------------[ KEEP(11)]-|
     | H006                        H005  |
     |--| |-----------------------       |
     | H005   0003                       |
0021 |--| |----| |------------[ KEEP(11)]-|
     | H001                        H006  |
     |--| |-----------------------       |
     | H006                   0501       |
0025 |--| |--------------------( )-------|
     | H001                   0502       |
0027 |--| |--------------------( )-------|
     | H003                   0503       |
0029 |--| |-+------------------( )-------|
     | H005 |                            |
     |--| |-+                            |
     | H002                   0504       |
0032 |--| |-+------------------( )-------|
     | H004 |                            |
     |--| |-+                            |
     |                                   |
0035 |------------------------[ END (01)]-|
```

MNEMONIC LIST

ADDRESS	MNEMONIC	OPERAND	
0000	LD	HR	006
0001	AND		0001
0002	AND		0000
0003	LD	HR	002
0004	KEEP (11)	HR	001
0005	LD	HR	001
0006	AND		0002
0007	LD	HR	003
0008	KEEP (11)	HR	002
0009	LD	HR	002
0010	AND		0004
0011	LD	HR	004
0012	KEEP (11)	HR	003
0013	LD	HR	003
0014	AND		0003
0015	LD	HR	005
0016	KEEP (11)	HR	004
0017	LD	HR	004
0018	AND		0004
0019	LD	HR	006
0020	KEEP (11)	HR	005
0021	LD	HR	005
0022	AND		0003
0023	LD	HR	001
0024	KEEP (11)	HR	006
0025	LD	HR	006
0026	OUT		0501
0027	LD	HR	001
0028	OUT		0502
0029	LD	HR	003
0030	OR	HR	005
0031	OUT		0503
0032	LD	HR	002
0033	OR	HR	004
0034	OUT		0504
0035	END (01)		

Figure 11-17. Ladder diagram and mnemonic list for starting conditions of original control problem 1. (before circuit is altered)

Figure 11-18. Altered circuit for control problem 1. (Circuit includes all changes as per control problem 1 specification.)

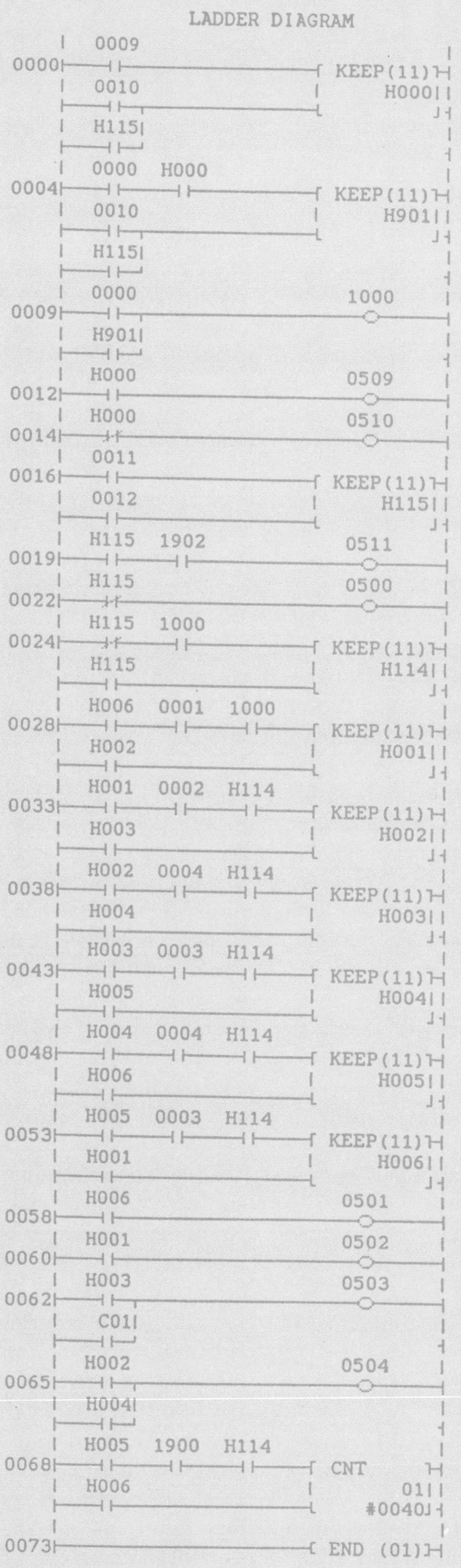

LADDER DIAGRAM

```
     | 0009                                     |
0000 |--| |-----------------------[ KEEP(11) ]--|
     | 0010                       |    H000|    |
     |--| |--+--------------------|            -|
     | H115|                                    |
     |--| |--+                                  |
     | 0000   H000                              |
0004 |--| |----| |----------------[ KEEP(11) ]--|
     | 0010                       |    H901|    |
     |--| |--+--------------------|            -|
     | H115|                                    |
     |--| |--+                                  |
     | 0000                          1000       |
0009 |--| |--+-----------------------( )--------|
     | H901|                                    |
     |--| |--+                                  |
     | H000                          0509       |
0012 |--| |--------------------------( )--------|
     | H000                          0510       |
0014 |--|/|--------------------------( )--------|
     | 0011                                     |
0016 |--| |-----------------------[ KEEP(11) ]--|
     | 0012                       |    H115|    |
     |--| |-----------------------|            -|
     | H115   1902                   0511       |
0019 |--| |----| |-----------------( )--------|
     | H115                          0500       |
0022 |--|/|--------------------------( )--------|
     | H115   1000                              |
0024 |--|/|----| |----------------[ KEEP(11) ]--|
     | H115                       |    H114|    |
     |--| |-----------------------|            -|
     | H006   0001   1000                       |
0028 |--| |----| |----| |---------[ KEEP(11) ]--|
     | H002                       |    H001|    |
     |--| |-----------------------|            -|
     | H001   0002   H114                       |
0033 |--| |----| |----| |---------[ KEEP(11) ]--|
     | H003                       |    H002|    |
     |--| |-----------------------|            -|
     | H002   0004   H114                       |
0038 |--| |----| |----| |---------[ KEEP(11) ]--|
     | H004                       |    H003|    |
     |--| |-----------------------|            -|
     | H003   0003   H114                       |
0043 |--| |----| |----| |---------[ KEEP(11) ]--|
     | H005                       |    H004|    |
     |--| |-----------------------|            -|
     | H004   0004   H114                       |
0048 |--| |----| |----| |---------[ KEEP(11) ]--|
     | H006                       |    H005|    |
     |--| |-----------------------|            -|
     | H005   0003   H114                       |
0053 |--| |----| |----| |---------[ KEEP(11) ]--|
     | H001                       |    H006|    |
     |--| |-----------------------|            -|
     | H006                          0501       |
0058 |--| |--------------------------( )--------|
     | H001                          0502       |
0060 |--| |--------------------------( )--------|
     | H003                          0503       |
0062 |--| |--+-----------------------( )--------|
     | C01|                                     |
     |--| |--+                                  |
     | H002                          0504       |
0065 |--| |--+-----------------------( )--------|
     | H004|                                    |
     |--| |--+                                  |
     | H005   1900   H114                       |
0068 |--| |----| |----| |---------[ CNT        ]-|
     | H006                       |       01|   |
     |--| |-----------------------|    #0040 -|
     |                                          |
0073 |----------------------------[ END (01) ]--|
```

ADDRESS	MNEMONIC	OPERAND
0000	LD	0009
0001	LD	0010
0002	OR	HR 115
0003	KEEP (11)	HR 000
0004	LD	0000
0005	AND	HR 000
0006	LD	0010
0007	OR	HR 115
0008	KEEP (11)	HR 901
0009	LD	0000
0010	OR	HR 901
0011	OUT	1000
0012	LD	HR 000
0013	OUT	0509
0014	LD NOT	HR 000
0015	OUT	0510
0016	LD	0011
0017	LD	0012
0018	KEEP (11)	HR 115
0019	LD	HR 115
0020	AND	1902
0021	OUT	0511
0022	LD NOT	HR 115
0023	OUT	0500
0024	LD NOT	HR 115
0025	AND	1000
0026	LD	HR 115
0027	KEEP (11)	HR 114
0028	LD	HR 006
0029	AND	0001
0030	AND	1000
0031	LD	HR 002
0032	KEEP (11)	HR 001
0033	LD	HR 001
0034	AND	0002
0035	AND	HR 114
0036	LD	HR 003
0037	KEEP (11)	HR 002
0038	LD	HR 002
0039	AND	0004
0040	AND	HR 114
0041	LD	HR 004
0042	KEEP (11)	HR 003
0043	LD	HR 003
0044	AND	0003
0045	AND	HR 114
0046	LD	HR 005
0047	KEEP (11)	HR 004
0048	LD	HR 004
0049	AND	0004
0050	AND	HR 114
0051	LD	HR 006
0052	KEEP (11)	HR 005
0053	LD	HR 005
0054	AND	0003
0055	AND	HR 114
0056	LD	HR 001
0057	KEEP (11)	HR 006
0058	LD	HR 006
0059	OUT	0501
0060	LD	HR 001
0061	OUT	0502
0062	LD	HR 003
0063	OR	CNT 01
0064	OUT	0503
0065	LD	HR 002
0066	OR	HR 004
0067	OUT	0504
0068	LD	HR 005
0069	AND	1900
0070	AND	HR 114
0071	LD	HR 006
0072	CNT	01
		# 0040

APPENDICES

APPENDIX 1
GLOSSARY

AC Alternating electrical current, normally alternating at 50 or 60 Hz.
Addend Number to be added to another. For example, in A + B, B is the addend.
Address The location in memory where data are stored. For data areas, an address consists of a two-letter data area designation and a number that designates the word or bit location. For the UM area, however, an address designates the instruction location (UM area); for the FM area, the block location (FM area) etc.
Allocation The process by which the PLC assigns certain bits or words in memory for various functions. This includes pairing I/O bits to I/O points on units.
Analogue input module A PLC module with terminals capable of receiving a continuous, varying, electrical value from an outside device such as a temperature, pressure or flow sensor.
Analogue output module A PLC module with terminals capable of furnishing a continuous, varying output voltage to an output device such as a proportional valve solenoid.
Analogue signal A continuous value between two limits. Can represent flow, pressure, temperature, position, voltage, angle or any electrical signal with a varying value.
Analogue timer unit A dedicated timer that interfaces through analogue signals externally and digital signals internally.
Analogue-to-digital converter A circuit, usually an input module on a PLC, for converting a varying analogue electrical signal to a corresponding representative binary number.
AND function A logic operation whereby the result is true if and only if both premises are true. In ladder-diagram programming the premises are usually "ON"/"OFF" states of bits or the logical combination of such states called execution conditions.
Arithmetic capability The ability of a PLC or computer to perform mathematical functions.
Arithmetic logic unit (ALU) A computer CPU subsystem that can perform arithmetic and logic gate operations.
ASCII American Standard Code for Information Interchange. A 7-bit code for representing letters, numbers and symbols appearing in written information output.
Astable relay A relay with only one stable logic state (monostable relay).
Augend Number to which another is added. For example, in A + B, A is the augend.

BASIC Beginner's All-purpose Symbolic Instruction Code. A high-level, procedural programming language with English-like statements that tells the computer what to do step by step.
Baud A rate of data transmission. Its rate is equal to the number of code elements per second that are transmitted.
BCD Binary-Coded Decimal is a numbering system in which each decimal digit from 0 to 9 is represented by a pattern of four binary bits.
BCD calculation An arithmetic calculation that uses numbers expressed in binary-coded decimal.
Binary A number system where all numbers are expressed in base 2. Although in a PLC all data are ultimately stored in binary form, binary is used to refer to data that are numerically equivalent to the binary value. It is not used to refer to binary-coded decimal. Each four binary bits are equivalent to one hexadecimal digit.
Binary-coded decimal See *BCD*.
Binary word A group of bits usually found in a single word or DM address location.
Bistable Having two stable states.
Bit The smallest unit of storage in a PLC. The status of a bit is either "ON" or "OFF" (logic 1 or logic 0). Four bits equal one digit; sixteen bits equal one word. Different bits are allocated to special purposes, such as holding the status input from external devices; other bits are available for general use in programming.
Bit address The location in memory where a bit of data is stored. A bit address must specify (sometimes by default) the data area and word that are being addressed as well as the number of the bit.
Bit designator An operand that is used to designate the bit or bits of a word to be used by an instruction.
Bit number A number that indicates the location of a bit within a word. Bit 00 is the rightmost (least significant) bit; bit 15 is the leftmost (most significant) bit.
Boolean algebra A shorthand notation that expresses logic functions in equation type expressions. The plus sign (+) denotes "OR" connection; the dot sign (•) denotes "AND" connection.
Boolean equation Expresses the relations between logic

functions in an equation to an equivalent logic written in Boolean algebra form.

Branch A parallel logic path within a user program rung of a ladder diagram.

Buffer A temporary storage area in a computer or printer used for intermediate storage of data. Typically receives data at one rate and then outputs the data at a different rate or in a different form.

Bus One or more conductors for transmitting data between destinations.

Bus bar The line leading down the left and sometimes the right side of a ladder diagram. Instruction execution follows down the bus bar, which is the starting point for all instruction lines.

Byte A sequence of four binary digits usually operated upon and retrieved as a unit.

Call A process by which instruction execution shifts from the main program to a subroutine. The subroutine may be called by an instruction or by an interrupt.

Carry flag A flag that is used with arithmetic operations to hold a carry from an addition or a multiplication operation or to indicate that the result is negative in a subtraction operation. The carry flag is also used with certain types of shift operation.

Cascading Placing two or more functions of the same kind in sequence. The purpose of cascading is to extend the number of operational steps beyond that of a single function.

Cassette recorder/player For PLCs, a device that can transfer information between PLC memory and magnetic tape. When recording, it makes a permanent record on tape of a program or data from a processor memory. In the play mode, the cassette recorder transfers the previously recorded program or data from the tape into the PLC memory.

Character One symbol of a set of standardised basic symbols, such as a digit, a punctuation mark or a letter.

Chip Another name for IC integrated circuit. A tiny piece of layered semiconductor material mounted in a small case with terminals. Contains a large number of transistors, resistors and capacitors in miniature.

CIM Computer Integrated Manufacturing. A manufacturing system controlled by an easily reprogrammable computer for manufacturing flexibility and speed of tool or product changeover.

Clear A command to remove data from one or more memory locations. Normally sets the memory location value to zero or resets the address counter to zero address.

Clock A circuit that generates timed pulses to synchronise the timing of computer operations or causes intermittent PLC-driven actions.

Clock pulse A pulse available at a certain bit in memory for use in timing operations. Various clock pulses are available with different pulse widths.

Clock pulse bit A bit in memory that supplies a pulse that can be used to time operations. Various clock pulse bits are available with different pulse widths.

Code A system of symbols or bits for representing data, ideas or characters.

Combinational control system A system or PLC program operating by combinational principles.

COMPARE function A PLC function that compares magnitudes to see if the compared magnitude is greater than, equal to or less than a reference magnitude.

Computer interface A device that communicates between various computers or computer and PLC.

Condition An "instruction" placed along an instruction line to determine how terminal instructions on the right side are to be executed. Each condition is assigned to a bit in memory that determines its status. The status of the bit assigned to each condition determines, in turn, the execution condition for each instruction up to a terminal instruction on the right side of the ladder diagram.

Constant An operand for which the actual numeric value (#) is entered by the programmer and in place of a data memory address or counter or timer PV.

Contact A switching device with two terminals. In PLCs, it is in a conducting or non-conducting state, depending on its corresponding coil's status and the coil's initial status, whether normally open or normally closed. In ladder diagrams contacts are embedded in the rungs.

Control bit A bit in a memory area that is set from either the program or a programming device to achieve a specific purpose, e.g., a restart bit is turned "ON" and "OFF" to restart a unit.

Control signal A signal sent from the PLC to affect the operation of the controlled system.

Control system All of the hardware and software components used to control a machine. A control system includes the PLC system, the PLC programs and all I/O devices that are used to control or obtain feedback from the controlled machine.

Controlled system The devices that are being controlled by a PLC system.

Counter Either a dedicated number of digits or words in memory used to count the number of times a specific process has occurred, or a location in memory addressed through a TC bit and used to count the number of times the status of a bit or an execution condition has changed from "OFF" to "ON".

CPU An acronym for Central Processing Unit. In a PLC system, the CPU executes the program, processes I/O signals, communicates with external devices etc.

CPU unit The CPU unit contains the CPU and provides a certain number of I/O points.

Cross reference In ladder diagrams, letters or numbers to the right of relays or functions. The letters or numbers indicate on what other ladder lines contacts depending on the relay or function are located. Normally closed contacts (N/C) are distinguished from normally open contacts (N/O) through the use of an asterisk (*) or by underlining.

Data area An area in the PLC's memory that is designed to hold a specific type of data, e.g., the SR area is designed to hold flags and control bits. Note, memory areas that hold programs are not considered data areas.
Data area boundary The highest address available in a data area. When designating an operand that requires multiple words, it is necessary that the highest address in the data area not be exceeded.
Data transfer A PLC operation that moves data from one DM area to another.
DC Direct Current. Electrical current flowing continuously in the same direction, usually at a fixed rate or value.
Debug Correcting mistakes in a program through various forms of analysis. Debugging includes both removal of syntax errors as well as fine-tuning of timing and co-ordination of control operations.
Decimal A number system where all numbers are expressed in base 10. Although in a PLC all data are ultimately stored in binary form, four binary bits are often used to represent one decimal digit, a system called *Binary-Coded Decimal*.
Decimal number system The base 10 system of counting. Digits are 0 to 9.
Decrement Decreasing a numeric value by 1.
Default A value assumed and automatically set by the PLC when a specific value is not entered by the user.
Definer A number used as an operand for an instruction but that serves to define the instruction itself rather than the data on which the instruction is to operate. Definers include jump numbers, subroutine numbers etc.
Delay In tracing, a value that specifies where tracing is to begin in relationship to the trigger. A delay can be either positive or negative, i.e., can designate an offset on either side of the trigger.
Destination The location where data of some sort in an instruction are to be placed as opposed to the location from which data are to be taken for use in the instruction. The location from which data are to be taken is called the source.
Detent A catch or restrainer that causes a switch or valve mechanism to stop striking. Mild force removes the catch and permits the mechanism to assume alternative positions.
Diagnose To find the cause of a fault in a control system.
Diagnostic program A program used to analyse faults in a PLC program or in a system's operation.
Differentiation instruction An instruction used to ensure that the operand bit is never turned "ON" for more than one scan after the execution condition goes either from "OFF" to "ON" for a DIFFERENTIATE UP instruction or from "ON" to "OFF" for a DIFFERENTIATE DOWN instruction.
Digit A unit of storage in memory that consists of four bits.
Digit designator An operand that is used to designate the digit or digits of a word to be used by an instruction.
Digital gate A device (electrical, pneumatic or mechanical) that analyses the digital states of its inputs and produces an appropriate logic output state.
Digital-to-analogue converter An electrical circuit, usually an output module on a PLC, that converts binary bits to a representative, continuous, analogue signal.
DIP switch A group of small, in-line, on–off switches. On PLCs used to indicate change of chip.
Discrete Having the characteristic of being clearly "ON" or "OFF".
Discrete input module A PLC module that processes input status information having two discrete states only, high or low.
Discrete output module A PLC module that produces only two states, "ON" or "OFF".
Diskette The flat, flexible disc on which a disc drive writes and reads. Synonymous with disc.
Distributed control An automation concept in which control of each portion of an automated system is located near the devices actually being controlled, i.e., control is decentralised and distributed over the system. Distributed control is a concept basic to PLC systems.
Dividend Number to be divided by divisor. For example, in A ÷ B, A is the dividend.
Divisor Number by which dividend is to be divided. For example, in A ÷ B, B is the divisor.
DM area A data area used to hold word data. A word in the DM area cannot be addressed by bit.
Documentation A logical, orderly, recorded or written document containing software or data listings, ladder diagrams, mnemonic lists, cross reference lists and machine commissioning instructions.
Double precision The system of using two addresses or words to display a number too large for one address or word. Allows the display of more significant figures since twice as many bits are used. Used in arithmetic operations.
Down counter A counter that starts from a specified number (SV) and decrements downward to zero.
Download The process of transferring a program or data from a higher-level computer to a lower-level computer or PLC or from a PLC to a printer buffer.
Drum switch Synonymous with *Sequencer.* Normally mechanical in nature, it operates through a multiple sequence of simultaneous "ON"–"OFF" states.

EEPROM Electrically Erasable Programmable Read Only Memory. A programmable integrated circuit chip

The program can be erased after use (all bits reset to zero) by applying an electrical current to two of its terminals.

Electrical noise Electrical static that can disturb electronic communications. The snow that can appear on a TV screen is an example of the effects of electrical noise.

Enable To allow a function to operate by energising a PLC ladder rung. If not enabled, the PLC function will not be active.

EPROM Erasable Programmable Read-Only Memory. Same as EEPROM, except that resetting is accomplished by exposing a small section under a "window" to ultraviolet light.

Error code A numeric code produced to indicate the existence of an error and something about its nature. Some error codes are generated by the system; other are defined in the program by the operator.

Examine off An instruction that is true only if the examined bit is "OFF" or logic 0.

Examine on An instruction that is true only if the examined bit is "ON" or logic 1.

Execution condition The "ON" or "OFF" status under which an instruction is executed. The execution condition is determined by the logical combination of conditions on the same instruction line and up to the instruction being executed.

Execution time The time required for the CPU to execute either an individual instruction or an entire program.

EXOR EXclusive OR gate. A digital gate in which the output is "ON" only when one of its two inputs, but not both, is "ON". (Also XOR.)

Expansion I/O unit An Expansion I/O unit is connected to increase the number of I/O points available.

Extended counter A counter created in a program that counts higher than any of the standard counters provided by the individual instructions.

Extended timer A timer created in a program that times longer than any of the standard timers provided by the individual instructions.

Factory intelligent terminal A programming device provided with advanced programming and debugging capabilities to facilitate PLC operation. The factory intelligent terminal also provides various interfaces for external devices, such as floppy disc drives.

Fail-safe A control situation that discontinues the operation of a process when the supply power source fails. A non-fail-safe operation requires the application of supply power to turn it off.

False Prescribed conditions are not met and the logic is therefore disabled.

Fatal error An error that will stop PLC operation and require correction before operation can be continued.

Feedback Information received from the machine controlled by the PLC.

FIT Short for *Factory Intelligent Terminal.*

Flag A dedicated bit in memory that is set by the system to indicate some type of operating status. Some flags, such as the carry flag, can also be set by the operator or program.

Flicker bit A bit that is programmed to turn "ON" and "OFF" at a specific interval.

Flip-flop A bistable PLC output relay requiring a distinct set and a distinct reset instruction; e.g. RS, flip-flop, JK flip-flop, D flip-flop.

Floppy disc A recording disc used with a computer disc drive for recording or retrieving data. The disc is flexible, not rigid.

FORCE function A keyboard function used to turn output elements of a PLC ladder diagram on and off. It overrides the normal input function status received through the input module.

Force reset The process of artificially turning "OFF" a bit from a programming device. Bits are usually turned "OFF" as a result of program execution.

Force set The process of artificially turning "ON" a bit from a programming device. Bits are usually turned "ON" as a result of program execution.

Function code A two-digit number used to enter an instruction into the PLC (e.g. FUN 11).

Fuse A device that rapidly interrupts electrical current. The interruption is achieved by the melting of a thin strip of metal or wire inside the fuse.

Gate Electronically, a logic element with several inputs and one output that makes logic decisions, depending on its input state: the gate's output is turned "ON" or "OFF".

GPC Short for *Graphic Programming Console.*

Grafcet A graphics-based, high-level, computer or PLC language used to program PLCs with less effort.

Graphic programming console A programming device provided with advanced programming and debugging capabilities to facilitate PLC operation. A graphic programming console is provided with a large display onto which ladder-diagram programs can be written directly in ladder-diagram symbols for entry into the PLC without conversion to mnemonic form.

GRAY code A special binary code where only one of its bits changes status when going sequentially from one number to the next. (See figure 10-9 on page 142 of the book *Pneumatic Control for Industrial Automation* by Peter Rohner and Gordon Smith.)

Ground An electrical connection made for safety from a PLC to ground potential.

Hand-held programmer A small, portable or detachable, programming keyboard, usually with a small LCD window on which portions of the entire ladder diagram or a mnemonic PLC instruction may be displayed.

Hard copy A printed copy of data or user programs.
Hard disc An inflexible recording disc used with a computer disc drive.
Hardware In contrast to software, the mechanical, electrical and electronic parts of a PLC.
Hardware error An error originating in the hardware structure of the PLC, as opposed to a software error, which originates in software (programs).
Hex An abbreviation for *Hexadecimal*.
Hexadecimal A numbering system with four binary bits, where all numbers are expressed in base 16 (radix). Represents 0 to 15 in the decimal system by using the digits 0 to 9 and the letters A to F. Although in a PLC all data are ultimately stored in binary form, programming devices often use hexadecimal to facilitate operation.
High A status representation of "ON" (1 or true).
Holding register A type of data storage area in the CPU for symbol or logic storage.
Holding relay A KEEP relay or flip-flop with a discrete set or reset state and memory capacity during power failure.
Host computer A computer that is used to transfer data or programs to or receive data or programs from a PLC in a *Host Link System*. The host computer is used for data management and overall system control. Host computers are generally small personal or business computers.
Host link system One or more *Host Computers* connected to one or more PLCs through host link units so that the host computer can be used to transfer data to and receive data from the PLC(s). Host link systems enable centralised management and control of a PLC system.
Host link unit An interface used to connect a PLC to a host computer in a *Host Link System*.
HR area A data area used to store and manipulate data and to preserve data when power to the PLC is turned off.
Hydraulic A system of control using a fluid.

I/O Input–output.
I/O capacity The number of inputs and outputs that a PLC is able to handle. This number ranges from around a hundred for smaller PLCs to two thousand for the largest ones.
I/O devices The devices to which terminals on I/O units are connected. I/O devices may be either part of the control system, if they function to help control other devices, or part of the controlled system.
I/O link Created in an optical remote I/O system to enable input or output of one or two IR words directly between PLCs. The words are transferred between the PLC controlling the master and a PLC connected to the remote I/O system through an I/O link unit or an I/O link rack.
I/O link unit A unit used with certain PLCs to create an I/O link in an optical remote I/O system.
I/O point The place at which an input signal enters the PLC system or an output signal leaves the PLC system. In physical terms, an I/O point corresponds to terminals or connector pins on a unit. In terms of programming, an I/O point corresponds to an I/O bit in the IR area.
I/O response time The time required for an output signal to be sent from the PLC in response to an input signal received from an external device.
I/O unit The most basic type of unit mounted to a backplane to create a rack. I/O units include input units and output units, each of which is available in a range of specifications. I/O units do not include special I/O units, link units etc.
I/O update time The time interval in milliseconds that it takes for a PLC to update the CPU image of all input and output modules.
I/O word A word in the IR area that is allocated to a unit in the PLC system.
Increment Increasing a numeric value by 1.
Inhibition function A logic function in which one input, if present, negates the output of the gate.
Initialise Part of the start-up process whereby some memory areas are cleared, system setup is checked and default values are set.
Initialisation error An error that occurs either in hardware or software before the PLC system has actually begun operation, i.e., during initialisation.
Input The signal coming from an external device into the PLC. Input often is used abstractly or collectively to refer to incoming signals.
Input bit A bit in the IR area that is allocated to hold the status of an input.
Input device An external device connected to the PLC input modules that sends signals into the PLC system. Switches, limit switches, push-buttons and electrical potentiometers are examples of input devices.
Input point The point at which an input enters the PLC system. An input point physically corresponds to terminals or connector pins.
Input register A PLC register associated where input device signals are recorded.
Input scan One of three parts of the PLC scan. During the input scan, input points are read and the input image table is updated accordingly.
Input signal A change in the status of a connection entering the PLC. Generally an input signal is said to exist when, for example, a connection point goes from low to high voltage or from a non-conductive to a conductive state.
Instruction A direction given in the program that tells the PLC an action to be carried out and the data to be used in carrying out the action. Instructions can simply turn a bit "ON" or "OFF", or they can perform much more complex actions, such as converting or transferring large blocks of data.

Instruction block A group of instructions that are logically related in a ladder-diagram program. Although any logically related group of instructions could be called an instruction block, the term is generally used to refer to blocks of instructions called *Logic Blocks* that require logic block instructions to relate them to other instructions or logic blocks.
Instruction execution time The time required to execute an instruction. The execution time for any one instruction can vary with the execution condition for the instruction and the operands used in it.
Instruction line A group of conditions that lie together on the same horizontal line of a ladder diagram. Instruction lines can branch or join to form instruction blocks.
Interface An interface is the conceptual boundary between systems or devices and usually involves changes in the way the communicated data are represented. Interface devices perform operations such as changing the coding, format or speed of the data.
Interlock A programming method used to treat a number of instructions as a group so that the entire group can be reset together when individual execution is not required. An interlocked program section is executed normally for an "ON" execution condition and partially reset for an "OFF" execution condition.
IR area A data area whose principal function is to hold the status of inputs entering the system and outputs leaving the system. Bits and words in the IR area that are used this way are called *I/O bits* and *I/O words*. The remaining bits in the IR area are *Work Bits*.

Jump A type of programming where execution moves directly from one point in a program to a separate point in the program without sequentially executing the instruction in between. Jumps are usually conditional on an execution condition.
Jump number A definer used with a jump that defines the points from which and to which a jump is to be made.

Keep relay See *Holding Relay*.
Keyboard The alphanumeric keypad on which the programmer enters instructions to the PLC (see *Programming Console*).

Label The means of identification of words, addresses, contacts and coils—normally in letters, numbers or alphanumerics.
Ladder diagram A system of successive horizontal rungs with contact symbols representing the logic operation of a control system. The symbols are drawn in relay logic or PLC-logic form. The control contacts are to the left and coils and functions are output to the right bus bar.
Ladder diagram program A form of program arising out of relay-based control systems that uses circuit-type diagrams to represent the logic flow of programming instructions. The appearance of the program suggests a ladder and thus the name.
Ladder diagram symbol A symbol used in a *Ladder-Diagram Program*.
Ladder instruction An instruction that represents the rung portion of a ladder-diagram program. The other instructions in a ladder diagram fall along the right side of the diagram and are called output or function instructions.
LAN Local Area Network. A system control network that controls devices relatively close to each other.
Language A group of letters and symbols used for intercommunication between people, computers or people and computers and PLCs.
Laptop computer A small, lightweight computer. Normally a laptop contains extensive computing power, a CPU, an LCD screen and a hard disc drive in one package. It may also be connected to a printer.
Latch An electronic or mechanical device that causes an energised relay to remain "ON" after its input signal is turned "OFF".
Latching relay A relay with a latching-type operation and two inputs, "ON" and "OFF".
LCD See *Liquid Crystal Display*.
LED Light-Emitting Diode. A type of small light used in combinations to give a visual display by emitting light. Used on PLCs to display active inputs and outputs.
Leftmost (bit or word) The highest numbered bit of a group of bits, generally of an entire word, or the highest numbered word of a group of words. Often called most significant bits or words.
Limit switch A mechanical device used to produce a machine feedback signal to the PLC. The electrical switch is actuated by depressing its protruding arm or cam roller.
Link A hardware or software connection formed between two units. "Link" can refer either to a part of the physical connection between two units or to a software connection created to data at another location (I/O links).
Liquid crystal display A low-power display, working off reflected light, used in hand-held PLC programmers and laptop computer monitors.
Load The processes of copying data either from an external device or from a storage area to an active portion of the system such as a display buffer. An output device connected to the PLC is also called a load.
Logic block A group of instructions that is logically related in a ladder-diagram program and that requires logic block instructions to relate it to other instructions or logic blocks.
Logic block instruction An instruction used to locally combine the execution condition resulting from a logic block with a current execution condition. The current

execution condition could be the result of a single condition or of another logic block. "AND LOAD" and "OR LOAD" are the two logic block instructions.
Loop control A control of a process or machine that uses feedback. An output status indicator modifies the input signal effect on the process control.
Low A state of being "OFF" (0 or false).
LR area A data area that is used in a PLC link system so that data can be transferred between two or more PLCs. If a PLC link system is not used, the LR area is available for use as work bits.

Magnetic tape A thin plastic tape covered with magnetic particles, used in tape recorders. The tape stores information by becoming magnetised at specific points as it passes through a fixed location. The stored information may be read from the tape on a subsequent pass through.
Main program All of a program except for the subroutines.
Master Short for *Remote I/O Master Unit*.
Memory In a PLC, the group of words or registers where information and programs are stored. Storage may be permanent, or temporary and erasable, depending on chips used.
Memory area Any of the areas in the PLC used to hold data or programs.
Microprocessor A computer on a chip containing functions normally found on many different chips. It has all of the capabilities of the digital computer.
Microsecond One millionth of a second.
Millisecond One thousandth of a second.
Minuend Number from which another is to be subtracted. For example, in A – B, A is the minuend.
Mnemonic The ability to memorise or retain information. The "m" is silent. (From Greek, mnhmonikoV = mindful.)
Mnemonic code A form of a ladder-diagram program that consists of a sequential list of the instructions without using a ladder diagram. Mnemonic code is required to enter a program into a PLC when using a programming console.
Mode The functional form in which a computer is operating; for example, run, program or monitor.
Module An electronic PLC functional device or subcircuit. As a device, it may be attached to or plugged into a *Bus* or *Rack*. The bus is connected electrically or electronically to other modules and to the CPU.
Monitor mode A mode of PLC operation in which normal program execution is possible but modification of data held in memory is still possible. Used for monitoring or debugging the PLC.
Monostable Having one stable state.
Multiplicand Number to be multiplied by the multiplier. For example, in A × B, A is the multiplicand.
Multiplier Number by which the multiplicand is multiplied. For example, in A × B, B is the multiplier.

NAND A digital gate whose output is "OFF" only when all of its inputs are "ON".
NC input An input that is normally closed, i.e., the input signal is considered to be present when the circuit connected to the input opens.
Nest Programming one jump within another jump; programming a call to a subroutine from within another subroutine etc.
Nesting In ladder diagrams, locating a series of contacts logically within another series of contacts (timer nesting, counter nesting, jump nesting).
Network A number of interconnected logic devices or contacts leading to an output.
No input An input that is normally open, i.e., the input signal is considered to be present when the circuit connected to the input closes.
Node A common electrical or logic point with two or more points of the circuit or diagram connected to it.
Noise interference Disturbances in signals caused by electrical noise.
Non-fatal error A hardware or software error that produces a warning but does not stop the PLC from operating.
Non-retentive Describes a PLC logic device (usually a flip-flop, timer, counter or relay) that loses its count of increments or status when the input or power goes "OFF" or low.
NOR A digital gate whose output is "OFF" when one or more of its inputs is "ON".
Normally closed condition A condition that produces an "ON" execution condition when the bit assigned to it is "OFF" and an "OFF" execution condition when the bit assigned to it is "ON".
Normally closed contact (NC) A contact that is conductive when its operating coil is not energised.
Normally open condition A condition that produces an "ON" execution condition when the bit assigned to it is "ON" and an "OFF" execution condition when the bit assigned to it is "OFF".
Normally open contact (NO) A contact that is non-conductive when its operating coil is not energised.
NOT A digital inverter gate. "ON" translates to "OFF" and "OFF" translates to "ON" through the gate.
NOT function A logic operation which inverts the status of the operand. For example "AND NOT" indicates an "AND" with the opposite of the actual status of the operand bit.

Octal A numbering system using three binary bits equivalent to a decimal 8. Digits used are 0 to 7.
"OFF" The status of an input or output when a signal is said not to be present. The "OFF" state is generally low

voltage or non-conductivity, but can be defined as the opposite of either.

Off delay The delay produced between the time at which turning "OFF" a signal is initiated (e.g., by an input device or PLC) and the time the signal reaches a state readable as an "OFF" signal (i.e., as no signal) by, for example, an output device or PLC.

Off-delay timer A timer that initiates an action at a specified time after being enabled.

"ON" The status of an input or output when a signal is said to be present. The "ON" state is generally high voltage or conductivity, but can be defined as the opposite of either.

On delay The delay produced between the time a signal is initiated (e.g., by an input device or PLC) and the time the signal reaches a state readable as an "ON" signal by, for example, an output device or PLC.

One-shot bit A bit that is turned "ON" for a specified time regardless of its enabling signal.

Operand Bit(s) or word(s) designated as the data to be used for an instruction. An operand can be entered as a constant expressing the actual numeric value to be used or as an address to express the location in memory of the data to be used.

Operand bit A bit designated as an operand for an instruction.

Operand word A word designated as an operand for an instruction.

Operating error An error that occurs during actual PLC operation, as opposed to an *Initialisation Error*, which occurs before actual operations can begin.

Optical isolation Electronic isolation of two parts of a circuit by using a small light beam between the two stages. One stage produces a light beam of appropriate varying intensity; the other receives and decodes the varying light's pattern.

Optical slave rack A slave rack connected through an optical *Remote I/O Slave Unit*.

OR function A logic operation whereby the result is true if either one or both of the premises is true. In ladder-diagram programming the premises are usually "ON"–"OFF" states of bits or the logical combination of such states called execution conditions.

OR gate A digital gate that is "ON" if any one or more of its inputs is "ON".

Output The signal sent from the PLC to an external device. Output often is used abstractly or collectively to refer to outgoing signals.

Output bit A bit in the IR area that is allocated to hold the status to be sent to an *Output Device*.

Output device An external device that receives signals from the PLC system.

Output module An electrical modular unit or circuit used to connect the PLC to outside devices that are to be turned "ON" or "OFF".

Output point The point at which an output leaves the PLC system. An output point physically corresponds to terminals or connector pins.

Output register A PLC register holding data for output devices.

Output scan One of three parts of the PLC scan. During the output scan, data associated with the output image table are transferred to the output terminals.

Output signal A change in the status of a connection leaving the PLC. Generally an output signal is said to exist when, for example, a connection point goes from low to high voltage or from a non-conductive to a conductive state.

Overseeing Part of the processing performed by the CPU that includes general tasks required to operate the PLC.

Overwrite Changing the content of a memory location so that the previous content is lost.

Parallel circuit An electrical circuit in which the opposite ends of two or more components or elements are each connected to the same nodes. A parallel circuit may make up the whole circuit or be a portion of an overall larger logic line.

PC Personal Computer.

PCB Printed Circuit Board.

Peripheral device Device connected to a PLC system to aid in system operation. Peripheral devices include printers, programming devices, external storage media etc.

PID Proportional-Integral-Derivative. A sophisticated, self-correcting, analogue control system for accurately and speedily controlling output parameters.

PLC See *Programmable Logic Controller*.

PLC system All of the units connected to the CPU unit up to, but not including, the I/O devices. The limits of the PLC system on the upper end are the PLC and the program in its CPU, and on the lower end are the I/O units, an I/O link unit etc.

Pneumatic A system run by air pressure (compressed air flow).

Pneumatic controller An industrial sequential control device using compressed air to control a machine.

Port In computers, a point of connection to a peripheral input or output device.

Power supply For PLCs, the device that converts line power, usually 115 or 240 V AC, to the power type required by the PLC and its attached devices.

Present value The current time left on a timer or the current count of a counter. Present value is abbreviated PV.

Printed circuit board A board onto which electrical circuits are printed for mounting into a computer or electrical device.

Program The list of instructions that tells the PLC the sequence of control actions to be carried out.

Program mode A mode of operation that allows for entering and debugging programs but that does not permit normal execution of the program.

Program scan One of three parts of the PLC scan. During the program scan, data in the input status table is applied to the user program, the program is executed and the output status table is updated appropriately.

Programmable logic controller A computerised device that can accept inputs from external devices and generate outputs to external devices according to a program held in memory. Programmable logic controllers are used to automate control of external devices. Abbreviated PLC.

Programmed alarm An alarm given as a result of execution of an instruction designed to generate the alarm in the program as opposed to one generated by the system.

Programmed error An error arising as a result of execution of an instruction designed to generate the error in the program as opposed to one generated by the system.

Programmed message A message generated as a result of execution of an instruction designed to generate the message in the program as opposed to one generated by the system.

Programming console The simplest form of programming device available for a PLC. Programming consoles are available as both hand-held and CPU-mounting models.

Programming device A peripheral device used to enter a program into a PLC or to alter or monitor a program already held in the PLC. There are dedicated programming devices, such as programming consoles, and there are non-dedicated devices, such as a host computer.

PROM Programmable Read-Only Memory. A ROM chip programmed at the factory for use with a given PLC. It is non-volatile.

PROM writer A peripheral device used to write programs and other data into a ROM for permanent storage and application.

Prompt A message or symbol that appears on a display to request input from the operator.

Proximity device A non-contact, input-indicating sensor for detecting the presence of an object associated with a process. Its emitted signal may be discrete or analogue, depending on the process being controlled.

PV See *Present Value.*

Rack A mechanical channel or chassis on which PLC input and output modules are mounted. May also include wiring channels and connectors.

Radix Number used as basis of a numbering system. For example, 10 is the radix of the decimal numbering system.

RAM Random Access Memory chip. A chip that has read and write capabilities.

Read/write memory A computer memory that can receive and store (read) information and can be used for retrieval of stored information (write). Stored information can be erased or replaced. Writing out information does not change the stored information that has been read: the stored information is duplicated in the write destination.

Refresh The process of updating output status sent to external devices so that it agrees with the status of output bits held in memory and of updating input bits in memory so that they agree with the status of inputs from external devices.

Register A location in a PLC's memory for storing information, usually in bit form.

Relay A device or PLC output function actuated by a voltage or signal.

Relay-based control The forerunner of PLCs. In relay-based control, groups of relays are wired to each other to form control circuits. In a PLC, these are replaced by programmable circuits.

Reliability The ability of a device to perform its function correctly over a period of time or through a number of actuations or operations. Can be expressed as a decimal or a percentage or in words, such as good or bad.

Remote I/O master unit The unit in a remote I/O system through which signals are sent to all other remote I/O units. The remote I/O master unit is mounted to an Omron C200H, C500, C1000H or C2000H CPU rack or to an expansion I/O rack connected to the CPU rack. Remote I/O master unit is generally abbreviated to simply "Master."

Remote I/O slave unit A unit mounted to an Omron C200H, C500, C1000H or C2000H backplane to form a slave rack. Remote I/O slave unit is generally abbreviated to simply "Slave."

Remote I/O system An Omron C200H, C500, C1000H or C2000H system in which remote I/O points are controlled through a master mounted to a CPU rack or through an expansion I/O rack connected to the CPU rack. K-type PLCs can be connected to remote I/O systems through I/O link units.

Remote I/O unit Any of the units in a remote I/O system. Remote I/O units include masters, slaves, optical I/O units, I/O link units and remote terminals.

Reset The process of turning a bit or signal "OFF" or of changing the present value of a timer or counter to its set value or to zero.

Retentive timer (or counter) A timer or counter that retains PV if the supply power to the PLC fails.

Return The process by which instruction execution shifts from a subroutine back to the point from which the subroutine was called. A return is automatic upon completion of the subroutine and the return is always to subroutine start.

Reversible counter A counter that can be both incremented and decremented depending on specified conditions.

Reversible shift register A shift register that can shift data in either direction depending on specified conditions.

Right-hand instruction Another term for terminal instruction.

Rightmost (bit or word) The lowest numbered bits of a group of bits, generally of an entire word, or the lowest numbered word of a group of words. Often called least significant bit or word.

ROM Read-Only Memory. An integrated circuit chip with unalterable information.

Run mode The operating mode used by the PLC for normal control operations.

Rung The horizontal ladder system that controls the output on the right-hand bus. May be on more than one horizontal control line.

RS-232 A standard plug used on PLCs.

Scan The process used to execute a ladder-diagram program. The program is examined sequentially from start to finish and each instruction is executed in turn based on execution conditions.

Scan time The time required for one complete sweep through the PLC's entire ladder-diagram program.

Schematic An electrical, pneumatic or hydraulic diagram symbolically showing components and their logic connections.

Seal See *Latch*.

Self-diagnosis A process whereby the system checks its own operation and generates a warning or error if an abnormality is discovered.

Self-maintaining bit A bit that is programmed to maintain either an "OFF" or "ON" status until set or reset by a specific condition different from the one that originally caused the bit to turn "OFF" or "ON".

Sensor A PLC input device (limit switch, pressure switch, flow switch etc.) that senses the process condition. Its status is fed to the PLC through an input module.

Sequence The order in which events take place.

Sequencer A control system that sequences machine actions through a fixed program.

Sequential control system A system or program causing a machine to operate its actuators in a defined sequential order.

Servicing The process whereby the PLC provides data to or receives data from external devices or a remote I/O, or otherwise handles data transactions for link systems.

Servomechanism A closed-loop control system that uses feedback for accuracy and process correction.

Set The process of turning a bit or signal "ON".

Set value The count from which a counter starts counting down (or, in the case of a reversible counter, the maximum count) or the time from which a timer starts timing. Set value is abbreviated to SV.

Shift register One or more words in which data are shifted in bit, digit or word units a specified number of units to the right or left.

Slave Short for *Remote I/O Slave Unit*.

Slave rack An Omron C200H, C500, C1000H or C2000H rack containing a remote I/O slave unit and controlled through a remote I/O master unit. Slave racks are generally located away from the CPU rack.

Software The programs that control a PLC or computer.

Software error An error that occurs in the execution of a program.

Software protects A software means of protecting data from being changed as opposed to a physical switch or other hardware setting.

Solenoid A magnetic coil that changes the "ON" or "OFF" status of an output device such as a pneumatic or hydraulic valve. The change is accomplished by the movement of an iron plunger, which may be a spring-return type (usually used as actuator for fluid power valves).

Solid state Made of semiconductor material. May be transistors, thyristors or complete circuits in the form of integrated circuits.

Source The location from which data are taken for use in an instruction as opposed to the location to which the result of an instruction is to be written, which is called the destination.

SR area A data area in a PLC used mainly for flags, control bits and other information provided about PLC operation. The status of only certain SR bits may be controlled by the operator, i.e., most SR bits can only be read.

Status The condition of a gate, flip-flop or contact, being "ON" or "OFF".

Step-action diagram A graphical method of representing motions and actions taking place in a sequential machine (actuation of pneumatic cylinders).

Step-counter A circuit design method used for sequential control systems.

Subroutine A group of instructions placed after the main program and executed only if called from the main program or activated by an interrupt.

Subroutine number A definer used to identify the subroutine that a subroutine call or interrupt activates.

Subtrahend Number to be subtracted. For example, in A – B, B is the subtrahend.

SV See *Set Value*.

Switching capacity The voltage or current that a relay can switch "ON" and "OFF".

Syntax error An error in the way in which a program is written. Syntax errors can include "spelling" mistakes (i.e., a function code that does not exist), mistakes in specifying operands within acceptable parameters (e.g., specifying unwritable SR bits as a destination), and mistakes in actual application of instructions (e.g., a call to a subroutine that does not exist).

System error An error generated by the system as opposed to one resulting from execution of an instruction designed to generate an error.

System error message An error message generated by the system as opposed to one resulting from execution of an instruction designed to generate a message.

TC area A data area that can be used only for timers and counters. Each bit in the TC area serves as the access point for the SV, PV and completion flag for the timer or counter defined with that bit.

TC number A definer that corresponds to a bit in the TC area and used to define the bit as either a timer or a counter.

Terminal instruction An instruction placed on the right side of a ladder diagram that uses the final execution condition on an instruction line.

Thumbwheel switch A series of small, adjacent, numbered (0 to 9) rotary wheels or ratchets that may be set to a given number. Their settings may be entered into a PLC for data control.

Timer A location in memory accessed through a TC bit and used to time down from the timer's set value. Timers are turned "ON" and "OFF" according to their execution conditions. They are used for monitoring or determining actions times.

TM area A memory area used to store the results of a trace.

Toggle switch A small, electrical, detent switch with a lever for manual actuation. Usually panel-mounted.

TR area A data area used to store execution conditions so that they can be reloaded later for use with other instructions.

Transfer The process of moving data from one location to another within the PLC or between the PLC and external devices. When data are transferred, generally a copy of the data is sent to the destination, i.e., the content of the source of the transfer is not changed.

Transmission distance The distance that a signal can be transmitted.

True "ON" logic condition.

Truth table A yes/no (I/O) matrix indicating the status of a logic function and how it depends on the status of its inputs.

UM area The memory area used to hold the active program, i.e., the program that is being currently executed (user memory).

Unit number A number assigned to some link units and special I/O units to assign words and sometimes other operating parameters to it.

Unlatch instruction A PLC command that turns a function "OFF" and keeps it "OFF", overriding any other subsequent instruction to turn it "ON".

Up counter An event counter that starts from 0 and counts up to the preset value.

User-friendly A term indicating that a PLC program can be designed, entered and run by a person with minimal training or instruction.

Volatile memory A memory of values or status that is lost when its supply power is turned "OFF". Memory locations are usually reset to zero at loss of power.

Watchdog timer A timer within the system that ensures that the scan time stays within specified limits. When limits are reached, either warnings are given or PLC operation is stopped, depending on the particular limit that is reached. Although a default value of 130ms is automatically set for the basic time limit, this value can be extended by the program.

Wired slave rack A slave rack connected through a wired remote I/O slave unit.

Word A unit of storage in memory that consists of 16 bits. All data areas consists of words. Some data areas can be addressed by words; others can be addressed by either words or bits.

Word address The location in memory where a word of data is stored. A word address must specify (sometimes by default) the data area and the number of the word that is being addressed.

Work bit A bit in a *Work Word*.

Work word A word that can be used for data calculation or other manipulation in programming, i.e., a "work space" in memory. A large portion of the IR area is always reserved for work words. Parts of other areas not required for special purposes may also be used as work words, e.g., I/O words not allocated to I/O units.

Write For PLCs, the programming instruction to insert a line of instruction into the user memory.

APPENDIX 2
BASIC PROGRAMMING INSTRUCTIONS FOR PLCs

Instruction	Symbol	Mnemonic	Operand
Load		LD | B	B: IR SR HR AR LR TC
Load Not		LD NOT | B	
And		AND | B	
And Not		AND NOT | B	
Or		OR | B	
Or Not		OR NOT | B	
And Load		AND LD | —	—
Or Load	+	OR LD | —	
Out	B	OUT | B	B: IR SR HR AR LR
Out Not	B	OUT NOT | B	
Timer	TIM	TIM | N — | SV	N: TC SV: IR HR AR LR DM *DM #
Counter	CP R CNT N SV	CNT | N — | SV	

LEGEND
B = bit
N = number for timer or counter
SV = set value # for timer or counter
IR = internal relay
SR = special relay
HR = holding relay
AR = auxiliary relay
LR = link relay
TC = timer or counter area
DM = data memory
CP = count pulse
R = reset
# = decimal number

APPENDIX 3
PROGRAMMING INSTRUCTIONS FOR AN OMRON C200H PLC

Code	Symbol	Mnemonic		Function	Operand
00		NOP(00)	—		—
01	END(01)	END(01)	—	Ends the program	
02	IL(02)	IL(02)	—	Causes program steps to be ignored and outputs cleared and timers reset depending on the result immediately before this instruction.	
03	ILC(03)	ILC(03)	—	Clears IL.	
04/05	JMP(04)	JMP(04)	—	Causes all the program steps between this instruction and JME to be ignored, or executed, according to the result immediately before this instruction.	N: 00 to 99
	JMP(05)	JME(05)	—		
	JMP(04)	JMP(04)	N		
	JMP(05)	JME(05)	N		
06	FAL(06)	FAL(06)	N	Indicates an error that does not stop the CPU.	N: 01 to 99
	@FAL(06)	@FAL(06)	N		
	FAL(06)	FAL(06)	00	Clears FAL area.	
	@FAL(06)	@FAL(06)	00		
07	FALS(07)	FALS(07)	N	Indicates an error that stops the CPU.	N: 01 to 99
08	STEP(08)	STEP(08)	—	Shows the section number. Must be included before each section.	—
		STEP(08)	N	Shows the end of the section.	N: IR HR AR LR
09	SNXT(09)	SNXT(09)	N	Resets the previous section, and activates the next section. Must be included before and after each section.	

Code	Symbol	Mnemonic	Function	Operands
10	I, P, R → SFT (10)	SFT(10) B — E	Shifts data in bit units. 15 0 15 0 E B ←IN	B/E: IR HR AR LR
11	St → KEEP B	KEEP(11) B	Causes data bit to become latching.	B: IR HR AR LR
12	II, DI, Rt → CNTR N SV	CNTR(12) N — SV	UP-DOWN (reversible) counter operation.	N: TC SV: IR HR AR LR DM *DM #
13	DIFU(13) P	DIFU(13) B	Causes the following instruction to operate for one scan time at the leading edge of the input signal.	B: IR HR AR LR
14	DIFD(14) B	DIFD(14) B	Causes the following instruction to operate for one scan time at the trailing edge of the input signal.	B: IR HR AR LR
15	TI → TIMH	TIMH(15) N — SV	High-speed, ON-delay timer operation. Set value: 0.01 to 99.99 s	N: TC SV: IR HR AR LR DM *DM #
16	WSFT(16) B E	WSFT(16) — — B — E	Shifts data in Ch units.	B/E: IR LR HR DM AR *DM
20	CMP(20) C1 C2 @CMP(20) C1 C2	CMP(20) — — C1 — C2 @CMP(20) — — C1 — C2	Compares one channel's data, or a 4-digit constant, against another channel's data. Note: C1 and C2 cannot both be constants.	C1/C2: IR SR HR AR LR TC DM *DM #

IR	SR	HR	AR	LR	TC	DM	#
00000 to 24615	24700 to 25507	0000 to 9915	Read: 0000 to 2715 Write: 0700 to 2215	0000 to 6315	000 to 511	Read: 0000 to 1999 Write: 0000 to 0999	0000 to 9999 0000 to FFFF

Code	Symbol	Mnemonic	Function	Operands	
21	MOV(21) S D	MOV(21) — — S — D	Transfers channel data, or a 4-digit constant, to a specified channel.	S: IR SR HR AR LR TC DM *DM #	D: IR HR AR LR DM *DM
	@MOV(21) S D	@MOV(21) — — S — D			
22	MVN(22) S D	MVN(22) — — S — D	Inverts channel data, or a 4-digit constant, and transfers it to a specified channel.	S: IR SR HR AR LR TC DM *DM #	D: IR HR AR LR DM *DM
	@MVN(22) S D	@MVN(22) — — S — D			
23	BIN(23) S R	BIN(23) — — S — R	Converts BCD data into binary data. 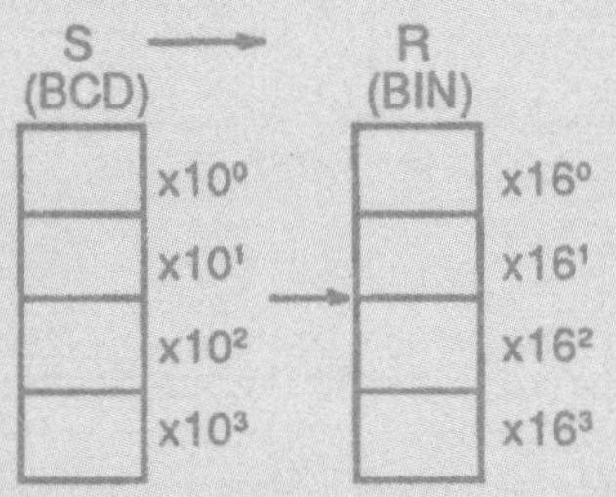 S (BCD) → R (BIN) x10⁰ x10¹ x10² x10³ → x16⁰ x16¹ x16² x16³	S: IR SR HR AR LR TC DM *DM	R: IR HR AR LR DM *DM
	@BIN(23) S R	@BIN(23) — — S — R			
24	BCD(24) S R	BCD(24) — — S — R	Converts binary data into BCD data. 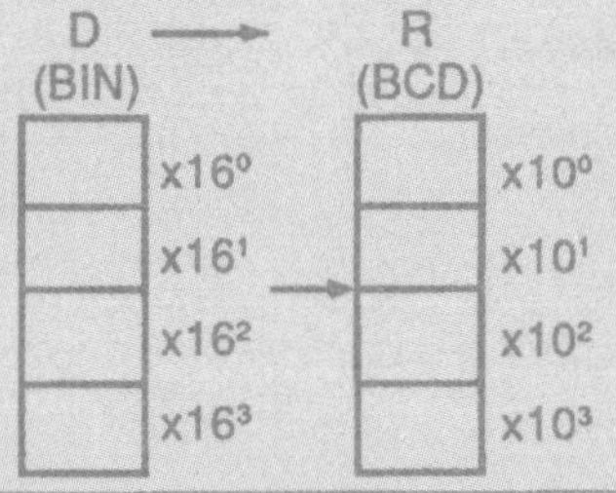 D (BIN) → R (BCD) x16⁰ x16¹ x16² x16³ → x10⁰ x10¹ x10² x10³	S: IR SR HR AR LR DM *DM	R: IR HR AR LR DM *DM
	@BCD(24) S R	@BCD(24) — — S — R			
25	ASL(25) Ch	ASL(25) — — Ch	Shifts Ch data left. 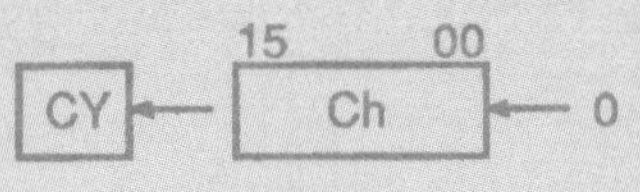 CY ← [15 Ch 00] ← 0	Ch: IR HR AR LR DM *DM	
	@ASL(25) Ch	@ASL(25) — — Ch			

Code	Symbol	Mnemonic	Function	Operand
26	ASR(26) / Ch	ASR(26) — / — Ch	Shifts Ch data right. 0 → Ch (15…00) → CY	Ch: IR, HR, AR, LR, DM, *DM
	@ASR(26) / Ch	@ASR(26) — / — Ch		
27	ROL(27) / Ch	ROL(27) — / — Ch	Rotates Ch left, with carry. Ch (15…00) ← CY (loop)	Ch: IR, HR, AR, LR, DM, *DM
	@ROL(27) / Ch	@ROL(27) — / — Ch		
28	ROR(28) / Ch	ROR(28) — / — Ch	Rotates Ch right, with carry. CY → Ch (15…00) (loop)	Ch: IR, HR, AR, LR, DM, *DM
	@ROR(28) / Ch	@ROR(28) — / — Ch		
29	COM(29) / Ch	COM(29) — / — Ch	Inverts Ch data Ch → Ch	Ch: IR, HR, AR, LR, DM, *DM
	@COM(29) / Ch	@COM(29) — / — Ch		
30	ADD(30) / Au / Ad / R	ADD(30) — / — Au / — Ad / — R	Performs BCD addition of one channel's data, or a 4-digit constant, and another channel's data. Au + Ad + CY → R CY	Au/Ad: IR, SR, HR, AR, LR, TC, DM, *DM, # R: IR, HR, AR, LR, DM, *DM
	@ADD(30) / Au / Ad / R	@ADD(30) — / — Au / — Ad / — R		

IR	SR	HR	AR	LR	TC	DM	#
00000 to 24615	24700 to 25507	0000 to 9915	Read: 0000 to 2715 Write: 0700 to 2215	0000 to 6315	000 to 511	Read: 0000 to 1999 Write: 0000 to 0999	0000 to 9999 0000 to FFFF

Code	Symbol	Mnemonic	Function	Operands	
31	SUB(31) Mi Su R	SUB(31) — — Mi — Su — R	Performs BCD subtraction of one channel's data, or a 4-digit constant, and another channel's data. Mi - Sv - [CY] → R [CY]	Mi/Su: IR SR HR AR LR TC DM *DM #	R: IR HR AR LR DM *DM
	@SUB(31) Mi Su R	@SUB(31) — — Mi — Su — R			
32	MUL(32) Md Mr R	MUL(32) — — Md — Mr — R	Performs BCD multiplication of one channel's data, or a 4-digit constant, and another channel's data. Md x Mr → [R + 1] [R]	Md/Mr: IR SR HR AR LR TC DM *DM #	R: IR HR AR LR DM *DM
	@MUL(32) Md Mr R	@MUL(32) — — Md — Mr — R			
33	DIV(33) Dd Dr R	DIV(33) — — Dd — Dr — R	Performs BCD division of one channel's data, or a 4-digit constant, and another channel's data. [R + 1] [R]	Dd/Dr: IR SR HR AR LR TC DM *DM #	R: IR HR AR LR DM *DM
	@DIV(33) Dd Dr R	@DIV(33) — — Dd — Dr — R			
34	ANDW(34) I1 I2 R	ANDW(34) — — I1 — I2 — R	Performs a logical AND operation between two channel's data. I1 ∧ I2 → R	I1/I2: IR SR HR AR LR TC DM *DM #	R: IR HR AR LR DM *DM
	@ANDW(34) I1 I2 R	@ANDW(34) — — I1 — I2 — R			

Code	Symbol	Mnemonic	Function	Operands	
35	ORW(35) / I1 / I2 / R	ORW(35) — / — I1 / — I2 / — R	Performs a logical OR operation between two channel's data. I1 V I2 → F	I1/I2: IR SR HR AR LR TC DM *DM #	R: IR HR AR LR DM *DM
	@ORW(35) / I1 / I2 / R	@ORW(35) — / — I1 / — I2 / — R			
36	XORW(36) / I1 / I2 / R	XORW(36) — / — I1 / — I2 / — R	Performs a logical exclusive OR operation between two channel's data.	I1/I2: IR SR HR AR LR TC DM *DM #	R: IR HR AR LR DM *DM
	@XORW(36) / I1 / I2 / R	@XORW(36) — / — I1 / — I2 / — R			
37	XNRW(37) / I1 / I2 / R	XNRW(37) — / — I1 / — I2 / — R	Performs a logical exclusive OR NOT operation between two channel's data.	I1/I2: IR SR HR AR LR TC DM *DM #	R: IR HR AR LR DM *DM
	@XNRW(37) / I1 / I2 / R	@XNRW(37) — / — I1 / — I2 / — R			
38	INC(38) / Ch	INC(38) — / — Ch	Increments a channel's data by 1.	Ch: IR HR AR LR DM *DM	
	@INC(38) / Ch	@INC(38) — / — Ch			

IR	SR	HR	AR	LR	TC	DM	#
00000 to 24615	24700 to 25507	0000 to 9915	Read: 0000 to 2715 Write: 0700 to 2215	0000 to 6315	000 to 511	Read: 0000 to 1999 Write: 0000 to 0999	0000 to 9999 0000 to FFFF

Code	Symbol	Mnemonic	Function	Operands
39	DEC(39) Ch	DEC(39) — — Ch	Decrements a channel's data by 1.	Ch: IR HR AR LR DM *DM
	@DEC(39) Ch	@DEC(39) — — Ch		
40	STC(40)	STC(40) —	Sets the carry flag (Ch) to "1". 1→ Ch	—
	@STC(40)	@STC(40) —		
41	CLC(41)	CLC(41) —	Resets the carry flag (Ch) to "0". 0 → Ch	—
	@CLC(41)	@CLC(41) —		

Code	Symbol	Mnemonic	Function	Operands	
46	MSG(46) B	MSG(46) — — B	Displays 8 channels of ASCII codes on the Programming Console or GPC starting from B. B Ch: A B B + 1 Ch: C D : B + 7 Ch: D P ↓ ABCD........DP	B: IR HR AR LR DM *DM	
	@MSG(46) B	@MSG(46) — — B			
50	ADB(50) Au Ad R	ADB(50) — — Au — Ad — R	Performs binary addition of one channel's data, or a 4-digit constant, and another channel's data. Au + Ad + CY ——— R CY	Au/Ad: IR SR HR AR LR TC DM *DM #	R: IR HR AR LR DM *DM
	@ADB(50) Au Ad R	@ADB(50) — — Au — Ad — R			

IR	SR	HR	AR	LR	TC	DM	#
00000 to 24615	24700 to 25507	0000 to 9915	Read: 0000 to 2715 Write: 0700 to 2215	0000 to 6315	000 to 511	Read: 0000 to 1999 Write: 0000 to 0999	0000 to 9999 0000 to FFFF

Code	Symbol	Mnemonic	Function	Operands	
51	SBB(51) / Mi / Su / R	SBB(51) — / — Mi / — Su / — R	Performs binary subtraction of one channel's data, or a 4-digit constant, and another channel's data. Mi − Su − CY = R, CY	**Mi/Su:** IR, SR, HR, AR, LR, TC, DM, *DM, #	**R:** IR, HR, AR, LR, DM, *DM
	@SBB(51) / Mi / Su / R	@SBB(51) — / — Mi / — Su / — R			
52	MLB(52) / Md / Mr / R	MLB(52) — / — Md / — Mr / — R	Performs binary multiplication of one channel's data, or a 4-digit constant, and another channel's data. Md X Mr; Quotient R; Remainder R + 1	**Md/Mr:** IR, SR, HR, AR, LR, TC, DM, *DM, #	**R:** IR, HR, AR, LR, DM, *DM
	@MLB(52) / Md / Mr / R	@MLB(52) — / — Md / — Mr / — R			
53	DVB(53) / Dd / Dr / R	DVB(53) — / — Dd / — Dr / — R	Performs binary division of one channel's data, or a 4-digit constant, and another channel's data. Dd ÷ Dr; Quotient R; Remainder R + 1	**Dd/Dr:** IR, SR, HR, AR, LR, TC, DM, *DM, #	**R:** IR, HR, AR, LR, DM, *DM
	@DVB(53) / Dd / Dr / R	@DVB(53) — / — Dd / — Dr / — R			
54	ADDL(54) / Au / Ad / D	ADDL(54) — / — Au / — Ad / — D	Performs BCD addition of one set of data (8 digits) and another set of data (8 digits). Au + 1, Au + Ad + 1, Ad + CY = CY, R + 1, R	**Au/Ad:** IR, SR, HR, AR, LR, TC, DM, *DM	**D:** IR, HR, AR, LR, DM, *DM
	@ADDL(54) / Au / Ad / D	@ADDL(54) — / — Au / — Ad / — D			

Code	Symbol	Mnemonic	Function	Operands	
55	SUBL(55) / Mi / Su / R	SUBL(55) — / — Mi / — Su / — R	Performs BCD subtraction of one set of data (8 digits) and another set of data (8 digits). Mi + 1, Mi — Su + 1, Su — CY CY, R + 1, R	Mi/Su: IR SR HR AR LR TC DM *DM	R: IR HR AR LR DM *DM
	@SUBL(55) / Mi / Su / R	@SUBL(55) — / — Mi / — Su / — R			
56	MULL(56) / Md / Mr / R	MULL(56) — / — Md / — Mr / — R	Performs BCD multiplication of one set of data (8 digits) and another set of data (8 digits). Md + 1, Md X Mr + 1, Mr R + 3, R + 2, R + 1, R	Md/Mr: IR SR HR AR LR TC DM *DM	R: IR HR AR LR DM *DM
	@MULL(56) / Md / Mr / R	@MULL(56) — / — Md / — Mr / — R			
57	DIVL(57) / Dd / Dr / R	DIVL(57) — / — Dd / — Dr / — R	Performs BCD division of one set of data (8 digits) and another set of data (8 digits). Dd + 1, Dd ÷ Dr + 1, Dr Quotient R + 1, R Remainder R + 3, R + 2	Dd/Dr: IR SR HR AR LR TC DM *DM	R: IR HR AR LR DM *DM
	@DIVL(57) / Dd / Dr / R	@DIVL(57) — / — Dd / — Dr / — R			
58	BINL(58) / S / R	BINL(58) — / — S / — R	Converts BCD data into binary data. S, S + 1 → R, R + 1	S: IR SR HR AR LR TC DM *DM	R: IR HR AR LR DM *DM
	@BINL(58) / S / R	@BINL(58) — / — S / — R			

IR	SR	HR	AR	LR	TC	DM	#
00000 to 24615	24700 to 25507	0000 to 9915	Read: 0000 to 2715 Write: 0700 to 2215	0000 to 6315	000 to 511	Read: 0000 to 1999 Write: 0000 to 0999	0000 to 9999 0000 to FFFF

Code	Symbol	Mnemonic	Function	Operands
59	BCDL(59) / S / R	BCDL(59) — ; — S ; — R	Converts binary data into BCD data. 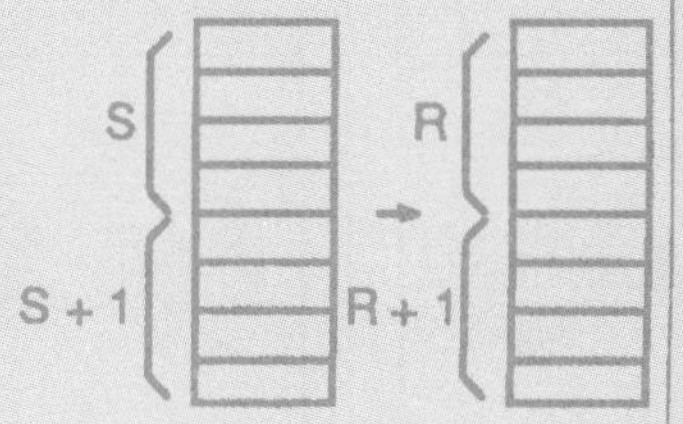 (S, S + 1 → R, R + 1)	S: IR SR HR AR LR TC DM *DM ; R: IR HR AR LR DM *DM
	@BCDL(59) / S / R	@BCDL(59) — ; — S ; — R		
67	FUN67 / N / SB / D	FUN67 — ; — N ; — SB ; — D	Counts the number of "1"s in channel data.	N: IR SR HR AR LR TC DM *DM # ; SB: IR SR HR AR LR DM *DM ; D: IR HR AR LR DM *DM
	@FUN67 / N / SB / D	@FUN67 — ; — N ; — SB ; — D		
68	BCMP(68) / CD / CB / R	BCMP(68) — ; — CD ; — CB ; — R	Outputs "1" to the specified bit of R when the data to be compared in CD is within the range of the compare table data (CB to CB + 31). 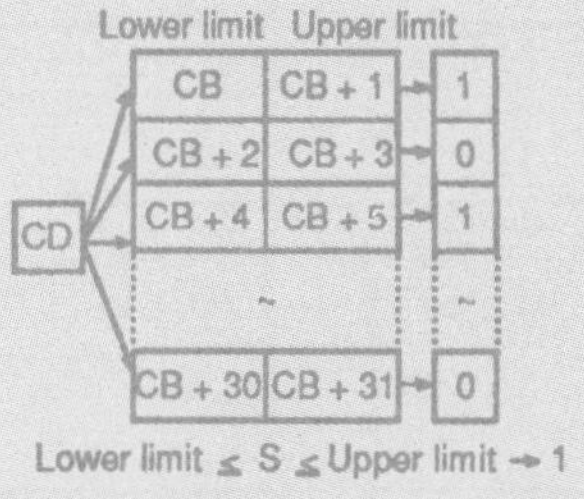 Lower limit / Upper limit: CB, CB + 1 → 1; CB + 2, CB + 3 → 0; CB + 4, CB + 5 → 1; ~; CB + 30, CB + 31 → 0. Lower limit ≤ S ≤ Upper limit → 1	CD: IR SR HR AR LR TC DM *DM # ; CB: IR HR LR TC DM *DM ; R: IR HR AR LR DM *DM
	@BCMP(68) / CD / CB / R	@BCMP(68) — ; — CD ; — CB ; — R		
70	XFER(70) / N / S / D	XFER(70) — ; — N ; — S ; — D	Moves the contents of several consecutive source channels to several consecutive destination channels. 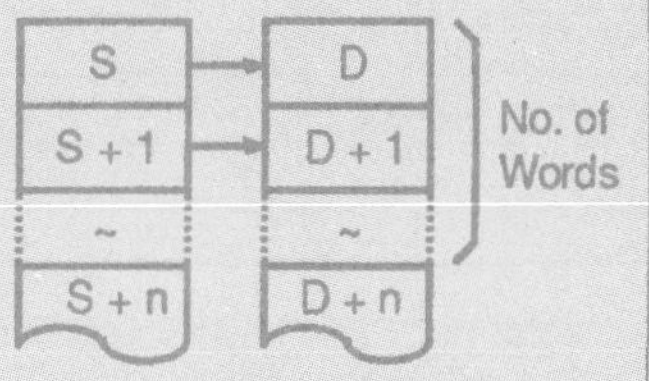 (S → D, S + 1 → D + 1, ~, S + n → D + n; No. of Words)	N/S: IR SR HR AR LR TC DM *DM # (N only) ; D: IR HR AR LR TC DM *DM
	@XFER(70) / N / S / D	@XFER(70) — ; — N ; — S ; — D		

Code	Symbol	Mnemonic	Function	Operands	
71	BSET(71) / S / B / E @BSET(71) / S / B / E	BSET(71) — / — S / — B / — E @BSET(71) — / — S / — B / — E	Sets the same data to the specified consecutive channels. S: Number of channels S → B ~ E	S: IR SR HR AR LR TC DM *DM #	B/E: IR HR AR LR TC DM *DM
72	ROOT(72) / Sq / R @ROOT(72) / Sq / R	ROOT(72) — / — Sq / — R @ROOT(72) — / — Sq / — R	Computes 8-digit BCD square root. √(Sq Sq) → R	Sq: IR HR AR LR TC DM *DM	R: IR HR AR LR DM *DM
73	XCHG(73) / E1 / E2 @XCHG(73) / E1 / E2	XCHG(73) — / — E1 / — E2 @XCHG(73) — / — E1 / — E2	Exchanges data between channels. E1 ↔ E2	E1/E2: IR HR AR LR TC DM *DM	
74	SLD(74) / B / E @SLD(74) / B / E	SLD(74) — / — B / — E @SLD(74) — / — B / — E	Shifts Ch data left in digit units (4 bits). B, B + 1, E ← 0	B/E: IR HR AR LR DM *DM	

IR	SR	HR	AR	LR	TC	DM	#
00000 to 24615	24700 to 25507	0000 to 9915	Read: 0000 to 2715 Write: 0700 to 2215	0000 to 6315	000 to 511	Read: 0000 to 1999 Write: 0000 to 0999	0000 to 9999 0000 to FFFF

Code	Symbol	Mnemonic	Function	Operands	
75	SRD(75) / B / E	SRD(75) — / — B / — E	Shifts Ch data right in digit units (4 bits). 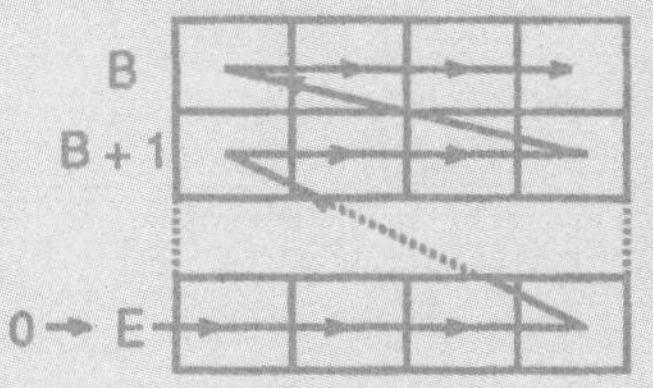	B/E: IR HR AR LR DM *DM	
	@SRD(75) / B / E	@SRD(75) — / — B / — E	B, B + 1, 0 → E		
76	MLPX(76) / S / Di / RB	MLPX(76) — / — S / — Di / — RB	Decodes 1-digit (4-bit) data into a bit position. 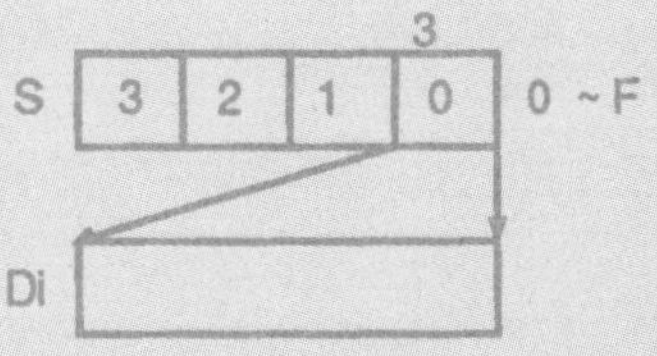	S: IR SR HR AR LR TC DM *DM	Di/RB: IR HR AR LR DM *DM TC (Di only)
	@MLPX(76) / S / Di / RB	@MLPX(76) — / — S / — Di / — RB	S 3 2 1 0 0 ~ F; Di		
77	DMPX(77) / SB / R / Di	DMPX(77) — / — SB / — R / — Di	Encodes the position of the highest bit that is ON into 1-digit (4-bit) data. 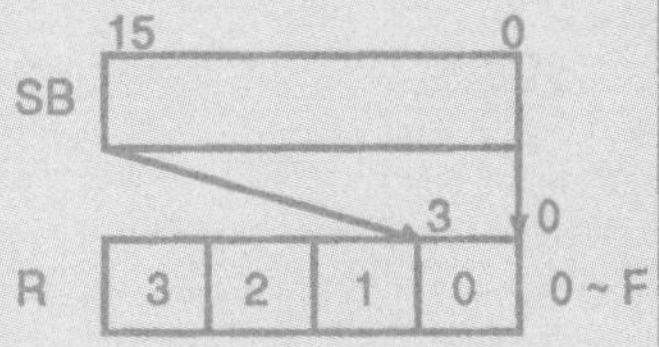	SB: IR SR HR AR LR TC DM *DM	R/Di: IR HR AR LR TC DM *DM TC (R only)
	@DMPX(77) / SB / R / Di	@DMPX(77) — / — SB / — R / — Di	SB 15 0; R 3 2 1 0 0 ~ F		
78	SDEC(78) / S / Di / DB	SDEC(78) — / — S / — Di / — DB	Decodes 1-digit (4 bits) of Ch data into data for 7-segment display. 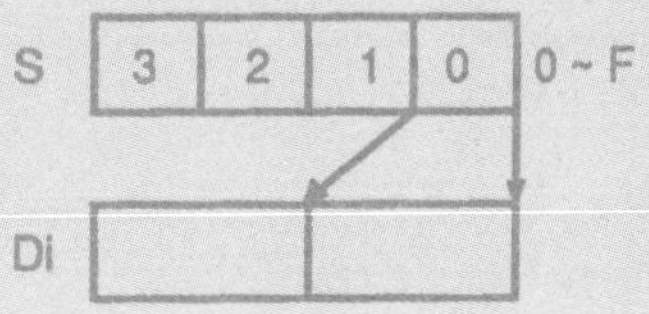	S: IR SR HR AR LR TC DM *DM	Di/DB: IR HR AR LR DM *DM TC (Di only)
	@SDEC(78) / S / Di / DB	@SDEC(78) — / — S / — Di / — DB	S 3 2 1 0 0 ~ F; Di		

Code	Symbol	Mnemonic	Function	Operands	
79	FDIV(79) / Dd / Dr / R @FDIV(79) / Dd / Dr / R	FDIV(79) — ; — Dd ; — Dr ; — R @FDIV(79) — ; — Dd ; — Dr ; — R	Performs floating point division between two 7-digit BCD data. 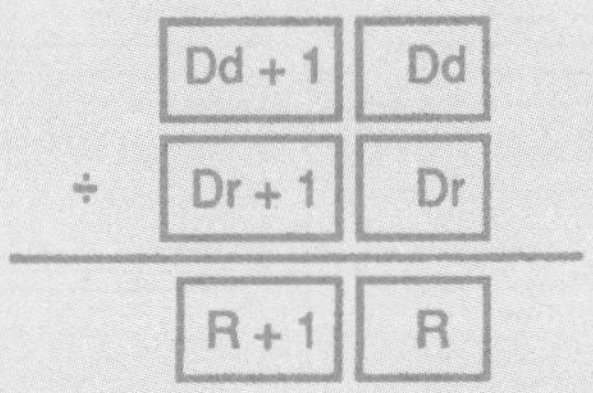 Dd + 1, Dd ÷ Dr + 1, Dr = R + 1, R	Dd/Dr: IR HR AR LR TC DM *DM	R: IR HR AR LR DM *DM
80	DIST(80) / S / DBs / Of @DIST(80) / S / DBs / Of	DIST(80) — ; — S ; — DBs ; — Of @DIST(80) — ; — S ; — DBs ; — Of	Transfers 16-bit data to a channel with address given by base plus offset. 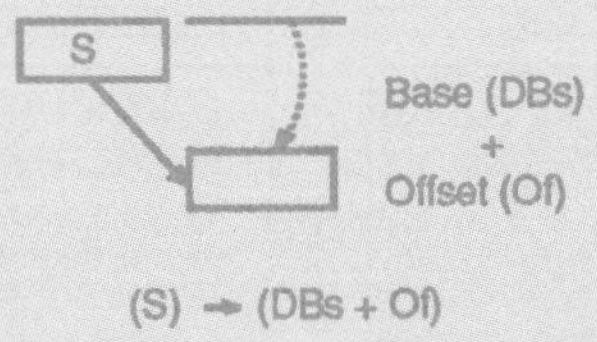 S → Base (DBs) + Offset (Of) (S) → (DBs + Of)	S: IR SR HR AR LR TC DM *DM #	DBs/Of: IR HR AR LR TC DM *DM
81	COLL(81) / SBs / Of / D @COLL(81) / SBs / Of / D	COLL(81) — ; — SBs ; — Of ; — D @COLL(81) — ; — SBs ; — Of ; — D	Extracts 16-bit data from a channel with address given by base plus offset and transfers the data to the specified channel. 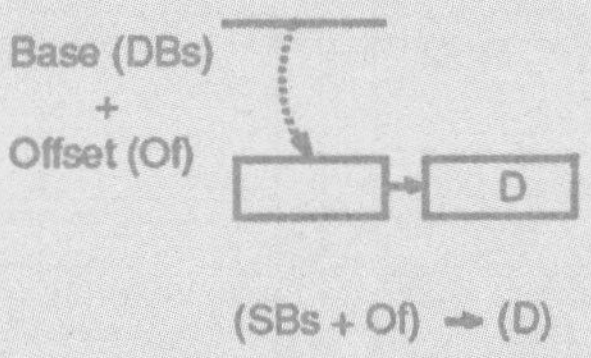 Base (DBs) + Offset (Of) → D (SBs + Of) → (D)	SBs: IR SR HR AR LR TC DM *DM #	Of/D: IR HR AR LR TC DM *DM

IR	SR	HR	AR	LR	TC	DM	#
00000 to 24615	24700 to 25507	0000 to 9915	Read: 0000 to 2715 Write: 0700 to 2215	0000 to 6315	000 to 511	Read: 0000 to 1999 Write: 0000 to 0999	0000 to 9999 0000 to FFFF

Code	Symbol	Mnemonic	Function	Operands
82	MOVB(82) / S / C / D	MOVB(82) — / — S / — C / — D	Transfers a specified bit in Ch S to a specified bit position in Ch D.	S: IR SR HR AR LR TC DM *DM # ; C/D: IR HR AR LR TC DM *DM
	@MOVB(82) / S / C / D	@MOVB(82) — / — S / — C / — D		
83	MOVD(83) / S / C / D	MOVD(83) — / — S / — C / — D	Transfers Ch data to the specified position in digit (4 bit) units.	S: IR SR HR AR LR TC DM *DM # ; C/D: IR HR AR LR TC DM *DM
	@MOVD(83) / S / C / D	@MOVD(83) — / — S / — C / — D		
84	SFTR(84) / C / B / E	SFTR(84) — / — C / — B / — E	Shifts the specified data one bit to the left or right, with carry.	C/B/E: IR HR AR LR DM *DM #
	@SFTR(84) / C / B / E	@SFTR(84) — / — C / — B / — E		
85	TCMP(85) / CD / CB / R	TCMP(85) — / — CD / — CB / — R	Compares 16-bit data with a table consisting of 16-channel data. 1: agreement 0: disagreement	CD: IR SR HR AR LR TC DM *DM # ; CB: IR HR AR LR TC DM *DM ; R: IR HR AR LR TC DM *DM
	@TCMP(85) / CD / CB / R	@TCMP(85) — / — CD / — CB / — R		

Code	Symbol	Mnemonic	Function	Operands
86	ASC(86) / S / Di / DB @ASC(86) / S / Di / DB	ASC(86) — / — S / — Di / — DB @ASC(86) — / — S / — Di / — DB	Converts 1-digit (4-bits) data of a channel into a 7-bit ASCII code. S: 3 2 1 0 (0 ~ F) → Di: 8-bit data	S: IR, SR, HR, AR, LR, TC, DM, *DM Di/DB: IR, HR, AR, LR, DM, *DM, TC (Di only)
89	FUN89 / CC / N / D @FUN89 / CC / N / D	FUN89 — / — CC / — N / — D @FUN89 — / — CC / — N / — D	Controls the scheduled interrupt.	D: IR, HR, AR, LR, DM, *DM, #
91	SBS(91) @SBS(91)	SBS(91) N @SBS(91) N	Calls a subroutine.	N: 00 to 99
92	SBN(92) @SBN(92)	SBN(92) N @SBN(92) N	Indicates the beginning of a subroutine.	N: 00 to 99
93	RET(93)	RET(93) —	Indicates the end of a subroutine.	—
94	WDT(94) @WDT(94)	WDT(94) N @WDT(94) N	Refreshes the watchdog timer the specified number of times.	N: 00 to 99

IR	SR	HR	AR	LR	TC	DM	#
00000 to 24615	24700 to 25507	0000 to 9915	Read: 0000 to 2715 Write: 0700 to 2215	0000 to 6315	000 to 511	Read: 0000 to 1999 Write: 0000 to 0999	0000 to 9999 0000 to FFFF

Code	Symbol	Mnemonic	Function	Operands
97	IORF(97) B E	IORF(97) — — B — E	Refreshes the specified I/O channels.	**B/E:** 000 to 049
	@IORF(97) B E	@IORF(97) — — B — E		

IR	SR	HR	AR	LR	TC	DM	#
00000 to 24615	24700 to 25507	0000 to 9915	Read: 0000 to 2715 Write: 0700 to 2215	0000 to 6315	000 to 511	Read: 0000 to 1999 Write: 0000 to 0999	0000 to 9999

APPENDIX 4
UNITS OF MEASUREMENT AND THEIR SYMBOLS

Mechanical oscillations are commonly expressed in cycles per unit of time, and rotational frequency in revolutions per unit of time. Since "cycle" and "revolution" are not units, they do not have internationally recognised symbols. However, they are often expressed as abbreviations, and in English the common expressions for them are r.p.m. (revolutions per minute), and c/nub (cycles per minute). The bar and Pascal are given equal status as pressure units. The American and Australian fluid power industries still do not agree on one preferred unit. However, the Pascal is the Sl unit for pressure. Thus, common current usage in the industry opts for the multiples kiloPascal (kPa) and megaPascal (MPa) and for the bar, which is equivalent to 100 kPa. The European fluid power industry predominantly uses the bar as the preferred unit, and both equipment and product information from European countries has its pressure ratings specified in bars.

PREFIXES FOR FRACTIONS AND MULTIPLE OF BASE UNITS

FRACTION	PREFIX	SYMBOL	
10–18	alto	a	
10–15	femento	f	
10–12	pico	p	
10–9	nano	n	
10–6	micro	μ	0.000001
10–3	milli	m	0.001
10–2	centi	c	0.01
10–1	deci	d	0.1

MULTIPLE	PREFIX	SYMBOL	
10^1	deca	da	10
10^2	hecto	h	100
10^3	kilo	k	1000
10^6	maga	M	1000000
10^9	giga	G	1000000000
10^{12}	lera	T	

COMMONLY USED AREA AND VOLUME CONVERSIONS

$1m^2$ = 10000 CM^2 = 1000000 MM^2

$1m^3$ = 1000 DM^3 = 1000 L = 1000000 CM^3 = 1000000000 MM^3

SYMBOLS USED IN THIS BOOK (BASE UNITS)

Quantity	Symbol	SI unit	Other recognised units
Area	A	m^2	mm^2, km^2, cm^2
Acceleration	a	m/s^2	
Displacement	V	m^3	mL, cm^3
Flow rate	Q	m^3/s	L/min
Force	F	N	kN, MN
Frequency	f	Hz	1/s
Circle constant	π	3.1416	
Length	I	m	mm, cm, km
Mass	m	kg	
Moment	M	Nm	kNm, MNm
Power	P	W	kW, MW
Pressure	p	Pa	kPa, MPa, BAR
Radius	r	m	mm, cm
Revolutions	n	1/s	1/min
Temperature	T	K	°C
Time	t	s	ms, min,h,d
Torque	m	Nm	kNm, MNm
Velocity	v	m/s	m/min, km/h
Viscocity (DYN)	n	Pa.s	$\frac{N.s}{m^2}$
Viscocity (KIN)	v	m^2/s	mm^2/s = 1cSt
Volume	V	m3	mL, cm^3
Work	W	J	kJ, mJ

EXAMPLES FOR USING FRACTIONS AND MULTIPLES OF BASE UNITS

FRACT./MULT.	SYMBOL	Pa	M	L	N	W
10^3	m		mm	mL		mW
10^{-2}	c		cm			
10^{-1}	d		dm*	dL*		
10^1	da				daN*	
10^2	h			hL		
10^3	k	kPa	km			kW
10^6	M	MPa			MN	MW

*UNITS USED IN EUROPE

APPENDIX 5
BOOLEAN LOGIC CONCEPTS

To understand the peculiarities of logic switching, whether this be applied to pneumatic or electric or electronic control makes no difference, one must understand some basic rules of Boolean algebra.

These rules are now presented in simple and abbreviated form and no attempts have been made to prove their origin or consecutive logic development.

Boolean postulates

$\overline{1} = 0$	$\overline{0} = 1$
$0 \cdot 0 = 0$	$0 + 0 = 0$
$0 \cdot 1 = 0$	$0 + 1 = 1$
$1 \cdot 0 = 0$	$1 + 0 = 1$
$1 \cdot 1 = 1$	$0 + 0 = 0$

Boolean theorems (for one input signal)

$A \cdot 0 = 0$	$A + 1 = 1$
$A \cdot 1 = A$	$A + 0 = A$
$A \cdot A = A$	$A + A = A$
$A \cdot \overline{A} = 0$	$A + \overline{A} = 1$

Theorems for more than one input signal

Commutative laws

$A \cdot B = B \cdot A$ $\qquad$ $A + B = B + A$

Associative laws

$(A \cdot B) \cdot C = A \cdot (B \cdot C) = A \cdot B \cdot C$

$(A + B) + C = A + (B + C) = A + B + C$

Distributive law

$(A \cdot B) + (C \cdot B) = B \cdot (A + C)$ or $(A + C) \cdot B$

$(A + B) \cdot (C + B) = B + (A \cdot C)$ or $(A \cdot C) + B$

Absorption law

$A + A \cdot B = A$

De Morgan's theorem

$\overline{A \cdot B \cdot C} = \overline{A} + \overline{B} + \overline{C}$ $\qquad$ $\overline{A + B + C} = \overline{A} \cdot \overline{B} \cdot \overline{C}$

APPENDIX 6
CHART OF ASCII CHARACTERS

	Upper Digit (four bits)												
Lower Digit (four bits)	**0、1、8、9**	**2**	**3**	**4**	**5**	**6**	**7**	**A**	**B**	**C**	**D**	**E**	**F**
0			0	@	P	`	p		ー	タ	ミ	α	p
1		!	1	A	Q	a	q	。	ア	チ	ム	ä	q
2		"	2	B	R	b	r	「	イ	ツ	メ	β	θ
3		#	3	C	S	c	s	」	ウ	テ	モ	ε	∞
4		$	4	D	T	d	t	、	エ	ト	ヤ	μ	Ω
5		%	5	E	U	e	u	・	オ	ナ	ユ	σ	ü
6		&	6	F	V	f	v	ヲ	カ	ニ	ヨ	ρ	Σ
7		'	7	G	W	g	w	ァ	キ	ヌ	ラ	g	π
8		(	8	H	X	h	x	ィ	ク	ネ	リ	√	$\bar{x}$
9		)	9	I	Y	i	y	ゥ	ケ	ノ	ル	$^{-1}$	y
A		*	:	J	Z	j	z	ェ	コ	ハ	レ	j	千
B		+	;	K	[	k	{	ォ	サ	ヒ	ロ	^{x}	万
C		,	<	L	¥	l	|	ャ	シ	フ	ワ	¢	円
D		-	=	M	]	m	}	ュ	ス	ヘ	ン	£	÷
E		.	>	N	^	n	→	ョ	セ	ホ	゛	ñ	
F		/	?	O	_	o	←	ッ	ソ	マ	゜	ö	█

APPENDIX 7
INTERNATIONALLY USED FLUID POWER SYMBOLS

The symbols are based on ISO 1219. Only the most common symbols and circuits are given.

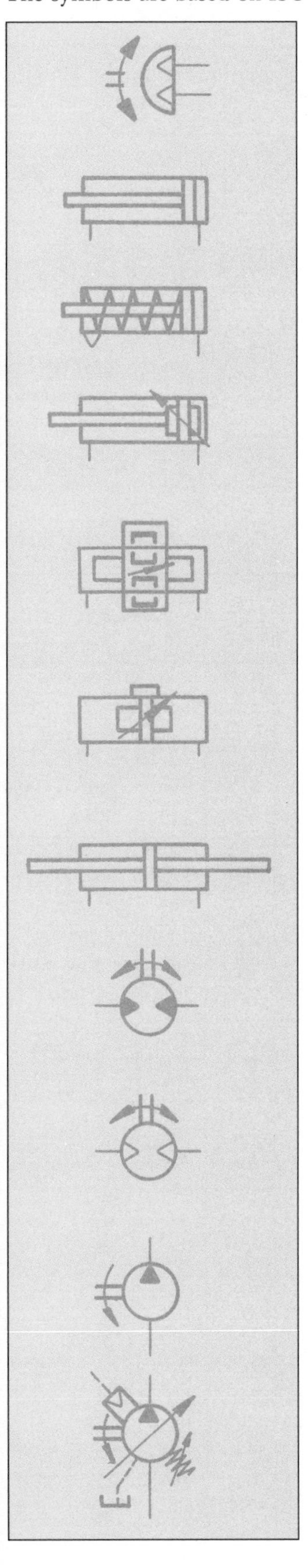

semi-rotary pneumatic actuator

double-acting linear actuator

single-acting, spring-return, linear actuator

linear actuator with adjustable end cushioning

rodless linear actuator with magnetic coupling

rodless linear actuator with zip-slot coupling

double-ended, rod-type, linear actuator

double-direction hydraulic motor

double-direction pneumatic motor

single-direction, fixed displacement hydraulic pump

single-direction, variable displacement hydraulic pump with control mechanism

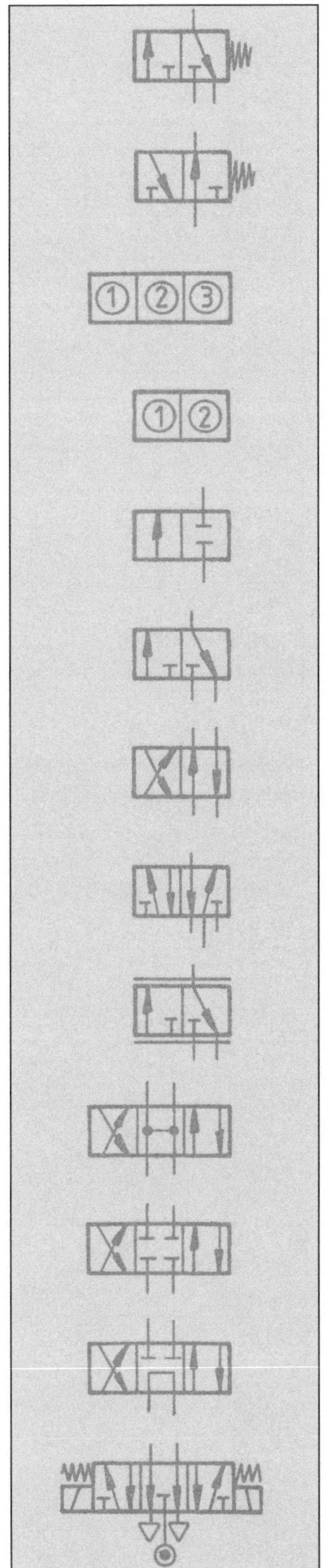

normally closed valve

normally open valve

directional control valve with three discrete positions

directional control valve with two discrete positions

two-position, two-port valve (2/2)

two-position, three-port valve (3/2)

two-position, four-port valve (4/2)

two-position, five-port valve (5/2)

valve with two discrete and an infinite number of intermediate throttling positions

three-position, four-port valve with open centre position

three-position, four-port valve with closed centre position

three-position, four-port valve with tandem flow centre position

three-position, four-port valve with 2 and 4 exhausted and 1 blocked (typical pneumatic valve)

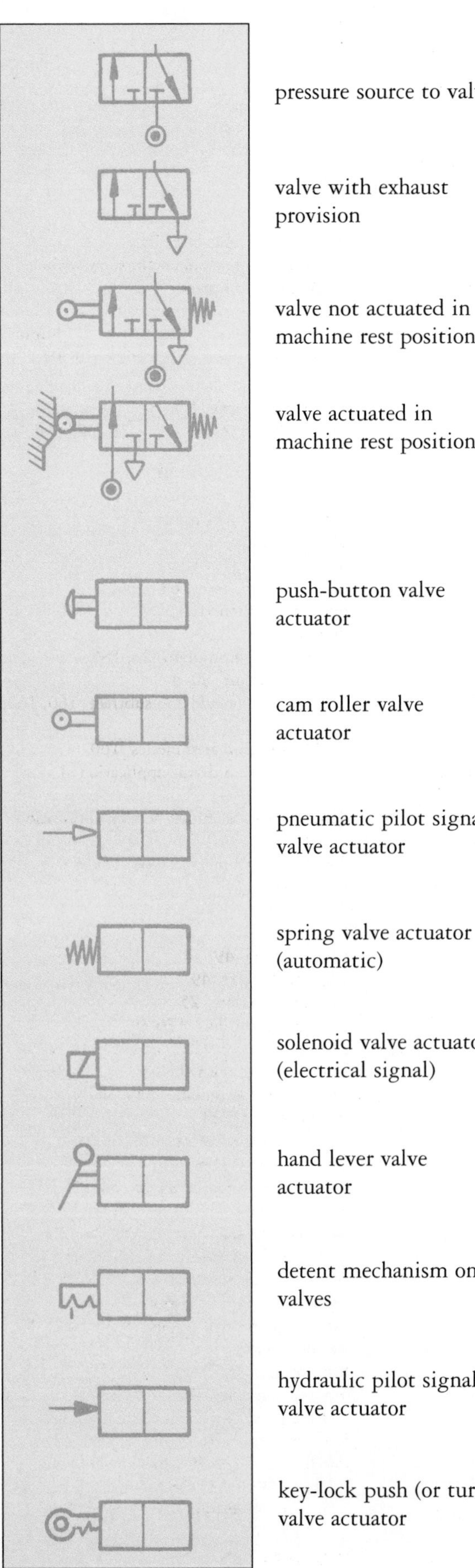

pressure source to valve

valve with exhaust provision

valve not actuated in machine rest position

valve actuated in machine rest position

push-button valve actuator

cam roller valve actuator

pneumatic pilot signal valve actuator

spring valve actuator (automatic)

solenoid valve actuator (electrical signal)

hand lever valve actuator

detent mechanism on valves

hydraulic pilot signal valve actuator

key-lock push (or turn) valve actuator

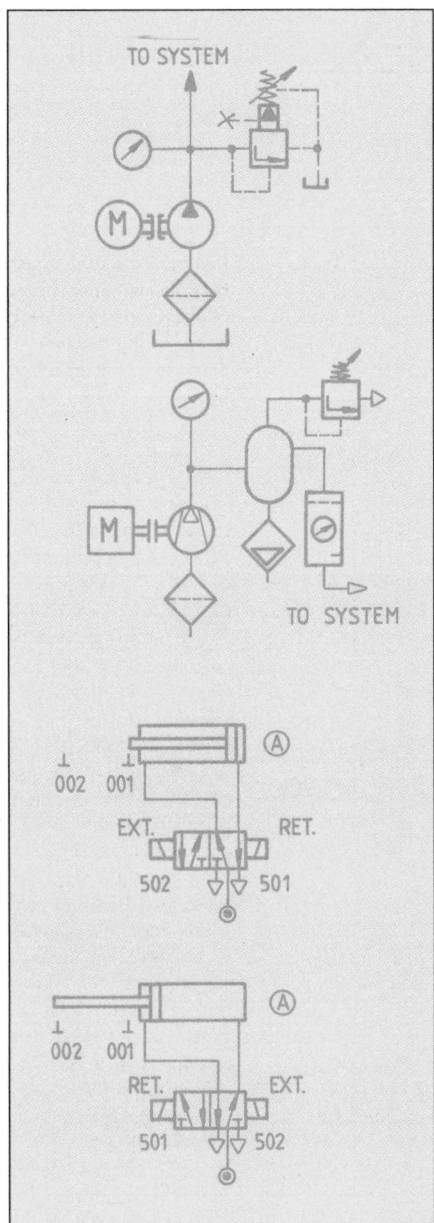

hydraulic pumping station (power unit) with relief valve, pressure gauge, pump, electric motor and suction filter

compressed-air pumping station with relief valve, air receiver, pressure gauge, compressor, internal combustion engine, suction filter, air service unit and automatic drain

linear, pneumatic, double-acting actuator with directional control valve. Actuator rest position is retracted

linear, pneumatic, double-acting actuator with directional control valve. Actuator rest position is extended

INDEX

ERRATA

Text correction to page 58

Figure 6-05. Cascaded counters to achieve extra-long counting tasks.

In figure 6-05, the counter flag CNT 02 is used to drive output relay 500, which announces the counting task as being completed. Cascaded counters, like all Omron counters, are memory-retentive, which means the accumulated value (or present value) is retained during PLC power failure.

Overstriking text correction to page 66

A BEFORE B CIRCUIT

In an assembly machine, where two products need to be attached to each other, and the assembling process must not start until both products are present but one product must arrive before the other, this circuit becomes an invaluable building block (figure 7-01).

If product "b" (input 015) is not present when product "a" arrives (input 014), the logic "AND" function consisting of a • $\overline{b}$ is established at the "AND" valve (014 • $\overline{015}$). This "AND" function creates the latch x (contact of relay 1001). When product b then arrives (input 015), the top "AND" valve switches to "ON" and produces output signal y (500). Thus, 500 is equal to 015 • latch, or in Boolean expression form:

1001 = $\overline{015}$ • 014 = first stage

1001 = 014 • 1001 = transition stage (latch)

500 = 015 • 1001 = final stage

A close investigation shows that input a needs to be ahead of input b for at least the time taken to establish the latch and hold the "AND" function while input $\overline{b}$ is disappearing and input b is being established. If the transition from $\overline{b}$ to b complementing is faster than the latch creation, output y (500) will not appear until b has been removed and a enters before b with sufficient time lag. This time lag is probably in the small millisecond range. (For complementing outputs see figure 4-02.)

Overstriking text correction to page 131

It is for these reasons that I suggest that the reset condition of the HR 114 flip-flop be used to conduct the 107 emergency stop signal ($\overline{\text{HR 114}}$) and not the artificial set condition. At machine commissioning stage, all HR flip-flops are reset, thus rendering $\overline{Q}$ as is shown in figures 4-02 and 4-03. If forcing of the artificial set condition (HR 114 being "ON") were neglected during machine commissioning, it could prevent the operator from starting the emergency program! Hence, the starting equation for step 201 is:

(TIM 01 + 107) • $\overline{\text{HR 114}}$ = SET HR 201.

(See also ladder diagram figure 9-12 and function plan figure 9-11.) Thus, HR 114 is set by HR 201 and reset by the reset push-button signal 108.

Overstriking text correction to page 170

Actuator Ⓐ will release a container into the weighing station only when the weighing station is empty ($\overline{0000}$), actuator Ⓑ gives free passage (003), a container is at the barrier (010) and the machine is started (1000).

Lamp output 109 is "ON" if HR 01 or HR 03 signals an incorrectly filled container being present between sensors 000 (weighing station) and 009 (exit for correctly filled containers). Lamp output 110 is "ON" if a container is weighed and swept off the weighing station, and the comparison function (FUN 20) has rendered a correct weighing result ($\overline{\text{HR 003}}$ • $\overline{\text{HR 001}}$). De Morgan's theorem is used here, which says:

$$\overline{\text{HR 003}} \bullet \overline{\text{HR 001}} = \overline{\text{HR 003} + \text{HR 001}}.$$

Internal relay 1001 is "ON" if no container is in the machine between sensor 000 and sensor 009 (weighing station and remotest exit point).

Flip-flop HR 001 and HR 003 can be reset only if internal relay 1001 is "ON".

The weighing process may be started only if actuator Ⓑ starts the sweeping motion and flip-flops HR 001 and HR 003 are reset.

```
     |  0010                              |
0000 |--| |--------------------------[ KEEP(11)]-|
     |  0011                              |   H000|
     |--| |--------------------------L         ]-|
     |  H000                       1000          |
0003 |--| |-+---------------------------( )------|
     |  0000|                                    |
     |--| |-+                                    |
     |  H000                       0510          |
0006 |--| |-----------------------------( )------|
     |  H000                       0511          |
0008 |--|/|-----------------------------( )------|
     |  H007   0002   0003   1000                |
0010 |--| |-----| |-----| |-----| |--[ KEEP(11)]-|
     |  H002                              |   H001|
     |--| |--------------------------L         ]-|
     |  H001   0001                              |
0016 |--| |-----| |-------------------[ KEEP(11)]-|
     |  H003                              |   H002|
     |--| |--------------------------L         ]-|
     |  H002   0004                              |
0020 |--| |-----| |-------------------[ KEEP(11)]-|
     |  H004                              |   H003|
     |--| |--------------------------L         ]-|
     |  H003   0003                              |
0024 |--| |-----| |-------------------[ KEEP(11)]-|
     |  H005                              |   H004|
     |--| |--------------------------L         ]-|
     |  H004   0004                              |
0028 |--| |-----| |-------------------[ KEEP(11)]-|
     |  H006                              |   H005|
     |--| |--------------------------L         ]-|
     |  H005   0003                              |
0032 |--| |-----| |-------------------[ KEEP(11)]-|
     |  H007                              |   H006|
     |--| |--------------------------L         ]-|
     |  H006   0004                              |
0036 |--| |-----| |-------------------[ KEEP(11)]-|
     |  H001                              |   H007|
     |--| |--------------------------L         ]-|
     |  H001                       0501          |
0040 |--| |-----------------------------( )------|
     |  T02                        0502          |
0042 |--| |-+---------------------------( )------|
     |  T03|                                     |
     |--| |-+                                    |
     |  H003                       0503          |
0045 |--| |-+---------------------------( )------|
     |  T01|                                     |
     |--| |-+                                    |
     |  T02|                                     |
     |--| |-+                                    |
     |  T03|                                     |
     |--| |-+                                    |
     |  H002                       0504          |
0050 |--| |-+---------------------------( )------|
     |  H004|                                    |
     |--| |-+                                    |
     |  H006|                                    |
     |--| |-+                                    |
     |  0014   H007                              |
0054 |--| |-----| |-------------------[ KEEP(11)]-|
     |  0015   H007                   |       H800|
     |--| |-----| |------------------L         ]-|
     |  H005                                     |
0059 |--| |---------------------------[ TIM     ]-|
     |                                |        01|
     |                                L    #0020]-|
     |  H800   H007                              |
0061 |--| |-----| |-------------------[ TIM     ]-|
     |                                |        02|
     |                                L    #0040]-|
     |  H800   H007                              |
0064 |--|/|-----| |-------------------[ TIM     ]-|
     |                                |        03|
                                      L    #0090]-|
```

ADDRESS	MNEMONIC		OPERAND	
0000	LD			0010
0001	LD			0011
0002	KEEP	(11)	HR	000
0003	LD		HR	000
0004	OR			0000
0005	OUT			1000
0006	LD		HR	000
0007	OUT			0510
0008	LD	NOT	HR	000
0009	OUT			0511
0010	LD		HR	007
0011	AND			0002
0012	AND			0003
0013	AND			1000
0014	LD		HR	002
0015	KEEP	(11)	HR	001
0016	LD		HR	001
0017	AND			0001
0018	LD		HR	003
0019	KEEP	(11)	HR	002
0020	LD		HR	002
0021	AND			0004
0022	LD		HR	004
0023	KEEP	(11)	HR	003
0024	LD		HR	003
0025	AND			0003
0026	LD		HR	005
0027	KEEP	(11)	HR	004
0028	LD		HR	004
0029	AND			0004
0030	LD		HR	006
0031	KEEP	(11)	HR	005
0032	LD		HR	005
0033	AND			0003
0034	LD		HR	007
0035	KEEP	(11)	HR	006
0036	LD		HR	006
0037	AND			0004
0038	LD		HR	001
0039	KEEP	(11)	HR	007
0040	LD		HR	001
0041	OUT			0501
0042	LD		TIM	02
0043	OR		TIM	03
0044	OUT			0502
0045	LD		HR	003
0046	OR		TIM	01
0047	OR		TIM	02
0048	OR		TIM	03
0049	OUT			0503
0050	LD		HR	002
0051	OR		HR	004
0052	OR		HR	006
0053	OUT			0504
0054	LD			0014
0055	AND		HR	007
0056	LD			0015
0057	AND		HR	007
0058	KEEP	(11)	HR	800
0059	LD		HR	005
0060	TIM			01
			#	0020
0061	LD		HR	800
0062	AND		HR	007
0063	TIM			02
			#	0040
0064	LD	NOT	HR	800
0065	AND		HR	007
0066	TIM			03
			#	0090
0067	END	(01)		

Figure 8-31. Ladder diagram and mnemonic list for control problem 2.

Circuit correction to page 116